Lecture Notes in Computer Science 10591

Commenced Publication in 1973
Founding and Former Series Editors:
Gerhard Goos, Juris Hartmanis, and Jan van Leeuwen

Wuyi Yue · Quan-Lin Li
Shunfu Jin · Zhanyou Ma (Eds.)

Queueing Theory and Network Applications

12th International Conference, QTNA 2017
Qinhuangdao, China, August 21–23, 2017
Proceedings

 Springer

Editors
Wuyi Yue
Konan University
Kobe
Japan

Quan-Lin Li
Yanshan University
Qinhuangdao
China

Shunfu Jin
Yanshan University
Qinhuangdao
China

Zhanyou Ma
Yanshan University
Qinhuangdao
China

ISSN 0302-9743   ISSN 1611-3349   (electronic)
Lecture Notes in Computer Science
ISBN 978-3-319-68519-9   ISBN 978-3-319-68520-5   (eBook)
https://doi.org/10.1007/978-3-319-68520-5

Library of Congress Control Number: 2017959630

LNCS Sublibrary: SL1 – Theoretical Computer Science and General Issues

Printed on acid-free paper

This Springer imprint is published by Springer Nature
The registered company is Springer International Publishing AG
The registered company address is: Gewerbestrasse 11, 6330 Cham, Switzerland

Preface

The International Conference on Queueing Theory and Network Applications aims to promote the knowledge and the development of high-quality research on queueing theory and its applications in networks and other related fields. It brings together researchers, scientists, and practitioners from the world and offers an open forum to share the latest important research accomplishments and challenging problems in the area of queueing theory and network applications.

This volume contains papers selected and presented at the 12th International Conference on Queueing Theory and Network Applications (QTNA 2017) held during August 21–23, 2017, in Qinhuangdao, China. The conference is organized by Yanshan University, China, and by the following sponsors: Academy of Mathematics and Systems Science, Chinese Academy of Sciences, National Natural Science Foundation of China, and Springer.

QTNA 2017 was a continuation of the series of successful QTNA conferences: QTNA 2006 (Seoul, Korea), QTNA 2007 (Kobe, Japan), QTNA 2008 (Taipei, Taiwan), QTNA 2009 (Singapore), QTNA 2010 (Beijing, China), QTNA 2011 (Seoul, Korea), QTNA 2012 (Kyoto, Japan), QTNA 2013 (Taichung, Taiwan), QTNA 2014 (Bellingham, USA), QTNA 2015 (Hanoi, Vietnam), and QTNA 2016 (Wellington, New Zealand).

We received 65 submissions from Belgium, China, Hong Kong, Japan, The Netherlands, Singapore, South Korea, Taiwan, USA, and UK. All papers were peer reviewed and evaluated on the quality, originality, soundness, and significance of their contributions by the members of the Technical Program Committee of QTNA 2017 and 19 papers were accepted as full papers appearing in this LNCS proceedings published by Springer.

In addition, 37 short papers were accepted to be presented at QTNA 2017. The papers presented disseminate the latest results covering up-to-date research fields such as performance modeling and analysis of telecommunication systems, retrial and vacation queueing models, optimization of queueing systems, modeling of social systems, and other application areas.

It is also our privilege to have had Profs. David D. Yao, Guy Latouche, Mu-Fa Chen, and Zhisheng Niu to give us keynote talks, and Profs. Minghua Chen, Longbo Huang, and Pengfei Guo to deliver invited talks at QTNA 2017.

The proceedings show that the potential of queueing theory is to be exploited, and this is a significant opportunity and a challenge for all of researchers, PhD and graduate students in queueing theory and its applications to share recent achievements and discoveries and create new friendships for future collaborative works.

We would like to thank the authors of papers for contributing their excellent technical contributions and new theoretical results to this book. Special thanks go to the co-chairs and members of the Technical Program Committee of QTNA 2017 for their contribution to keeping the high quality of the selected papers. We would also like to express our gratitude to the co-chairs and members of the local Organizing Committee for their hard work throughout the process from planning to holding the conference. Finally, we cordially thank Springer for support in publishing this volume.

August 2017   Wuyi Yue
                                                                    Quan-Lin Li
                                                                    Shunfu Jin
                                                                    Zhanyou Ma

Organization

General Chairs

Quan-Lin Li   Yanshan University, China
Wuyi Yue   Konan University, Japan

Steering Committee

Co-chairs

Bong Dae Choi   Korea University, Korea
Yutaka Takahashi   Kyoto University, Japan
Wuyi Yue   Konan University, Japan

Members

Hsing Paul Luh, Taiwan
Winston K. G. Seah, New Zealand
Hideaki Takagi, Japan
Y. C. Tay, Singapore
Kuo-Hsiung Wang, Taiwan
Jinting Wang, China
Dequan Yue, China
Zhe George Zhang, USA

Technical Program Committee

Co-chairs

Shunfu Jin   Yanshan University, China
Shoji Kasahara   NAIST, Japan
Yutaka Takahashi   Kyoto University, Japan
Naishuo Tian   Yanshan University, China

Members

Azam Asanjarani, Australia
Herwig Bruneel, Belgium
Kuo-Hwa Chang, Taiwan
Wai-Ki Ching, Hong Kong
Bong Dae Choi, Korea
Wanyang Dai, China
Tien V. Do, Hungary
Alexander Dudin, Belarus

Antonis Economou, Greece
Pengfei Guo, Hong Kong
Yongjiang Guo, China
Guangyue Han, Hong Kong
Qi-Ming He, Canada
Ibrahim Hokelek, USA
Ganguk Hwang, Korea
Konosuke Kawashima, Japan
Jau-Chuan Ke, Taiwan
Jyh-Bin Ke, Taiwan
Bara Kim, Korea
Masahiro Kobayashi, Japan
Achyutha Krishnamoorthy, India
Yanfei Lan, China
Ho Woo Lee, Korea
Se Won Lee, Korea
Na Li, China
Shiyong Li, China
Zhaotong Lian, Macau
Chih-Chin Liang, Taiwan
Bin Liu, China
Zaiming Liu, China
Hsing Paul Luh, Taiwan
Yong-Hua Mao, China
Hiroyuki Masuyama, Japan
Agassi Melikov, Azerbaijan
Zhisheng Niu, China
Rein D. Nobel, The Netherlands
Yoshikuni Onozato, Japan
Tuan Phung-Duc, Japan
Wouter Rogiest, Belgium
Poompat Saengudomlert, Thailand
Zsolt Saffer, Hungary
Yutaka Sakuma, Japan
Winston K. G. Seah, New Zealand
Andreev Sergey, Finland
Yang Woo Shin, Korea
Jie Song, China
Wei Sun, China
Zhankun Sun, Hong Kong
János dr Sztrik, Hungary
Hideaki Takagi, Japan
Lixin Tang, China
Yinghui Tang, China
Y. C. Tay, Singapore
Miklos Telek, Hungary

Koen De Turck, France
Chia-Hung Wang, China
Jinting Wang, China
Kuo-Hsiung Wang, Taiwan
Sabine Wittevrongel, Belgium
Jingui Xie, China
Li Xia, China
Deju Xu, China
Dacheng Yao, China
Hengqing Ye, Hong Kong
Xue-Ming Yuan, Singapore
Dequan Yue, China
Zhe George Zhang, USA
Yigiang Q. Zhao, Canada
Wenhui Zhou, China

Local Organizing Committee

Co-chairs

Chunling Li   Yanshan University, China
Xiuli Xu   Yanshan University, China
Dequan Yue   Yanshan University, China

Members

Wanyang Dai, China
Yongjiang Guo, China
Yuanyuan Liu, China
Zhanyou Ma, China
Jie Song, China
Jinbiao Wu, China
Li Xia, China
Jingui Xie, China
Dacheng Yao, China
Yuan Zhao, China

Contents

Queueing Models II

Queueing Applications II

Queueing Models I

Equilibrium Analysis of the M/M/1 Queues with Setup Times Under N-Policy

Yaqian Hao, Jinting Wang[(✉)], Mingyu Yang, and Ruoyu Wang

Department of Mathematics, Beijing Jiaotong University, Beijing 100044, China
{14271053,jtwang,14271068,14274078}@bjtu.edu.cn

Abstract. Chen et al. (2015) studied the equilibrium threshold balking strategies for the fully observable and fully unobservable single-server queues with threshold policy and setup times. The server shuts down whenever the system becomes empty, and is only resumed when the number of customers reaches to a given threshold. Customers decide whether to join or to balk the system based on their observations of the queue length and status of the server at arrival instants. This paper aims to study the partially observable case and the unobservable case. The stationary probability distribution, the mean queue length and the social welfare are derived. The equilibrium strategies for the customers and the system performance under these strategies are analyzed.

Keywords: M/M/1 queue · Setup times · N-policy · Balking Queue length

1 Introduction

During the last decades, the game-theoretic analysis of queueing systems with strategic customers has been paid considerable attention since the pioneer work by Naor (1969). In general, to reflect customers' desire for service and their unwillingness to wait, some reward-cost structures are imposed on the system. Arriving customers can make decisions to decide whether to join or not, based on different levels of information of the system at their arrival, to maximize their utility. These customers take into account that the other customers have the same objective to maximize their benefit, so the situation can be regarded as a game among them. In these studies, the characterization and computation of individual and social optimal strategies is the fundamental problem.

Studies on customers decentralized behavior as well as socially optimal control of customers' arrivals was pioneered by Naor (1969) with a single-server system in an observable framework, i.e., upon arrival, a customer is informed about the length of queue before his decision is made to join. Edelson and Hildebrand (1975) considered the unobservable case. There is more related work

This work is supported by the National Natural Science Foundation of China (Grant Nos. 71571014 and 71390334).

W. Yue et al. (Eds.): QTNA 2017, LNCS 10591, pp. 3–17, 2017.
https://doi.org/10.1007/978-3-319-68520-5_1

by Hassin and Haviv (2003) in their survey book. Burnetas and Economou (2007) assumed $N = 1$ and an exponential setup time when the server starts a new busy period. They considered the strategic behavior of customers under different levels of information. In particular, if only the queue length is known and the set-up time is of considerable length, the "Follow-The-Crowd" behavior of customers is observed.

The pioneering work on queues with N-policy can go back to Yadin and Naor (1963) for an $M/M/1$ queue with multiple vacations. The server is immediately turned on whenever N $(N \geq 1)$ or more customers are present in the system and is switched off once there are no customers in the system. When the server shuts down, the server can not operate until N customers are present in the system. Guo and Hassin (2011) first considered customers' strategic behavior and social optimization in a single Markovian queue with N-policy, in which the server is activated only if there are N customers in the system and turned off once there are no customers in the system. They concluded that a customer can induce positive externalities in the fully observable and the fully unobservable cases. Some recent papers that deal with the strategic behavior of customers in various queueing systems can be found in Economou and Kanta (2008a, b), Economou et al. (2011), Hassin (2007), Sun et al. (2010), Wang et al. (2017), Wang and Zhang (2011), Zhang et al. (2015), among others.

Evidently, frequent setups increase the operating cost, and it is crucial for the server to decide when to start service in practice. In principle, an appropriate value N can be determined by avoiding excessively frequent setups and the associated cost. For instance, to reduce the operating cost, in a Make-To-Order (MTO) system, the firm will set up the machines when the quantity of orders reaches a threshold. Another example is energy-saving issues arising from wireless sensor networks (WSNs), the N-policy is actually used in switching the sensor's on-off states for prolonging the lifetime of the WSN system. Furthermore, the threshold-type control policy could be applied on optimizing elevator configuration, increasing the defense effectiveness of a missile defense system, and improving the connectivity of communications network.

The main objective of our work is to investigate the customers' equilibrium balking strategies for both the partial observable single-server queues with N-policy and setup times which make our model more practical and valuable. In the present paper, we assume that customers are aware of the service policy, specifically the threshold N, and react to it in a strategic way. We also consider the social optimization problems. The model under consideration can be viewed as an $M/M/1$ queue in a Markovian environment. Also, customer equilibrium strategies are studied in each case, along with the system's performance and the social welfare. Our model has potential in many practical applications.

2 Model Description

We consider a single server Markovian queue with infinite waiting room under the FCFS discipline, where customers arrive according to a Poisson process with rate λ. We are interested in customers' strategic behavior when they can decide whether to join or balk the system based on available information upon their arrival. The server works with service rate μ and it shuts down once there is no customer upon completion of a service. After the server shuts down, the server can not work until N customers are presented in the queue and then a setup process begins, and we assume that the setup time is exponentially distributed with rate θ. We suppose that arrival times, service times, setup times are mutually independent. More specifically, every customer gets a reward of R units for completing service, however there exists a waiting cost of C per time unit when waiting in the queue or in service. Customers are risk neutral and they want to maximize their expected net benefit. We assume that $R > \frac{C}{\mu}$, which enables that a customer joins in the queue when he finds the system empty, because the profit for service definitely surpasses the expected cost. Finally, the decisions are irrevocable, which means that retrials of balking customers and reneging of entering customers are not allowed.

The state of the system can be represented by a pair $(N(t), I(t))$ at time t, where $N(t)$ denotes the number of customers in the system, and $I(t)$ stands for the state of the server. More specifically, the state $(0, n)$, $0 \leq n \leq N-1$, implies that the system is down with n customers in the system; The state $(1, n)$, $n \geq N$, means the system is in a setup process with n customers in the system; And the state $(2, n)$, $n \geq 1$, implies that server is busy with n customers in the system. It is easy to see that the process $\{N(t), I(t), t \geq 0\}$ is a two-dimensional continuous time Markov chain with state space $\mathbb{S} = \{(n, i)|n = 0, 1, 2, \ldots; i = 0, 1, 2\}$ and non-zero transition rates are given by:

$$q_{(1,2)(0,0)} = \mu; \tag{1}$$

$$q_{(N-1,0)(N,1)} = \lambda; \tag{2}$$

$$q_{(n,i)(n+1,i)} = \lambda, \quad n = 1, 2, \ldots \ldots, \quad i = 0, 1, 2; \tag{3}$$

$$q_{(n+1,2)(n,2)} = \mu, \quad n = 1, 2, \ldots; \tag{4}$$

$$q_{(n,1)(n,2)} = \theta, \quad n = N, N+1 \ldots \tag{5}$$

In the next sections we will investigate the stationary probability distribution in the queueing system. In this paper, as mentioned above, we focus on two different levels of information that are available to customers before their decisions are made. We follow the notations in Burnetas and Economou (2007), i.e.

(1) Almost observable case: Customers are informed only about the queue length $N(t)$;
(2) Almost unobservable case: Customers are informed only about the server state $I(t)$.

For convenience, throughout this paper we denote S_{ao} by the social benefit per time unit in almost observable case, and S_{au} by the social benefit per time unit in almost unobservable case.

3   Almost Observable Case

In this section we proceed to the almost observable case where the arriving customers observe the number of the present customers in the system upon their arrival, but not the state of the server. The transition diagram is depicted in Fig. 1.

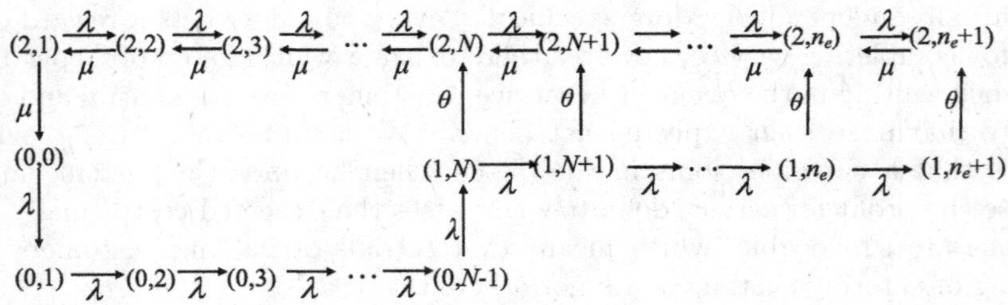

Fig. 1. Transition rate diagram for the n_e equilibrium threshold strategy in the almost observable model

To this end, it is necessary to obtain the stationary distribution of the system when the customers follow a given pure threshold strategy. In general, the strategy of never joining is always an equilibrium when $N > 1$, since if all customers adopt this strategy, the server is never active. We concentrate on the existence of other equilibrium strategies in which the server could be reactivated.

A threshold strategy with a threshold n_e is a strategy where customers join if and only if they find at most n_e customer in the system upon arrival. Thus the maximum number of customers in the system at any time is $n_e + 1$. We analyze the mean queue length and social welfare under this maximum number of customers. Firstly, we need to consider the condition for a server to be active in queues with setup time and N-policy which is different from the work of Guo and Hassin (2011). As mentioned, the server can only be active when the number of customers in the system reaches N. We know that, the longest expected waiting time for an arriving customer who arrives at state 0 is when there are n $(0 < n < N)$ customers in the queue, thus the longest expected waiting time W can be reached when there are 0 or $N - 1$ customers in the queue, thus W is shown as follows:

$$W = \begin{cases} \frac{N-1}{\lambda} + \frac{1}{\mu} + \frac{1}{\theta} & n = 0, \\ \frac{1}{\theta} + \frac{N}{\mu} & n = N - 1. \end{cases}$$

By assumptions, an incoming customer always joins if his net benefit U is non-negative. The sufficient condition for a server to be active is given as follows.

$$R - CW = max\left\{R - C(\frac{N-1}{\lambda} + \frac{1}{\mu} + \frac{1}{\theta}), R - C(\frac{1}{\theta} + \frac{N}{\mu})\right\} > 0.$$

Now we assume that the stability condition is satisfied. Obviously, a customer who arrives at state $(n, 0)$ has a higher expected waiting time than one who arrives at state $(n, 2)$. Thus, all arriving customers join the queue if there are no more than $N > 1$ customers in the system. The equilibrium strategy is therefore characterized by a threshold value $n_e > N$.

Lemma 1. *In the almost observable M/M/1 queue with N policy and setup time where the customers enter the system according to a threshold strategy: While arriving at time t, observe N(t); enter if N(t) $\leq n_e$ and balk otherwise. The stationary distribution $(p_{ao}(n, i)$:$(n, i) \in \{(0, 0)\} \cup \{1, \ldots, N - 1\} \times \{0, 2\} \cup \{N, \ldots, n_e + 1\} \times \{1, 2\})$ is given as follows:*

$$p(n, 0) = p(0, 0), \qquad n = 1, \ldots, N - 1, \tag{6}$$

$$p(N, 1) = \sigma p(0, 0), \tag{7}$$

$$p(n, 1) = \sigma^{n-N+1} p(0, 0), \qquad n = N + 1, \ldots, n_e, \tag{8}$$

$$p(n_e + 1, 1) = (1 - \sigma)\sigma^{n_e - N} p(0, 0), \tag{9}$$

$$p(1, 2) = \rho p(0, 0), \tag{10}$$

$$p(2, 2) = \frac{\rho}{1 - \rho}(1 - \rho^2)p(0, 0), \tag{11}$$

$$p(n, 2) = \frac{\rho}{1 - \rho}(1 - \rho^n)p(0, 0), \qquad n = 1, \ldots, N - 1, \tag{12}$$

$$p(n, 2) = \left(\frac{(1 - \sigma)\rho^{n-N+2} + (1 - \rho)\rho\sigma^{n-N+1}}{(1 - \sigma)(\sigma - \rho)} - \frac{\rho^{n+1}}{1 - \rho}\right)p(0, 0), n = N, \ldots, n_e, \tag{13}$$

where

$$p(0, 0) = ((N - 1) + \frac{\sigma^2 - \sigma^{n_e - N + 4}}{(1 - \sigma)(\sigma - \rho)} + \frac{\rho^{n_e + 3} - N\rho^2 + (N - 1)\rho}{(1 - \rho)^2}$$

$$+ \frac{(\sigma - 1)(\rho^2 - \rho^{n_e - N + 4})}{(1 - \rho)^2(\sigma - \rho)})^{-1}, \tag{14}$$

and $\rho = \frac{\lambda}{\mu}$, $\sigma = \frac{\lambda}{\lambda + \theta}$.

Proof. The corresponding stationary distribution is obtained as the unique positive normalized solution of the following system of balance equations:

$$\lambda p(0, 0) = \mu p(1, 2), \tag{15}$$

$$\lambda p(n, 0) = \lambda p(n - 1, 0) \qquad n = 1, 2, \ldots, N - 1, \tag{16}$$

$$(\lambda + \theta)p(N, 1) = \lambda p(N - 1, 0), \tag{17}$$

$$(\lambda + \theta)p(n, 1) = \lambda p(n - 1, 1), \qquad n = N + 1, \ldots, n_e, \tag{18}$$

$$\theta p(1, n_e + 1) = \lambda p(n_e, 1), \tag{19}$$

$$(\lambda + \mu)p(1, 2) = \mu p(2, 2), \tag{20}$$

$$(\lambda + \mu)p(n, 2) = \mu p(n + 1, 2) + \lambda p(n - 1, 2), \qquad n = 2, 3 \ldots, N - 1, \tag{21}$$

$$(\lambda + \mu)p(n, 2) = \mu p(n + 1, 2) + \lambda p(n - 1, 2) \qquad n = N + 1, \ldots, n_e. \tag{22}$$

By iterating (16), we can get

$$p(n, 0) = p(0, 0).$$

By (17) and iterating (18) we can get

$$p(1,n) = \sigma^{n-N+1}p(0,0) \qquad n = N+1, ..., n_e.$$

By (19) we can get $p(n_e + 1, 1)$ as follows

$$p(1, n_e + 1) = (1 - \sigma)\sigma^{n_e - N}p(0,0).$$

On the other hand, by (21), we can get

$$\mu p(n+1,2) - (\lambda + \mu)p(n,2) + \lambda p(n-1,2) = 0, \qquad n = 2, 3..., N-1. \quad (23)$$

In the following, we use a rather standard method to solve this type of equation by solving a linear difference equation with constant coefficients as

$$\mu x^2 - (\lambda + \mu)x + \lambda = 0. \tag{24}$$

It is readily seen that the above equation has two roots 1 and ρ and the common root of the homogeneous transformation Eq. (21) is

$$\begin{cases} x_n^{\text{hom}} = A1^n + B\rho^n, & \rho \neq 1; \\ x_n^{\text{hom}} = A1^n + Bn1^n, & \rho = 1. \end{cases} \tag{25}$$

Since we assume $\rho \neq 1$, thus the solution of Eq. (21) is

$$x_n = A + B\rho^n \qquad n = 1, 2, 3, ..., N-1. \tag{26}$$

Now, we need to know the values of A and B for the purpose of getting the expression of $x_n(n = 1, 2, ..., N-1)$.

Letting $n = 1$ and $n = 2$, we can get

$$\begin{cases} A + B\rho = p(1,2), \\ A + B\rho^2 = p(2,2). \end{cases} \tag{27}$$

We get $p(1,1)$ and $p(2,1)$ from (15) and (20):

$$\begin{cases} p(1,2) = \rho p(0,0), \\ p(2,2) = \frac{\rho}{1-\rho}(1 - \rho^2)p(0,0). \end{cases} \tag{28}$$

Solving the Eq. (29), we can get A and B as follows

$$\begin{cases} A = \frac{1-\rho}{\rho}p(0,0); \\ B = (-1)\frac{1-\rho}{\rho}p(0,0). \end{cases} \tag{29}$$

Next, we consider $p(n,2)(n = N, N+1, ..., n_e+1)$. Similarly the general solution of Eq. (24) is $x_n^{gen} = x_n^{\text{hom}} + x_n^{spec}$, where x_n^{spec} is a special root of the Eq. (22).

We want to find a special root of Eq. (24) to replace x_n^{spec}, and find the special root is like $D\sigma^n$ (when $\sigma \neq 1$ and $\sigma \neq \rho$), or like $Dn\sigma^n$ (when $\sigma = 1$ or $\sigma = \rho$), or like $Dn^2\sigma^n$ (when $\sigma = 1 = \rho$). According to the discussion on the root solution given by Burnetas and Economou (2007), we need only consider the common situation. That is, find the special root is like $D\sigma^n$ for the regular case $\sigma \neq 1$ and $\sigma \neq \rho$. Therefore, by simple computation, the solution of the Eq. (22) is given by:

$$x_n^{gen} = A1^n + B\rho^n + D\sigma^n, \quad n = 1, 2, 3, \ldots, n_e - 1. \tag{30}$$

Letting $x_n = D\sigma^n$ and take (24) into account, we can get the value of D as follows.

$$D = \frac{-\theta\sigma^{2-N}}{(\mu\sigma^2 - (\lambda + \mu)\sigma + \lambda)}p(0,0) = \frac{\rho\sigma^{1-N}}{(\sigma - \rho)}p(0,0).$$

Now, we need to know the values of A and B for the purpose of getting the expression of x_n^{gen}. Letting $n = N$ and $n = N + 1$, using (33), we can get

$$\begin{cases} A + B\rho^N + D\sigma^N = p(N, 2); \\ A + B\rho^{N+1} + D\sigma^{N+1} = p(N+1, 2). \end{cases} \tag{31}$$

So we can get the expression of $p(N, 2)$, $p(N + 1, 2)$ by taking (8) and (12) into (22):

$$\begin{cases} p(N, 2) = \frac{\rho}{1-\rho}(1 - \rho^N)p(0,0); \\ p(N+1, 2) = (\frac{\rho}{1-\rho}(1 - \rho^{N+1}) - \rho(1 - \sigma))p(0,0). \end{cases} \tag{32}$$

Solving the Eq. (32), we can get A and B:

$$\begin{cases} A = 0; \\ B = \frac{\sigma - 1 - \rho^{N-1}(\sigma-\rho)}{(1-\rho)(\sigma-\rho)}\rho^{2-N}p(0,0). \end{cases} \tag{33}$$

With the help of known values of A, B, D, we can obtain (13). Consequently, we can get the expression of $p(n_e + 1, 2)$ by taking (13) into (22):

$$p(n_e + 1, 2) = \rho p(n_e, 2) + \frac{1 - \sigma}{\sigma}p(n_e + 1, 1). \tag{34}$$

Based on the above results, we can conclude that all probabilities involved can be expressed via $p(0,0)$. Finally, we can get the expression of $p(0,0)$ by normalization equation:

$$\sum_{n=0}^{N-1} p(n, 0) + \sum_{n=N}^{n_e+1} p(n, 1) + \sum_{n=1}^{n_e+1} p(n, 2) = 1, \tag{35}$$

which reaches the result (14). This completes the proof of this lemma. $\qquad\square$

Next, we will proceed to study the profit net of the almost observable case. In this case, the arriving customers can only observe the number of customers.

$T(n, i)(i = 0, 1, 2)$ represents the sojourn time of an arriving customer when he finds n customers in front of him and the state of the sever $I = i(i = 0, 1, 2)$:

$$T(n, 2) = \frac{n+1}{\mu},$$

$$T(n, 1) = \frac{1}{\theta} + \frac{n+1}{\mu},$$

$$T(n, 0) = \frac{1}{\theta} + \frac{n+1}{\mu} + \frac{N - (n+1)}{\lambda}. \tag{36}$$

So for an arriving customer if he finds $n(n > N - 1)$ customers in front of him and decides to enter the system, the profit for this customer is given by

$$R - C(T(n, 1) \Pr(I(t) = 1|N(t) = n) + T(n, 2) \Pr(I(t) = 2|N(t) = n)), \tag{37}$$

where $\Pr(I(t) = 1|N(t) = n)$ is the conditional probability that the server is on setup when the system have n customers waiting, and $\Pr(I(t) = 2|N(t) = n)$ is the conditional probability that the server is working when the system have n customers waiting.

To find the equilibrium strategies of threshold type, we should compute $\Pr(I^- = i|N^- = n) (i = 1, 2)$ as follow.

$$\Pr(I(t) = i \,|N(t) = n) = \frac{\lambda p_{ao}(n, i)}{\lambda p_{au}(n, 1) + \lambda p_{au}(n, 0) I\{n \geq N\}},$$

$$n = N, N + 1 \ldots, n_e + 1, \tag{38}$$

where

$$I\{n \geq N\} = \begin{cases} 0, & n < N; \\ 1, & n \geq N. \end{cases}$$

Taking Eqs. (8), (13) and (36) into Eq. (38), we can get the profit of the customer as follows.

$$U = R - C\left\{(\frac{1}{\theta} + \frac{n+1}{\mu}) \Pr(I^- = 1|N^- = n) + \frac{n+1}{\mu} \Pr(I^- = 2|N^- = n)\right\}. \tag{39}$$

Theorem 1. *In the almost observable M/M/1 queue with N policy and setup time where the customers enter the system according to a threshold strategy 'While arriving at time t, observe $N(t)$; enter if $N(t) \leq n_e$ and balk otherwise', we conclude that there exists unique equilibrium strategy of threshold n_e^* if $\mu > \lambda + \theta$.*

Proof. Take the expression of U into consideration and we can get:

$$U = R - C\left(\frac{n+1}{\mu} + (\frac{\sigma^{n-N+1}(1 - \rho)(\sigma - \rho)}{(1 - \rho)\sigma^{n-N+1} + (\sigma - 1)\rho^{n-N+2} - (\sigma - \rho)\rho^{n+1}})\frac{1}{\theta}\right). \tag{40}$$

More specifically, taking transformation of the formula

$$\frac{n+1}{\mu} + \left(\frac{\sigma^{n-N+1}(1-\rho)(\sigma-\rho)}{(1-\rho)\sigma^{n-N+1} + (\sigma-1)\rho^{n-N+2} - (\sigma-\rho)\rho^{n+1}} \right) \frac{1}{\theta}$$

$$= \frac{n+1}{\mu} + \frac{1}{1 + \frac{\rho}{\sigma-\rho} + (\frac{\sigma-1}{(1-\rho)(\sigma-\rho)(\frac{\rho}{\sigma})^{N-1}} - \frac{\rho}{1-\rho}\sigma^{N-1})(\frac{\rho}{\sigma})^n},$$

which is strictly increasing when $\rho < \sigma$, equally $\mu > \lambda + \theta$, therefore $U(n)$ is strictly decreasing. So, there exists a unique threshold, denoted by $n_e^* = max\{n|U(n) \geq 0\}$. $\qquad\square$

Lemma 2. *In the observable M/M/1 queue with N policy and setup time where the arriving customers know the number of customers in the system, the social welfare per time unit SW_{ao} is given below:*

$$SW_{ao} = R\lambda(1 - p_{ao}(n_e, 1) - p_{ao}(n_e, 2)) - CL_{ao},$$

Proof. The mean sojourn time of customer is $E[W_{ao}]$ and the mean queue length is L_{ao}.

$$E[W_{ao}] = L_{ao}\lambda(1 - p_{ao}(n_e, 1) - p_{ao}(n_e, 2)),$$

where $p_{ao}(n_e, i)(i = 1, 2)$ is the steady state probability that the queue is at its maximum size, and $\lambda(1 - p_{ao}(n_e, 1) - p_{ao}(n_e, 2))$ is the efficient arrival rate of customer.

The mean queue length is shown below:

$$L_{ao} = \sum_{n=0}^{N-1} n(p(n, 0) + p(n, 2)) + \sum_{n=N}^{n_e+1} n(p(n, 1) + p(n, 2))$$

$$= \frac{1}{1-\rho} \left[\frac{N(N-1)}{2} + \frac{\rho^2 - N\rho^{N+1} + (N-1)\rho^{N+1}}{(1-\rho)^2} \right] p(0, 0)$$

$$+ \frac{N\rho^2 + (1-N)\rho^3 - (n_e + 1)\rho^{n_e-N+1} + n_e\rho^{n_e-N+4}}{(\sigma-\rho)(1-\rho)^2} p(0, 0)$$

$$+ \frac{\sigma(1-\rho) + \rho(\sigma-\rho)}{(1-\sigma)^3(\sigma-\rho)}(N\sigma + (1-N)\sigma^2 - (n_e + 1)\sigma^{n_e-N} + n_e\sigma^{n_e-N+3})p(0, 0)$$

$$- \frac{N\rho^{N+1} + (1-N)\rho^{N+2} - (n_e + 1)\rho^{n_e+2} + n_e\rho^{n_e+3}}{(1-\rho)^3} p(0, 0)$$

$$+ (n_e + 1)(1-\sigma)\sigma^{n_e-N}p(0, 0) + \frac{n_e + 1}{\mu} \left[\frac{\rho^{n_e-N+1}(1-\sigma)(\rho(\lambda+\mu) - \lambda)}{(1-\sigma)(\sigma-\rho)} \right.$$

$$\left. - \frac{\rho_e^n(\rho(\lambda+\mu) - \lambda)}{1-\rho} \right] p(0, 0). \qquad\square$$

4   Almost Unobservable Case

We now turn our interest to the unobservable cases where the customers have no information on the queue length when they make their decision to join or balk. Two cases, almost unobservable case and fully unobservable case, will be studied separately regarding whether the customers can observe the state of the server or not at their arrival instants. We will prove that there exist equilibrium mixed strategies.

We begin with the almost unobservable case in which the customers are informed about the state of the server before their decision is made to join upon arrival. Now the optimal decision of a customer has to take into account the strategies of the other customers.

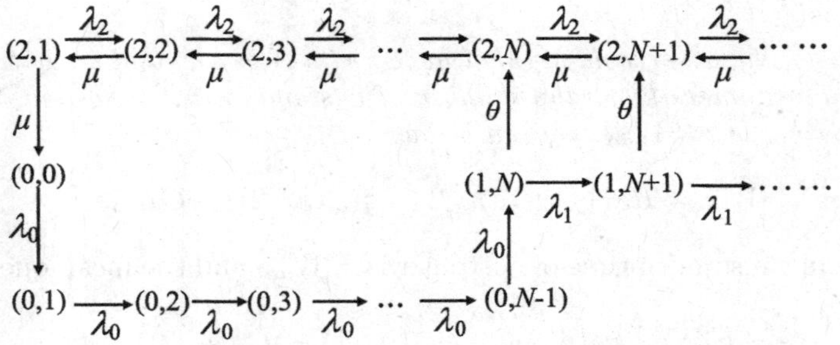

Fig. 2. Transition rate diagram for the $(q(0), q(1), q(2))$ mixed strategy in the almost unobservable model

Since all customers are assumed indistinguishable, we can consider the situation as a symmetric game among them. In the present model, there are only two pure strategies, to join or to balk. And a mixed strategy is specified by the joining probability of an arriving customer that finds the server is on working vacation or not. Our goal is to identify the Nash equilibrium mixed balking strategies. Suppose that all customers follow a mixed strategy $(q(0), q(1), q(2))$, where $q(i)$ is the probability of joining when the server is in state i. Then, the system behaves as the original, but with arrival rate $\lambda_i = \lambda q(i)$ for states where the server is in state i instead of λ. The transition diagram is shown in Fig. 2.

Lemma 1. *In the almost unobservable queue with N policy and setup time in which all customers adopt a mixed balking strategy $(q(0), q(1), q(2))$, where $q(i)$ is the probability of joining when the server is in state i, the stationary distribution is given as follows*

$$p(n, 0) = p(0, 0), \qquad n = 1, ..., N - 1, \qquad (41)$$

$$p(N, 1) = \frac{\lambda_0}{\lambda_1 + \theta} p(0, 0), \qquad (42)$$

$$p(n, 1) = \left(\frac{\lambda_0}{\lambda_1 + \theta} \right) \sigma^{n-N} p(0, 0), \quad n = N + 1, ..., \qquad (43)$$

$$p(1, 2) = \frac{\lambda_0}{\mu} p(0, 0), \qquad (44)$$

$$p(n,2) = \frac{\lambda_0}{\mu}\frac{1-\rho^n}{(1-\rho)}p(0,0), \qquad n = 2,3,...,N-1, \tag{45}$$

$$p(n,2) = \frac{\lambda_0}{\mu}\frac{\rho^n(\sigma-\rho)+(1-\sigma)\rho^{n-N+1}-(1-\rho)\sigma^{n-N+1}}{(\rho-1)(\sigma-\rho)}, n = N, N+1,..., \tag{46}$$

where

$$p(0,0) = \left(N + \frac{\lambda_0}{\lambda_1+\theta}\frac{1}{1-\sigma} + \frac{\lambda_0}{\mu}\frac{N(1-\sigma)+\sigma}{(1-\rho)(1-\sigma)}\right)^{-1}, \tag{47}$$

and $\rho = \frac{\lambda_2}{\mu}$, $\sigma = \frac{\lambda_1}{\lambda_1+\theta}$.

Proof. The corresponding stationary distribution is obtained as solution of the following system of balance equations:

$$\lambda_0 p(0,n) = \mu p(n-1,0), \qquad n = 1,2,...N-1, \tag{48}$$
$$(\lambda_1+\theta)p(1,N) = \lambda_1 p(N-1,1), \qquad n = N+1..., \tag{49}$$
$$(\lambda_1+\theta)p(1,n) = \lambda_1 p(n-1,1), \qquad n = N+1..., \tag{50}$$
$$\mu p(2,1) = \lambda_0 p(0,0), \tag{51}$$
$$(\lambda_2+\mu)p(2,1) = \mu p(2,2), \tag{52}$$
$$(\lambda_2+\mu)p(2,n) = \mu p(n+1,2) + \lambda_2 p(n-1,2), n = 2,...,N-1, \tag{53}$$
$$(\lambda_2+\mu)p(2,n) = \mu p(2,n+1) + \lambda_2 p(n-1,2) + \theta p(n,1), n = N, N+1,.... \tag{54}$$

By iterating (49) we can obtain

$$p(n,1) = (\frac{\lambda_0}{\lambda_1+\theta})\sigma^{n-N}p(0,0), \quad n = N+1,....$$

From (53) and (54) it follows that $p(n,2)$ is a solution of the nonhomogeneous linear difference equation with constant coefficients, i.e.,

$$\mu x_{n+1} - (\lambda_2+\mu)x_n + \lambda_2 x_{n-1} = -\theta(\frac{\lambda_0}{\lambda_1+\theta})\sigma^{n-N}p(0,0).$$

By using the same approach as used in the proof of Lemma 1, we can get (44). Solving equations with respect to $p(n,0)$ and substituting in

$$\sum_{n=0}^{N-1} p(0,n) + \sum_{n=N}^{\infty} p(1,n) + \sum_{n=1}^{\infty} p(2,n) = 1, \tag{55}$$

and we can obtain $p(0,0)$. □

Next, we consider an arriving customer who finds the server is at state $i(i = 0,1,2)$ and we will give the expected sojourn time of a customer that decides to enter given that the others follow the same mixed strategy $(q(0), q(1), q(2))$.

Case 1. When the server is at state 0, the expected sojourn time is

$$T_{au}(0) = \frac{E[N|0]+1}{\mu} + \frac{1}{\theta} + \frac{N-(E[N|0]+1)}{\lambda_0}. \tag{56}$$

Case 2. When the server is at state 1, the expected sojourn time is

$$T_{au}(1) = \frac{E[N|1]+1}{\mu} + \frac{1}{\theta}. \tag{57}$$

Case 3. When the server is at state 2, the expected sojourn time is

$$T_{au}(2) = \frac{E[N|2]+1}{\mu}. \tag{58}$$

To get the $T_{au}(i), (i = 0, 1, 2)$, we first give the probability that the server in idle, setup, busy steady state as follows

$$P(i = 0) = \sum_{n=0}^{N-1} p(n, 0) = Np(0, 0),$$

$$P(i = 1) = \sum_{n=N}^{\infty} p(n, 1) = (\frac{\lambda_0}{\lambda_1 + \theta})(\frac{1}{1 - \sigma})p(0, 0), \tag{59}$$

$$P(i = 2) = \sum_{n=1}^{\infty} p(n, 2) = \sum_{n=1}^{N-1} p(n, 2) + \sum_{n=N}^{\infty} p(n, 2),$$

$$= \frac{\lambda_0}{\mu} \frac{N(1 - \sigma) + \sigma}{N(1 - \rho)(1 - \sigma) + \rho^2(1 - \sigma) + \sigma(1 - \rho)}.$$

Using Eq. (59) we can get:

$$P(n|0) = \frac{p(n, 0)}{P(i = 0)} = \frac{1}{N}, \quad n = 1, 2, ..., N - 1;$$

$$P(n|1) = \frac{p(n, 1)}{P(i = 1)} = (1 - \sigma)\sigma^{n-N}, \quad n = N, N + 1, ...;$$

$$P(n|2) = \frac{p(n, 2)}{P(i = 2)} \tag{60}$$

$$= \begin{cases} \frac{(1-\rho)(1-\rho^n)(1-\sigma)}{N(1-\rho)(1-\sigma)+\rho^2(1-\sigma)+\sigma(1-\rho)}, & n = 1, 2..., N - 1; \\ \\ \frac{(1-\sigma)(\rho-1)(\rho^n(\sigma-\rho)+(1-\sigma)^{n-N+1})+(\rho-1)\sigma n-N+1}{(\sigma-\rho)(N(1-\rho)(1-\sigma)+\rho^2(1-\sigma)+\sigma(1-\rho))}, & n = N, N + 1, ... \end{cases}$$

With the help of Eqs. (59) and (60), we can compute $E[N|i]$ and get that

$$E[N|0] = \sum_{n=0}^{N-1} nP(n|0) = \frac{N-1}{2}, \tag{61}$$

$$E[N|1] = \sum_{n=N}^{\infty} nP(n|1) = N + \frac{\sigma}{1-\sigma}, \tag{62}$$

$$E[N|2] = \sum_{n=1}^{N-1} nP(n|2) + \sum_{n=N}^{\infty} nP(n|2)$$

$$= \frac{1}{N+(1-N)\sigma}\left(\frac{N(N-1)(1-\rho)(1-\sigma)}{2}\right.$$

$$+ \frac{(1-\sigma)((1-N)\rho^{N+2} + (2N-1)\rho^{N+1} - (N+1)\rho^N + (2-\rho)(1-\rho)\rho^2)}{(1-\rho)^2}$$

$$\left. - \frac{(1-\sigma)^2(N\rho+(1-N)\rho^2)}{(1-\rho)^2(\sigma-\rho)} + \frac{(1-\rho)(N\rho+(1-N)\sigma^2)}{\sigma-\rho}\right). \tag{63}$$

Taking Eqs. (61), (62), and (63) into Eqs. (56), (57) and (58), we can derive the $T_{au}(0)$ and $T_{au}(1)$ and $T_{au}(2)$ as follows:

$$T_{au}(0) = \frac{1}{\theta} + \frac{N-1}{2\lambda_0} + \frac{N+1}{2\mu},$$

$$T_{au}(1) = \frac{1}{\theta} + \frac{1}{\mu}(N + \frac{\sigma}{1-\sigma}) = \frac{1}{\theta} + \frac{N+1}{\mu} + \frac{\lambda_1}{\mu\theta},$$

$$T_{au}(2) = \frac{E[N|2]+1}{\mu}.$$

Based on the reward-cost structure, the expected benefit for an arriving customer who is informed the server is at state i is given as follows.

$$S_{au}(0) = R - C(\frac{1}{\theta} + \frac{N-1}{2\lambda_0} + \frac{N+1}{2\mu}),$$

$$S_{au}(1) = R - C(\frac{1}{\theta} + \frac{N+1}{\mu} + \frac{\lambda_1}{\mu\theta}),$$

$$S_{au}(2) = R - C(\frac{E[N|2]+1}{\mu}).$$

According to the above assumptions, an incoming customer always joins when the state of sever I is 0 as long as his net benefit U is non-negative. The sufficient condition for the system stability is

$$\frac{1}{\theta} + \frac{N-1}{2\lambda_0} + \frac{N+1}{2\mu} < \frac{R}{C}, \tag{64}$$

which means $S_{au}(0) > 0$. We now assume that the stability condition is satisfied. Obviously, a customer who arrives at state $(n,0)$ suffers a higher expected waiting time than who arrives at state $(N,1)$. Thus, all arriving customers join the queue if there are less than N customers in the system and we can know $q_e(0) = 1$.

Next we consider $q_e(1)$. To this end, we tag an arriving customer when the server is at state 1. His expected benefit is given as follows:

$$S_{au}(1) = R - C(\frac{1}{\theta} + \frac{N+1}{\mu} + \frac{\lambda_1}{\mu\theta}) = 0.$$

By solving the above equation, we can get $q_e(1) = \frac{\mu\theta}{\lambda}(\frac{R}{C} - \frac{1}{\theta} - \frac{N+1}{\mu})$. We have the following lemma.

Lemma 2. *In the almost observable M/M/1 queue with N policy and setup time where the arriving customers know the number of customers in the system, the social welfare per time unit SW_{au} is given below:*

$$SW_{ao} = R\lambda(1 - p_{ao}(n_e, 1) - p_{ao}(n_e, 2)) - CL_{ao},$$

Proof. The mean sojourn time of customer is $E[W_{au}]$ and the mean queue length is L_{au}.

$$E[W_{au}] = L_{au}\lambda.$$

The mean queue length is shown below:

$$L_{au} = \sum_{n=0}^{N-1} n(p(n,0) + p(n,2)) + \sum_{n=N}^{\infty} n(p(n,1) + p(n,2))$$

$$= \frac{N(N-1)}{2}\frac{\mu - \lambda_2 + \lambda_0}{\mu - \lambda_2}p(0,0) + \frac{\lambda_0}{\mu}\frac{(N-1)\rho^{N+1} - N\rho^N + \rho}{(1-\rho)^3}p(0,0)$$

$$+ \frac{\lambda_0(N + (1-N)\sigma)}{(\lambda_1 + \theta)(1-\sigma)^2}p(0,0) + \frac{\lambda_0}{\mu}\left(\frac{(N-1)\rho^{N+1} - N\rho^N}{(1-\rho)^3}\right.$$

$$+ \frac{(1-\sigma)(N-1)\rho^2 - N\rho}{(\sigma - \rho)(1-\rho)^3} + \left.\frac{N\sigma + (1-N)\sigma^2}{(\sigma - \rho)(1-\sigma)^2}\right)p(0,0).$$

This completes the proof.   □

5   Conclusions

In this paper we analyzed the strategic behavior of the customers and social optimization in a single server queueing system with N-policy and setup time. Arriving customers decide whether to join or to balk the system. Specifically, two different cases with respect to the levels of information provided to arriving customers have been investigated extensively. The customers' strategies have been analyzed and the expressions of the social welfare function of customers for two cases were derived. For future research, analyzing a model in which the setup time is generally distributed is worthy of further investigation.

Acknowledgments. The authors would like to thank the anonymous reviewers for their valuable comments and constructive suggestions that help to improve the presentation of this paper.

References

Burnetas, A., Economou, A.: Equilibrium customer strategies in a single server Markovian queue with setup times. Queueing Syst. **56**, 213–228 (2007)

Chen, P., Zhou, W.H., Zhou, Y.: Equilibrium customer strategies in the queue with threshold policy and setup times. Math. Probl. Eng. Article ID 361901, 11 (2015)

Economou, A., Kanta, S.: On balking strategies and pricing for the single server Markovian queue with compartmented waiting space. Queueing Syst. **59**, 237–269 (2008a)

Economou, A., Kanta, S.: Equilibrium balking strategies in the observable single-server queue with breakdowns and repairs. Oper. Res. Lett. **36**, 696–699 (2008b)

Economou, A., Gómez-Corral, A., Kanta, S.: Optimal balking strategies in single-server queues with general service and vacation times. Perform. Eval. **68**, 967–982 (2011)

Edelson, N.M., Hildebrand, K.: Congestion tolls for poisson queueing processes. Econometrica **43**, 81–92 (1975)

Guo, P., Hassin, R.: Strategic behavior and social optimization in Markovian vacation queues. Oper. Res. **59**, 986–997 (2011)

Hassin, R.: Information and uncertainty in a queueing system. Probab. Eng. Inf. Sci. **21**, 361–380 (2007)

Hassin, R., Haviv, M.: To Queue or Not to Queue: Equilibrium Behavior in Queueing Systems. Kluwer Academic Publishers, Boston (2003)

Naor, P.: The regulation of queue size by levying tolls. Econometrica **37**, 15–24 (1969)

Sun, W., Guo, P., Tian, N.: Equilibrium threshold strategies in observable queueing systems with setup/closedown times. Central Eur. J. Oper. Res. **18**, 241–268 (2010)

Wang, J., Zhang, X., Huang, P.: Stategic behavior and social optimization in a constant retrial queue with the N-policy. Eur. J. Oper. Res. **256**, 841–849 (2017)

Wang, J., Zhang, F.: Equilibrium analysis of the observable queues with balking and delayed repairs. Appl. Math. Comput. **218**, 716–2729 (2011)

Yadin, M., Naor, P.: Queueing systems with a removable service station. J. Oper. Res. Soc. **14**, 393–405 (1963)

Zhang, X., Wang, J., Do, T.V.: Threshold properties of the M/M/1 queue under T-policy with applications. Appl. Math. Comput. **261**, 284–301 (2015)

Waiting and Service Time of a Unique Customer in an M/M/m Preemptive LCFS Queue with Impatient Customers

Hideaki Takagi[✉]

University of Tsukuba, Tsukuba Science City 305-8573, Japan
takagi@sk.tsukuba.ac.jp

Abstract. We study an M/M/m preemptive last-come, first-served queue with impatient customers without priority classes. We focus on the probability of service completion and abandonment as well as the waiting and service times of a *unique customer* who has the mean service and patience times that are different from those of all other customers in the steady state. The problem is formulated as the first passage times in a combination of two one-dimensional birth-and-death processes each with two absorbing states. We provide explicit expressions in terms of Laplace-Stieltjes transform of the distribution function for the time to service completion or abandonment, which is decomposed into the waiting and service times of the unique customer. A numerical example is presented in order to demonstrate the computation of theoretical formulas.

Keywords: M/M/m preemptive LCFS queue · Impatient customers · First passage time

1 Introduction

We are concerned with an M/M/m queueing system with impatient customers without exogenous priority classes. Customers arrive according to a Poisson process at rate λ. The service time of each customer is exponentially distributed with mean $1/\mu$. There are m servers and a waiting room of infinite capacity. At any time, each customer present in the system is either being served or staying in the waiting room. Each customer in the waiting room leaves the system (abandons the waiting process) with probability $\theta \Delta t$ within a short time interval $(t, t + \Delta t)$. That is to say, the patience time for each customer is exponentially distributed with mean $1/\theta$. Customers never leave the system while being served before the service is completed.

It is assumed that the service to each customer is started immediately upon arrival. If all servers are busy, the arriving customer preempts the ongoing service to the customer who arrived first among those who are being served. The

---

This work is supported by the Grant-in-Aid for Scientific Research (C) No. 26330354 from the Japan Society for the Promotion of Science (JSPS) in the academic year 2016.

© Springer International Publishing AG 2017
W. Yue et al. (Eds.): QTNA 2017, LNCS 10591, pp. 18–35, 2017.
https://doi.org/10.1007/978-3-319-68520-5_2

customer whose service is preempted is placed at the head of the queue in the waiting room. When one of the servers becomes available, a customer at the head of the queue, if any, is called in for service to be resumed. This discipline is equivalent to the one called "preemptive *last-in, first-out* (LIFO)" for an M/G/1 queue by Wolff [4, p. 456].

In our previous work [2,3], we studied the time interval from arrival to either service completion or abandonment, whichever occurs first, of an arbitrary customer in steady state. The problem was formulated as a combination of two one-dimensional birth-and-death processes, each with two absorbing states, for the behavior of a tagged customer. We provided explicit expressions in terms of Laplace-Stieltjes transform (LST) of the distribution function (DF) for the first passage time to service completion or abandonment, which is decomposed into the waiting time and the received service time.

In the present paper, we turn our attention to the waiting and service time of a *unique customer* who has the mean service time $1/\mu_0$ and mean patience time $1/\theta_0$ that may be different from $1/\mu$ and $1/\theta$, respectively, of other customers. It is assumed that such a customer arrives during the steady state of an M/M/m queueing system with otherwise uniformly impatient customers. We are interested in the waiting and service time of the unique customer. The analysis technique is similar to the one in [3]. Through a numerical example, we compare the probability of service completion and abandonment as well as the mean waiting and service time of the unique customer to those of other customers. For a more patient customer, we find that (i) the probability of service completion is higher, (ii) the mean time spent in the system is longer whether he abandons waiting or he gets served, and (iii) the received service time is not much different from that of other customers.

2   First Passage Time to Service Preemption or Completion from State $k, 0 \leq k \leq m - 1$

We focus on a unique customer in *state k*, signifying that there are k other customers who compete with him for service at any given time in the steady state, where $k = 0, 1, 2, \ldots$. They are the customers who arrived after the unique one and have been staying in the system until that time. According to the preemptive LCFS discipline, an arriving unique customer always joins the system at state $k = 0$.

We first consider a birth-and-death process of state transitions for the unique customer in state k, $0 \leq k \leq m - 1$, in which he is being served. The service to this customer, with probability one, is eventually either preempted by another customer who arrives after him or completed without preemption.

2.1   Behavior of a Unique Customer Until Service Preemption or Completion

The state transition diagram for the discrete-time, one-dimensional birth-and-death process modeling the behavior of a unique customer in service is shown in

Fig. 1. This process has m transient states $\{0, 1, 2, \ldots, m-1\}$ and two absorbing states denoted by "Pr" (state m) and "Sr", representing service preemption and service completion, respectively. The state transition probabilities and the LST of the DF for the time spent by the unique customer in state k are given by

$$\alpha_k = \frac{k\mu}{\lambda + k\mu + \mu_0}; \quad \beta_k = \frac{\mu_0}{\lambda + k\mu + \mu_0}; \quad B_k^*(s) = \frac{\lambda + k\mu + \mu_0}{s + \lambda + k\mu + \mu_0}.$$

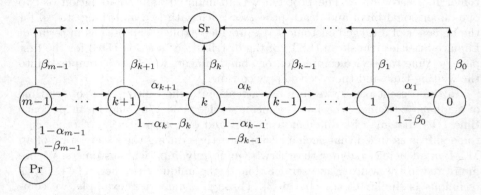

Fig. 1. State transitions for a unique customer until service preemption or completion.

2.2 LST of the DF for the Time to Service Preemption or Completion

By $H_k^*(s, \text{Pr})$, we denote the joint probability of service preemption and the LST of the DF for the first passage time from state k to state m ("Pr") without reaching state "Sr". Moreover, we denote by $H_k^*(s, \text{Sr})$ the joint probability of service completion and the LST of the DF for the first passage time from state k to state "Sr" without reaching state "Pr".

Applying the *first step analysis* for the discrete-time Markov chain, we have the following finite sets of equations for $\{H_k^*(s, \text{Pr}); 0 \le k \le m-1\}$ and $\{H_k^*(s, \text{Sr}); 0 \le k \le m-1\}$:

$$(s + \lambda + \mu_0)H_0^*(s, \text{Pr}) = \lambda H_1^*(s, \text{Pr}),$$
$$(s + \lambda + k\mu + \mu_0)H_k^*(s, \text{Pr}) = k\mu H_{k-1}^*(s, \text{Pr}) + \lambda H_{k+1}^*(s, \text{Pr})$$
$$1 \le k \le m-2,$$
$$[s + \lambda + (m-1)\mu + \mu_0]H_{m-1}^*(s, \text{Pr}) = (m-1)\mu H_{m-2}^*(s, \text{Pr}) + \lambda.$$

$$(s + \lambda + \mu_0)H_0^*(s, \text{Sr}) = \mu_0 + \lambda H_1^*(s, \text{Sr}),$$
$$(s + \lambda + k\mu + \mu_0)H_k^*(s, \text{Sr}) = k\mu H_{k-1}^*(s, \text{Sr}) + \mu_0 + \lambda H_{k+1}^*(s, \text{Sr})$$
$$1 \le k \le m-2,$$
$$[s + \lambda + (m-1)\mu + \mu_0]H_{m-1}^*(s, \text{Sr}) = (m-1)\mu H_{m-2}^*(s, \text{Sr}) + \mu_0.$$

In addition, we let $H_m^*(s, \mathrm{Pr}) \equiv 1$ and $H_m^*(s, \mathrm{Sr}) \equiv 0$. The solution can be obtained in terms of functions $\{h_k^*(s); 0 \le k \le m\}$ in the form

$$H_k^*(s, \mathrm{Pr}) = \frac{h_k^*(s)}{h_m^*(s)}; \quad H_k^*(s, \mathrm{Sr}) = \frac{\mu_0}{s + \mu_0}\left[1 - \frac{h_k^*(s)}{h_m^*(s)}\right] \qquad 0 \le k \le m.$$

2.3   Solution for $\{h_k^*(s); 0 \le k \le m\}$

A finite set of equations for $\{h_k^*(s); 0 \le k \le m\}$ is given by

$$h_0^*(s) = 1; \quad s + \lambda + \mu_0 = \lambda h_1^*(s),$$
$$(s + \lambda + k\mu + \mu_0)h_k^*(s) = k\mu h_{k-1}^*(s) + \lambda h_{k+1}^*(s) \qquad 1 \le k \le m - 1,$$

which can be written as the following set of recurrence relations:

$$h_k^*(s) = \frac{s + \lambda + (k-1)\mu + \mu_0}{\lambda} h_{k-1}^*(s) - \frac{(k-1)\mu}{\lambda} h_{k-2}^*(s) \qquad 2 \le k \le m.$$

The solution is given by *Cramer's formula* as the determinant of the $k \times k$ tridiagonal matrix

$$h_k^*(s) = (-1)^k$$

$$\times \begin{vmatrix} -\frac{s+\lambda+\mu_0}{\lambda} & 1 & 0 & 0 & 0 & \cdots & 0 \\ \frac{\mu}{\lambda} & -\frac{s+\lambda+\mu+\mu_0}{\lambda} & 1 & 0 & 0 & \cdots & 0 \\ 0 & \frac{2\mu}{\lambda} & -\frac{s+\lambda+2\mu+\mu_0}{\lambda} & 1 & 0 & \cdots & 0 \\ 0 & 0 & \frac{3\mu}{\lambda} & -\frac{s+\lambda+3\mu+\mu_0}{\lambda} & 1 & \cdots & 0 \\ \vdots & \vdots & \vdots & \vdots & \ddots & \ddots & \vdots \\ 0 & 0 & 0 & 0 & \cdots & -\frac{s+\lambda+(k-2)\mu+\mu_0}{\lambda} & 1 \\ 0 & 0 & 0 & 0 & \cdots & \frac{(k-1)\mu}{\lambda} & -\frac{s+\lambda+(k-1)\mu+\mu_0}{\lambda} \end{vmatrix}$$

for $1 \le k \le m$. Note that $h_k^*(s)$ is a kth-degree polynomial in s, the coefficient of s^k being $(1/\lambda)^k$. Thus, we obtain the probability of service preemption and completion

$$p_k\{\mathrm{Pr}\} := H_k^*(0, \mathrm{Pr}) = \frac{h_k^*(0)}{h_m^*(0)}; \quad p_k\{\mathrm{Sr}\} := H_k^*(0, \mathrm{Sr}) = 1 - \frac{h_k^*(0)}{h_m^*(0)}$$
$$0 \le k \le m.$$

In particular, we have $p_0\{\mathrm{Pr}\} = 1/h_m^*(0)$ and $p_m\{\mathrm{Pr}\} = 1$.

3   First Passage Time to Service Resumption or Abandonment from State k, $k \geq m$

We next consider another birth-and-death process of state transitions for a unique customer in state k, $k \geq m$, in which he is staying in the waiting room. With probability one, this customer, eventually, either is called in to resume his service or abandons waiting.

3.1   Behavior of a Unique Customer Until Service Resumption or Abandonment

The state transition diagram for the discrete-time, one-dimensional birth-and-death process modeling the behavior of a unique customer in the waiting room is shown in Fig. 2. The process has an infinite number of transient states $\{m, m + 1, \ldots\}$ and two absorbing states denoted by "Rs" (state $m - 1$) and "Ab", representing service resumption and abandonment, respectively. The state transition probabilities and the LST of the DF for the time spent by the unique customer in state k are given by

$$\alpha'_k = \frac{m\mu + (k - m)\theta}{\lambda + m\mu + (k - m)\theta + \theta_0}; \quad \beta'_k = \frac{\theta_0}{\lambda + m\mu + (k - m)\theta + \theta_0},$$

$$B'^*_k(s) = \frac{\lambda + m\mu + (k - m)\theta + \theta_0}{s + \lambda + m\mu + (k - m)\theta + \theta_0}.$$

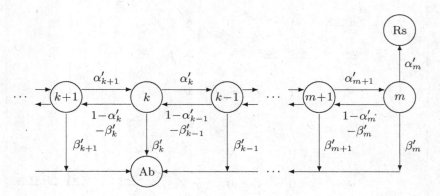

Fig. 2. State transitions for a unique customer until service resumption or abandonment.

3.2   LST of the DF for the Time to Service Resumption or Abandonment

By $W^*_k(s, \mathrm{Rs})$, we denote the joint probability of service resumption and the LST of the DF for the first passage time from state k to state $m - 1$ ("Rs") without reaching state "Ab". Moreover, we denote by $W_k^*(s, \text{Ab})$ the joint probability of abandonment and the LST of the DF for the first passage time from state k to state "Ab" without reaching state "Rs".

Infinite sets of equations for $\{W_k^*(s, \text{Rs}); k \geq m\}$ and $\{W_k^*(s, \text{Ab}); k \geq m\}$ are given by

$$(s + \lambda + m\mu + \theta_0)W_m^*(s, \text{Rs}) = m\mu + \lambda W_{m+1}^*(s, \text{Rs}),$$
$$[s + \lambda + m\mu + (k - m)\theta + \theta_0]W_k^*(s, \text{Rs})$$
$$= [m\mu + (k - m)\theta]W_{k-1}^*(s, \text{Rs}) + \lambda W_{k+1}^*(s, \text{Rs}) \qquad k \geq m + 1.$$
$$(s + \lambda + m\mu + \theta_0)W_m^*(s, \text{Ab}) = m\mu + \theta_0 + \lambda W_{m+1}^*(s, \text{Ab}),$$
$$[s + \lambda + m\mu + (k - m)\theta + \theta_0]W_k^*(s, \text{Ab})$$
$$= [m\mu + (k - m)\theta]W_{k-1}^*(s, \text{Ab}) + \theta_0 + \lambda W_{k+1}^*(s, \text{Ab}) \qquad k \geq m + 1.$$

The solution can be obtained in terms of functions $\{G_k^*(s); k \geq m\}$ in the form

$$W_k^*(s, \text{Rs}) = G_k^*(s + \theta_0); \quad W_k^*(s, \text{Ab}) = \frac{\theta_0}{s + \theta_0}[1 - G_k^*(s + \theta_0)] \qquad k \geq m.$$

Thus the probability of service preemption and abandonment is given by

$$p_k\{\text{Rs}\} := W_k^*(0, \text{Rs}) = G_k^*(\theta_0); \quad p_k\{\text{Ab}\} := W_k^*(0, \text{Ab}) = 1 - G_k^*(\theta_0)$$
$$k \geq m.$$

3.3 Busy Period

A *busy period* started with k ($\geq m$) customers in an M/M/m queue is the time interval, denoted by $\mathcal{G}_k$, from the instant at which there are k customers in the system (all servers are busy and $k - m$ customers are waiting) to the first instant at which any one of the servers becomes available. Let us denote by $f_{W_k}(t, \text{Rs})$ and $f_{W_k}(t, \text{Ab})$ the density functions of the time until service resumption and the time until abandonment, respectively, for a customer in state k, $k \geq m$. They are related with the density function $f_{\mathcal{G}_k}(t)$ for $\mathcal{G}_k$ and the probability $P\{\mathcal{G}_k > t\}$ as follows:

$$f_{W_k}(t, \text{Rs}) = e^{-\theta_0 t}f_{\mathcal{G}_k}(t); \quad f_{W_k}(t, \text{Ab}) = \theta_0 e^{-\theta_0 t}P\{\mathcal{G}_k > t\}.$$

The function $G_k^*(s)$ introduced in Sect. 3.2 is the LST of the DF for $\mathcal{G}_k$, $k \geq m$. The set of equations for $\{G_k^*(s), k \geq m\}$ is given by

$$(s + \lambda + m\mu)G_m^*(s) = \lambda G_{m+1}^*(s) + m\mu,$$
$$[s + \lambda + m\mu + (k - m)\theta]G_k^*(s) = [m\mu + (k - m)\theta]G_{k-1}^*(s) + \lambda G_{k+1}^*(s)$$
$$k \geq m + 1.$$

Iravani and Balcıoğlu [1] provides the LST of the DF for the duration of the busy period in an M/M/m queue with an exponentially distributed service time with mean $1/(m\mu)$ as follows:

$$G_k^*(s)$$

$$= \frac{\dfrac{m\mu}{s+m\mu} + \sum_{i=1}^{\infty}(-1)^i\psi_{i,k-m}(\lambda/\theta)\left[\prod_{j=0}^{i-1}\left(1-\dfrac{m\mu}{s+m\mu+j\theta}\right)\right]\dfrac{m\mu}{s+m\mu+i\theta}}{1+\sum_{i=1}^{\infty}\dfrac{(\lambda/\theta)^i}{i!}\prod_{j=0}^{i-1}\left(1-\dfrac{m\mu}{s+m\mu+j\theta}\right)}$$

$$k \geq m,$$

where we have defined

$$\psi_{i,k}(x) := \sum_{j=\max\{0,i-k\}}^{i}\frac{(-x)^j}{j!}\binom{k}{i-j} \qquad i \geq 1,\ k \geq 0.$$

In particular, since $\psi_{i,0}(x) = (-x)^i/i!$, we have

$$G_m^*(s) = \frac{\dfrac{m\mu}{s+m\mu} + \sum_{i=1}^{\infty}\dfrac{(\lambda/\theta)^i}{i!}\left[\prod_{j=0}^{i-1}\left(1-\dfrac{m\mu}{s+m\mu+j\theta}\right)\right]\dfrac{m\mu}{s+m\mu+i\theta}}{1+\sum_{i=1}^{\infty}\dfrac{(\lambda/\theta)^i}{i!}\prod_{j=0}^{i-1}\left(1-\dfrac{m\mu}{s+m\mu+j\theta}\right)}.$$

4   Joint Distribution of the Waiting and Service Time

We are now in a position to consider the distribution of the time until departure (either by abandonment or service completion) for a unique customer in a combination of two birth-and-death processes whose state transitions are shown in Figs. 1 and 2. We note that state "Pr" in Fig. 1 is identical to state m in Fig. 2, whereas state "Rs" in Fig. 2 is identical to state $m-1$ in Fig. 1.

The time until departure consists of the *waiting time* (the time that the customer spends staying in the waiting room) and the *service time* (the time during which the customer is being served). These are not independent. Therefore, we will derive the joint LST of the DF for the waiting and service time for a unique customer who abandons waiting, denoted by $\mathcal{T}_k^*(s, s', \mathrm{Ab})$, and for a unique customer who gets served until completion, denoted by $\mathcal{T}_k^*(s, s', \mathrm{Sr})$. Then, we obtain the probability of abandonment and service completion, the marginal LST of the DF for the waiting time, the service time, and the total time spent in the system as follows:

$$\begin{aligned}
\mathcal{P}_k\{\mathrm{Ab}\} &:= \mathcal{T}_k^*(0,0,\mathrm{Ab}) \quad &;& \quad \mathcal{P}_k\{\mathrm{Sr}\} := \mathcal{T}_k^*(0,0,\mathrm{Sr}),\\
\mathcal{W}_k^*(s,\mathrm{Ab}) &:= \mathcal{T}_k^*(s,0,\mathrm{Ab}) \quad &;& \quad \mathcal{H}_k^*(s,\mathrm{Ab}) := \mathcal{T}_k^*(0,s,\mathrm{Ab}),\\
\mathcal{W}_k^*(s,\mathrm{Sr}) &:= \mathcal{T}_k^*(s,0,\mathrm{Sr}) \quad &;& \quad \mathcal{H}_k^*(s,\mathrm{Sr}) := \mathcal{T}_k^*(0,s,\mathrm{Sr}),\\
\mathcal{T}_k^*(s,\mathrm{Ab}) &:= \mathcal{T}_k^*(s,s,\mathrm{Ab}) \quad &;& \quad \mathcal{T}_k^*(s,\mathrm{Sr}) := \mathcal{T}_k^*(s,s,\mathrm{Sr}).
\end{aligned}$$

4.1 Waiting and Service Time Until Abandonment

We first consider the waiting and service time until abandonment for a unique customer who abandons waiting.

(1) For the unique customer being served in state k, $0 \le k \le m-1$, the first passage to abandonment ("Ab") consists of the following passages:
  (i) the initial passage from state k to state "Pr" in Fig. 1,
  (ii) several repetitions of the transition from state m to state "Rs" in Fig. 2, followed by the transition from state $m-1$ back to state "Pr" in Fig. 1, and
  (iii) the final passage from state m to state "Ab" in Fig. 2.

Owing to the Markovian property of state transitions, the times to take these passages in succession are independent of each other. Therefore, we get

$$
\begin{aligned}
T_k^*(s, s', \mathrm{Ab}) &= H_k^*(s', \mathrm{Pr})W_m^*(s, \mathrm{Ab}) \\
&\quad + H_k^*(s', \mathrm{Pr})[W_m^*(s, \mathrm{Rs})H_{m-1}^*(s', \mathrm{Pr})]W_m^*(s, \mathrm{Ab}) \\
&\quad + H_k^*(s', \mathrm{Pr})[W_m^*(s, \mathrm{Rs})H_{m-1}^*(s', \mathrm{Pr})]^2 W_m^*(s, \mathrm{Ab}) + \cdots \\
&= H_k^*(s', \mathrm{Pr})W_m^*(s, \mathrm{Ab}) \sum_{n=0}^{\infty}[W_m^*(s, \mathrm{Rs})H_{m-1}^*(s', \mathrm{Pr})]^n \\
&= \frac{H_k^*(s', \mathrm{Pr})W_m^*(s, \mathrm{Ab})}{1 - W_m^*(s, \mathrm{Rs})H_{m-1}^*(s', \mathrm{Pr})} \\
&= \frac{\theta_0}{s+\theta_0} \cdot \frac{h_k^*(s')[1 - G_m^*(s+\theta_0)]}{h_m^*(s') - h_{m-1}^*(s')G_m^*(s+\theta_0)}.
\end{aligned}
$$

This joint distribution leads to the marginal distributions

$$
\begin{aligned}
\mathcal{W}_k^*(s, \mathrm{Ab}) &= \frac{p_k\{\mathrm{Pr}\}W_m^*(s, \mathrm{Ab})}{1 - p_{m-1}\{\mathrm{Pr}\}W_m^*(s, \mathrm{Rs})} \\
&= \frac{\theta_0}{s+\theta_0} \cdot \frac{h_k^*(0)[1 - G_m^*(s+\theta_0)]}{h_m^*(0) - h_{m-1}^*(0)G_m^*(s+\theta_0)}, \\
\mathcal{H}_k^*(s, \mathrm{Ab}) &= \frac{p_m\{\mathrm{Ab}\}H_k^*(s, \mathrm{Pr})}{1 - p_m\{\mathrm{Rs}\}H_{m-1}^*(s, \mathrm{Pr})} = \frac{h_k^*(s)[1 - G_m^*(\theta_0)]}{h_m^*(s) - h_{m-1}^*(s)G_m^*(\theta_0)}, \\
\mathcal{T}_k^*(s, \mathrm{Ab}) &= \frac{H_k^*(s, \mathrm{Pr})W_m^*(s, \mathrm{Ab})}{1 - W_m^*(s, \mathrm{Rs})H_{m-1}^*(s, \mathrm{Pr})} \\
&= \frac{\theta_0}{s+\theta_0} \cdot \frac{h_k^*(s)[1 - G_m^*(s+\theta_0)]}{h_m^*(s) - h_{m-1}^*(s)G_m^*(s+\theta_0)}.
\end{aligned}
$$

Then we get the probability of abandonment

$$
\mathcal{P}_k\{\mathrm{Ab}\} = \frac{p_k\{\mathrm{Pr}\}p_m\{\mathrm{Ab}\}}{1 - p_m\{\mathrm{Rs}\}p_{m-1}\{\mathrm{Pr}\}} = \frac{h_k^*(0)[1 - G_m^*(\theta_0)]}{h_m^*(0) - h_{m-1}^*(0)G_m^*(\theta_0)},
$$

and the mean waiting and service time until abandonment

$$E[\mathcal{W}_k, \text{Ab}] = \frac{1}{\theta_0}\mathcal{P}_k\{\text{Ab}\} + \frac{h_k^*(0)[h_m^*(0) - h_{m-1}^*(0)]G_m'(\theta_0)}{[h_m^*(0) - h_{m-1}^*(0)G_m^*(\theta_0)]^2},$$

$$E[\mathcal{H}_k, \text{Ab}] = [1 - G_m^*(\theta_0)]\left\{\frac{h_k^*(0)[h_m'(0) - h_{m-1}'(0)G_m^*(\theta_0)]}{[h_m^*(0) - h_{m-1}^*(0)G_m^*(\theta_0)]^2}\right.$$
$$\left. - \frac{h_k'(0)}{h_m^*(0) - h_{m-1}^*(0)G_m^*(\theta_0)}\right\},$$

$$E[\mathcal{W}_k\mathcal{H}_k, \text{Ab}] = \frac{1}{\theta_0}E[\mathcal{H}_k, \text{Ab}]$$
$$+ G_m'(\theta_0)\left\{\frac{2h_k^*(0)[h_m^*(0) - h_{m-1}^*(0)][h_m'(0) - h_{m-1}'(0)G_m^*(\theta_0)]}{[h_m^*(0) - h_{m-1}^*(0)G_m^*(\theta_0)]^3}\right.$$
$$\left. - \frac{h_k^*(0)[h_m'(0) - h_{m-1}'(0)] + h_k'(0)[h_m^*(0) - h_{m-1}^*(0)]}{[h_m^*(0) - h_{m-1}^*(0)G_m^*(\theta_0)]^2}\right\},$$

where $h_k'(0) := [dh_k^*(s)/ds]_{s=0}$, $0 \leq k \leq m$, and $G_m'(\theta_0) := [dG_m^*(s)/ds]_{s=\theta_0}$.

(2) For the unique customer waiting in state k, $k \geq m$, the first passage to abandonment ("Ab") is either

(a) a direct passage from state k to state "Ab" in Fig. 2, or

(b) a sequence of the following passages:

    (i) the initial passage from state k to state "Rs" in Fig. 2,

    (ii) several repetitions of the transition from state $m-1$ to state "Pr" in Fig. 1, followed by the transition from state m to state "Rs" in Fig. 2, and

    (iii) the passage from state $m-1$ to state "Pr" in Fig. 1, followed by the final passage from state m to state "Ab" in Fig. 2.

Therefore, from the Markovian property of state transitions, we get

$$\mathcal{T}_k^*(s, s', \text{Ab}) = W_k^*(s, \text{Ab}) + W_k^*(s, \text{Rs})H_{m-1}^*(s', \text{Pr})W_m^*(s, \text{Ab})$$
$$+ W_k^*(s, \text{Rs})[H_{m-1}^*(s', \text{Pr})W_m^*(s, \text{Rs})]H_{m-1}^*(s', \text{Pr})W_m^*(s, \text{Ab})$$
$$+ W_k^*(s, \text{Rs})[H_{m-1}^*(s', \text{Pr})W_m^*(s, \text{Rs})]^2 H_{m-1}^*(s', \text{Pr})W_m^*(s, \text{Ab})$$
$$+ \cdots$$
$$= W_k^*(s, \text{Ab}) + W_k^*(s, \text{Rs})H_{m-1}^*(s', \text{Pr})W_m^*(s, \text{Ab})$$
$$\times \sum_{n=0}^{\infty}[W_m^*(s, \text{Rs})H_{m-1}^*(s', \text{Pr})]^n$$
$$= W_k^*(s, \text{Ab}) + \frac{W_k^*(s, \text{Rs})H_{m-1}^*(s', \text{Pr})W_m^*(s, \text{Ab})}{1 - W_m^*(s, \text{Rs})H_{m-1}^*(s', \text{Pr})}$$
$$= \frac{\theta_0}{s + \theta_0}\left\{1 - \frac{[h_m^*(s') - h_{m-1}^*(s')]G_k^*(s + \theta_0)}{h_m^*(s') - h_{m-1}^*(s')G_m^*(s + \theta_0)}\right\}.$$

This joint distribution leads to the marginal distributions

$$\mathcal{W}_k^*(s,\mathrm{Ab}) = W_k^*(s,\mathrm{Ab}) + \frac{p_{m-1}\{\mathrm{Pr}\}W_k^*(s,\mathrm{Rs})W_m^*(s,\mathrm{Ab})}{1 - p_{m-1}\{\mathrm{Pr}\}W_m^*(s,\mathrm{Rs})}$$

$$= \frac{\theta_0}{s+\theta_0}\left\{1 - \frac{[h_m^*(0)-h_{m-1}^*(0)]G_k^*(s+\theta_0)}{h_m^*(0)-h_{m-1}^*(0)G_m^*(s+\theta_0)}\right\},$$

$$\mathcal{H}_k^*(s,\mathrm{Ab}) = p_k\{\mathrm{Ab}\} + \frac{p_k\{\mathrm{Rs}\}p_m\{\mathrm{Ab}\}H_{m-1}^*(s,\mathrm{Pr})}{1 - p_m\{\mathrm{Rs}\}H_{m-1}^*(s,\mathrm{Pr})}$$

$$= 1 - \frac{[h_m^*(s)-h_{m-1}^*(s)]G_k^*(\theta_0)}{h_m^*(s)-h_{m-1}^*(s)G_m^*(\theta_0)},$$

$$\mathcal{T}_k^*(s,\mathrm{Ab}) = W_k^*(s,\mathrm{Ab}) + \frac{W_k^*(s,\mathrm{Rs})H_{m-1}^*(s,\mathrm{Pr})W_m^*(s,\mathrm{Ab})}{1 - W_m^*(s,\mathrm{Rs})H_{m-1}^*(s,\mathrm{Pr})}$$

$$= \frac{\theta_0}{s+\theta_0}\left\{1 - \frac{[h_m^*(s)-h_{m-1}^*(s)]G_k^*(s+\theta_0)}{h_m^*(s)-h_{m-1}^*(s)G_m^*(s+\theta_0)}\right\}.$$

Then we get the probability of abandonment

$$\mathcal{P}_k\{\mathrm{Ab}\} = p_k\{\mathrm{Ab}\} + \frac{p_k^*\{\mathrm{Rs}\}p_{m-1}\{\mathrm{Pr}\}p_m\{\mathrm{Ab}\}}{1 - p_{m-1}^*\{\mathrm{Pr}\}p_m\{\mathrm{Rs}\}}$$

$$= 1 - \frac{[h_m^*(0)-h_{m-1}^*(0)]G_k^*(\theta_0)}{h_m^*(0)-h_{m-1}^*(0)G_m^*(\theta_0)}$$

and the mean waiting and service time until abandonment

$$E[\mathcal{W}_k,\mathrm{Ab}] = \frac{1}{\theta_0}\mathcal{P}_k\{\mathrm{Ab}\} + [h_m^*(0)-h_{m-1}^*(0)]$$

$$\times \left\{\frac{G_k'(\theta_0)}{h_m^*(0)-h_{m-1}^*(0)G_m^*(\theta_0)} + \frac{G_k^*(\theta_0)h_{m-1}^*(0)G_m'(\theta_0)}{[h_m^*(0)-h_{m-1}^*(0)G_m^*(\theta_0)]^2}\right\},$$

$$E[\mathcal{H}_k,\mathrm{Ab}] = \frac{G_k^*(\theta_0)[h_m'(0)h_{m-1}^*(0)-h_m^*(0)h_{m-1}'(0)][1-G_m^*(\theta_0)]}{[h_m^*(0)-h_{m-1}^*(0)G_m^*(\theta_0)]^2},$$

$$E[\mathcal{W}_k\mathcal{H}_k,\mathrm{Ab}] = \frac{1}{\theta_0}E[\mathcal{H}_k,\mathrm{Ab}] + [h_m'(0)h_{m-1}^*(0)-h_{m-1}'(0)h_m^*(0)]$$

$$\times \left\{\frac{G_k^*(\theta_0)G_m'(\theta_0)[h_m^*(0)-2h_{m-1}^*(0)+h_{m-1}^*(0)G_m^*(\theta_0)]}{[h_m^*(0)-h_{m-1}^*(0)G_m^*(\theta_0)]^3}\right.$$

$$\left. - \frac{G_k'(\theta_0)[1-G_m^*(\theta_0)]}{[h_m^*(0)-h_{m-1}^*(0)G_m^*(\theta_0)]^2}\right\},$$

where $G_k'(\theta_0) := [dG_k^*(s)/ds]_{s=\theta_0}, k \geq m$.

4.2 Waiting and Service Time Until Service Completion

We next consider the waiting and service time until service completion for a unique customer who gets served.

(1) For the unique customer being served in state k, $0 \le k \le m-1$, we have

$$\mathcal{T}_k^*(s, s', \mathrm{Sr}) = H_k^*(s', \mathrm{Sr}) + \frac{H_k^*(s', \mathrm{Pr})W_m^*(s, \mathrm{Rs})H_{m-1}^*(s', \mathrm{Sr})}{1 - W_m^*(s, \mathrm{Rs})H_{m-1}^*(s', \mathrm{Pr})}$$

$$= \frac{\mu_0}{s' + \mu_0}\left\{1 - \frac{h_k^*(s')[1 - G_m^*(s + \theta_0)]}{h_m^*(s') - h_{m-1}^*(s')G_m^*(s + \theta_0)}\right\}.$$

This joint distribution leads to the marginal distributions

$$\mathcal{W}_k^*(s, \mathrm{Sr}) = p_k\{\mathrm{Sr}\} + \frac{p_k\{\mathrm{Pr}\}p_{m-1}\{\mathrm{Sr}\}W_m^*(s, \mathrm{Rs})}{1 - p_{m-1}\{\mathrm{Pr}\}W_m^*(s, \mathrm{Rs})}$$

$$= 1 - \frac{h_k^*(0)[1 - G_m^*(s + \theta_0)]}{h_m^*(0) - h_{m-1}^*(0)G_m^*(s + \theta_0)},$$

$$\mathcal{H}_k^*(s, \mathrm{Sr}) = H_k^*(s, \mathrm{Sr}) + \frac{p_m\{\mathrm{Rs}\}H_k^*(s, \mathrm{Pr})H_{m-1}^*(s, \mathrm{Sr})}{1 - p_m\{\mathrm{Rs}\}H_{m-1}^*(s, \mathrm{Pr})}$$

$$= \frac{\mu_0}{s + \mu_0}\left\{1 - \frac{h_k^*(s)[1 - G_m^*(\theta_0)]}{h_m^*(s) - h_{m-1}^*(s)G_m^*(\theta_0)}\right\},$$

$$\mathcal{T}_k^*(s, \mathrm{Sr}) = H_k^*(s, \mathrm{Sr}) + \frac{H_k^*(s, \mathrm{Pr})W_m^*(s, \mathrm{Rs})H_{m-1}^*(s, \mathrm{Sr})}{1 - W_m^*(s, \mathrm{Rs})H_{m-1}^*(s, \mathrm{Pr})}$$

$$= \frac{\mu_0}{s + \mu_0}\left\{1 - \frac{h_k^*(s)[1 - G_m^*(s + \theta_0)]}{h_m^*(s) - h_{m-1}^*(s)G_m^*(s + \theta_0)}\right\}.$$

Then we get the probability of service completion

$$\mathcal{P}_k\{\mathrm{Sr}\} = p_k\{\mathrm{Sr}\} + \frac{p_k\{\mathrm{Pr}\}p_{m-1}\{\mathrm{Sr}\}p_m\{\mathrm{Rs}\}}{1 - p_{m-1}\{\mathrm{Pr}\}p_m\{\mathrm{Rs}\}}$$

$$= 1 - \frac{h_k^*(0)[1 - G_m^*(\theta_0)]}{h_m^*(0) - h_{m-1}^*(0)G_m^*(\theta_0)} = 1 - \mathcal{P}_k\{\mathrm{Ab}\}$$

and the mean waiting and service time until service completion

$$E[\mathcal{W}_k, \mathrm{Sr}] = \frac{h_k^*(0)[h_{m-1}^*(0) - h_m^*(0)]G_m'(\theta_0)}{[h_m^*(0) - h_{m-1}^*(0)G_m^*(\theta_0)]^2},$$

$$E[\mathcal{H}_k, \mathrm{Sr}] = \frac{1}{\mu_0}\mathcal{P}_k\{\mathrm{Sr}\} - [1 - G_m^*(\theta_0)]$$

$$\times \left\{\frac{h_k^*(0)[h_m'(0) - h_{m-1}'(0)G_m^*(\theta_0)]}{[h_m^*(0) - h_{m-1}^*(0)G_m^*(\theta_0)]^2} - \frac{h_k'(0)}{h_m^*(0) - h_{m-1}^*(0)G_m^*(\theta_0)}\right\},$$

$$E[\mathcal{W}_k\mathcal{H}_k, \mathrm{Sr}] = \frac{1}{\mu_0}E[\mathcal{W}_k, \mathrm{Sr}] - G_m'(\theta_0)$$

$$\times \left\{\frac{2h_k^*(0)[h_m^*(0) - h_{m-1}^*(0)][h_m'(0) - h_{m-1}'(0)G_m^*(\theta_0)]}{[h_m^*(0) - h_{m-1}^*(0)G_m^*(\theta_0)]^3}\right.$$

$$\left. - \frac{h_k^*(0)[h_m'(0) - h_{m-1}'(0)] + h_k'(0)[h_m^*(0) - h_{m-1}^*(0)]}{[h_m^*(0) - h_{m-1}^*(0)G_m^*(\theta_0)]^2}\right\}.$$

(2) For the unique customer waiting in state k, $k \geq m$, we have

$$\mathcal{T}_k^*(s, s', \mathrm{Sr}) = \frac{W_k^*(s, \mathrm{Rs})H_{m-1}^*(s', \mathrm{Sr})}{1 - W_m^*(s, \mathrm{Rs})H_{m-1}^*(s', \mathrm{Pr})}$$

$$= \frac{\mu_0}{s' + \mu_0} \cdot \frac{[h_m^*(s') - h_{m-1}^*(s')]G_k^*(s + \theta_0)}{h_m^*(s') - h_{m-1}^*(s')G_m^*(s + \theta_0)}.$$

This joint distribution leads to the marginal distributions

$$\mathcal{W}_k^*(s, \mathrm{Sr}) = \frac{p_{m-1}\{\mathrm{Sr}\}W_k^*(s, \mathrm{Rs})}{1 - p_{m-1}\{\mathrm{Pr}\}W_m^*(s, \mathrm{Rs})} = \frac{[h_m^*(0) - h_{m-1}^*(0)]G_k^*(s + \theta_0)}{h_m^*(0) - h_{m-1}^*(0)G_m^*(s + \theta_0)},$$

$$\mathcal{H}_k^*(s, \mathrm{Sr}) = \frac{p_k\{\mathrm{Rs}\}H_{m-1}^*(s, \mathrm{Sr})}{1 - p_m\{\mathrm{Rs}\}H_{m-1}^*(s, \mathrm{Pr})} = \frac{\mu_0}{s + \mu_0} \cdot \frac{[h_m^*(s) - h_{m-1}^*(s)]G_k^*(\theta_0)}{h_m^*(s) - h_{m-1}^*(s)G_m^*(\theta_0)},$$

$$\mathcal{T}_k^*(s, \mathrm{Sr}) = \frac{W_k^*(s, \mathrm{Rs})H_{m-1}^*(s, \mathrm{Sr})}{1 - W_m^*(s, \mathrm{Rs})H_{m-1}^*(s, \mathrm{Pr})}$$

$$= \frac{\mu_0}{s + \mu_0} \cdot \frac{[h_m^*(s) - h_{m-1}^*(s)]G_k^*(s + \theta_0)}{h_m^*(s) - h_{m-1}^*(s)G_m^*(s + \theta_0)}.$$

Then we get the probability of service completion

$$\mathcal{P}_k\{\mathrm{Sr}\} = \frac{p_{m-1}\{\mathrm{Sr}\}p_k\{\mathrm{Rs}\}}{1 - p_{m-1}\{\mathrm{Pr}\}p_m\{\mathrm{Rs}\}} = \frac{[h_m^*(0) - h_{m-1}^*(0)]G_k^*(\theta_0)}{h_m^*(0) - h_{m-1}^*(0)G_m^*(\theta_0)} = 1 - \mathcal{P}_k\{\mathrm{Ab}\},$$

and the mean waiting and service time until service completion

$$E[\mathcal{W}_k, \mathrm{Sr}] = [h_{m-1}^*(0) - h_m^*(0)]$$

$$\times \left\{ \frac{G_k'(\theta_0)}{h_m^*(0) - h_{m-1}^*(0)G_m^*(\theta_0)} + \frac{h_{m-1}^*(0)G_k^*(\theta_0)G_m'(\theta_0)}{[h_m^*(0) - h_{m-1}^*(0)G_m^*(\theta_0)]^2} \right\},$$

$$E[\mathcal{H}_k, \mathrm{Sr}] = \frac{1}{\mu_0}\mathcal{P}_k\{\mathrm{Sr}\}$$

$$- \frac{G_k^*(\theta_0)[h_m'(0)h_{m-1}^*(0) - h_{m-1}'(0)h_m^*(0)][1 - G_m^*(\theta_0)]}{[h_m^*(0) - h_{m-1}^*(0)G_m^*(\theta_0)]^2},$$

$$E[\mathcal{W}_k\mathcal{H}_k, \mathrm{Sr}] = \frac{1}{\mu_0}E[\mathcal{W}_k, \mathrm{Sr}] - [h_m'(0)h_{m-1}^*(0) - h_{m-1}'(0)h_m^*(0)]$$

$$\times \left\{ \frac{G_k^*(\theta_0)G_m'(\theta_0)[h_m^*(0) - 2h_{m-1}^*(0) + h_{m-1}^*(0)G_m^*(\theta_0)]}{[h_m^*(0) - h_{m-1}^*(0)G_m^*(\theta_0)]^3} \right.$$

$$\left. - \frac{G_k'(\theta_0)[1 - G_m^*(\theta_0)]}{[h_m^*(0) - h_{m-1}^*(0)G_m^*(\theta_0)]^2} \right\}.$$

4.3 Waiting and Service Time Until Departure

We finally consider the waiting and service time until departure (either abandonment or service completion) for a unique customer in state k ($k \geq 0$). Let

$$\mathcal{T}_k^*(s, s') := \mathcal{T}_k^*(s, s', \mathrm{Ab}) + \mathcal{T}_k^*(s, s', \mathrm{Sr}) \qquad k \geq 0$$

be the unconditional joint LST of the DF for the waiting and service time for the unique customer in state k. Then, we obtain the marginal LSTs of the DF for the waiting time, the service time, and the total time spent in the system as follows:

$$\mathcal{W}_k^*(s) := \mathcal{T}_k^*(s,0); \quad \mathcal{H}_k^*(s) := \mathcal{T}_k^*(0,s); \quad \mathcal{T}_k^*(s) := \mathcal{T}_k^*(s,s) \qquad k \geq 0.$$

(1) For the unique customer being served in state k, $0 \leq k \leq m-1$, we have

$$\mathcal{T}_k^*(s,s') = \frac{\mu_0}{s'+\mu_0} + \left(\frac{\theta_0}{s+\theta_0} - \frac{\mu_0}{s'+\mu_0} \right) \frac{h_k^*(s')[1 - G_m^*(s+\theta_0)]}{h_m^*(s') - h_{m-1}^*(s')G_m^*(s+\theta_0)}.$$

This joint distribution leads to the marginal distributions

$$\mathcal{W}_k^*(s) = \mathcal{W}_k^*(s, \mathrm{Ab}) + \mathcal{W}_k^*(s, \mathrm{Sr})$$
$$= 1 - \frac{s}{s+\theta_0} \cdot \frac{h_k^*(0)[1 - G_m^*(s+\theta_0)]}{h_m^*(0) - h_{m-1}^*(0)G_m^*(s+\theta_0)},$$
$$\mathcal{H}_k^*(s) = \mathcal{H}_k^*(s, \mathrm{Ab}) + \mathcal{H}_k^*(s, \mathrm{Sr})$$
$$= \frac{\mu_0}{s+\mu_0} + \frac{s}{s+\mu_0} \cdot \frac{h_k^*(s)[1 - G_m^*(\theta_0)]}{h_m^*(s) - h_{m-1}^*(s)G_m^*(\theta_0)},$$
$$\mathcal{T}_k^*(s) = \mathcal{T}_k^*(s, \mathrm{Ab}) + \mathcal{T}_k^*(s, \mathrm{Sr})$$
$$= \frac{\mu_0}{s+\mu_0} + \left(\frac{\theta_0}{s+\theta_0} - \frac{\mu_0}{s+\mu_0} \right) \frac{h_k^*(s)[1 - G_m^*(s+\theta_0)]}{h_m^*(s) - h_{m-1}^*(s)G_m^*(s+\theta_0)}.$$

The mean waiting time, service time, and total time until departure are given by

$$E[\mathcal{W}_k] = \frac{h_k^*(0)[1 - G_m^*(\theta_0)]}{\theta_0[h_m^*(0) - h_{m-1}^*(0)G_m^*(\theta_0)]},$$
$$E[\mathcal{H}_k] = \frac{1}{\mu_0} \left\{ 1 - \frac{h_k^*(0)[1 - G_m^*(\theta_0)]}{h_m^*(0) - h_{m-1}^*(0)G_m^*(\theta_0)} \right\},$$
$$E[\mathcal{T}_k] = E[\mathcal{W}_k] + E[\mathcal{H}_k] = \frac{1}{\mu_0} + \left(\frac{1}{\theta_0} - \frac{1}{\mu_0} \right) \frac{h_k^*(0)[1 - G_m^*(\theta_0)]}{h_m^*(0) - h_{m-1}^*(0)G_m^*(\theta_0)}.$$

We also have

$$E[\mathcal{W}_k \mathcal{H}_k] = E[\mathcal{W}_k \mathcal{H}_k, \mathrm{Ab}] + E[\mathcal{W}_k \mathcal{H}_k, \mathrm{Sr}] = \frac{1}{\theta_0} E[\mathcal{H}_k, \mathrm{Ab}] + \frac{1}{\mu_0} E[\mathcal{W}_k, \mathrm{Sr}]$$
$$= - \frac{h_k^*(0)[h_m^*(0) - h_{m-1}^*(0)]G_m'(\theta_0)}{\mu[h_m^*(0) - h_{m-1}^*(s)G_m^*(\theta_0)]^2} - \frac{1 - G_m^*(\theta_0)}{\theta}$$
$$\times \left\{ \frac{h_k'(0)}{h_m^*(0) - h_{m-1}^*(s)G_m^*(\theta_0)} - \frac{h_k^*(0)[h_m'(0) - h_{m-1}'(0)G_m^*(\theta_0)]}{[h_m^*(0) - h_{m-1}^*(s)G_m^*(\theta_0)]^2} \right\}.$$

(2) For the unique customer waiting in state k, $k \geq m$, we have

$$\mathcal{T}_k^*(s,s') = \frac{\theta_0}{s+\theta_0} + \left(\frac{\mu_0}{s'+\mu_0} - \frac{\theta_0}{s+\theta_0} \right) \frac{[h_m^*(s') - h_{m-1}^*(s')]G_k^*(s+\theta_0)}{h_m^*(s') - h_{m-1}^*(s')G_m^*(s+\theta_0)}.$$

This joint distribution leads to the marginal distributions

$$W_k^*(s) = \frac{\theta_0}{s + \theta_0} + \frac{s}{s + \theta_0} \cdot \frac{[h_m^*(0) - h_{m-1}^*(0)]G_k^*(s + \theta_0)}{h_m^*(0) - h_{m-1}^*(0)G_m^*(s + \theta_0)},$$

$$\mathcal{H}_k^*(s) = 1 - \frac{s}{s + \mu_0} \cdot \frac{[h_m^*(s) - h_{m-1}^*(s)]G_k^*(\theta_0)}{h_m^*(s) - h_{m-1}^*(s)G_m^*(\theta_0)},$$

$$\mathcal{T}_k^*(s) = \frac{\theta_0}{s + \theta_0} + \left(\frac{\mu_0}{s + \mu_0} - \frac{\theta_0}{s + \theta_0}\right) \frac{[h_m^*(s) - h_{m-1}^*(s)]G_k^*(s + \theta_0)}{h_m^*(s) - h_{m-1}^*(s)G_m^*(s + \theta_0)}.$$

The mean waiting time, service time, and total time until departure are given by

$$E[\mathcal{W}_k] = \frac{1}{\theta_0} \left\{1 - \frac{h_m^*(0) - h_{m-1}^*(0)]G_k^*(\theta_0)}{h_m^*(0) - h_{m-1}^*(0)G_m^*(\theta_0)}\right\},$$

$$E[\mathcal{H}_k] = \frac{[h_m^*(0) - h_{m-1}^*(0)]G_k^*(\theta_0)}{\mu_0[h_m^*(0) - h_{m-1}^*(0)G_m^*(\theta_0)]},$$

$$E[\mathcal{T}_k] = E[\mathcal{W}_k] + E[\mathcal{H}_k] = \frac{1}{\theta_0} + \left(\frac{1}{\mu_0} - \frac{1}{\theta_0}\right) \frac{h_m^*(0) - h_{m-1}^*(0)]G_k^*(\theta_0)}{h_m^*(0) - h_{m-1}^*(0)G_m^*(\theta_0)}.$$

We also have

$$E[\mathcal{W}_k\mathcal{H}_k] = E[\mathcal{W}_k\mathcal{H}_k, \mathrm{Ab}] + E[\mathcal{W}_k\mathcal{H}_k, \mathrm{Sr}] = \frac{1}{\theta_0}E[\mathcal{H}_k, \mathrm{Ab}] + \frac{1}{\mu_0}E[\mathcal{W}_k, \mathrm{Sr}]$$

$$= -\frac{[h_m^*(0) - h_{m-1}^*(0)]G_k'(\theta_0)}{\mu[h_m^*(0) - h_{m-1}^*(0)G_m^*(\theta_0)]}$$

$$- \frac{G_k^*(\theta_0)[h_m^*(0) - h_{m-1}^*(0)]h_{m-1}^*(0)G_m'(\theta_0)}{\mu[h_m^*(0) - h_{m-1}^*(0)G_m^*(\theta_0)]^2}$$

$$+ \frac{G_k^*(\theta_0)[h_m'(0)h_{m-1}^*(0) - h_m^*(0)h_{m-1}'(0)][1 - G_m^*(\theta_0)]}{\theta[h_m^*(0) - h_{m-1}^*(0)G_m^*(\theta_0)]^2}.$$

(3) Recursive relations among moments of distribution for the waiting and service time.

From the explicit expressions for $\mathcal{T}_k^*(s, s', \mathrm{Ab})$, $\mathcal{T}_k^*(s, s', \mathrm{Sr})$, and $\mathcal{T}_k^*(s, s')$ given above, it can be shown that the unconditional and conditional joint LST of the DF for the waiting and service time until departure for a unique customer in state k satisfies the following relation in both cases $0 \le k \le m-1$ and $k \ge m$:

$$\mathcal{T}_k^*(s, s') = 1 - \frac{s}{\theta_0}\mathcal{T}_k^*(s, s', \mathrm{Ab}) - \frac{s'}{\mu_0}\mathcal{T}_k^*(s, s', \mathrm{Sr}) \qquad k \ge 0.$$

This yields the recursive relations among unconditional and conditional moments

$$E[\mathcal{W}_k^\ell\mathcal{H}_k^{\ell'}] = \frac{\ell}{\theta_0}E[\mathcal{W}_k^{\ell-1}\mathcal{H}_k^{\ell'}, \mathrm{Ab}] + \frac{\ell'}{\mu_0}E[\mathcal{W}_k^\ell\mathcal{H}_k^{\ell'-1}, \mathrm{Sr}] \qquad \ell, \ell' = 2, 3, \dots .$$

In particular, we have

$$E[\mathcal{W}_k] = \frac{1}{\theta_0}\mathcal{P}_k\{\text{Ab}\}; \quad E[\mathcal{H}_k] = \frac{1}{\mu_0}\mathcal{P}_k\{\text{Sr}\},$$

$$E[\mathcal{W}_k\mathcal{H}_k] = \frac{1}{\theta_0}E[\mathcal{H}_k, \text{Ab}] + \frac{1}{\mu_0}E[\mathcal{W}_k, \text{Sr}],$$

$$E[\mathcal{W}_k^\ell] = \frac{\ell}{\theta_0}E[\mathcal{W}_k^{\ell-1}, \text{Ab}]; \quad E[\mathcal{H}_k^\ell] = \frac{\ell}{\mu_0}E[\mathcal{H}_k^{\ell-1}, \text{Sr}] \quad \ell = 2, 3, \dots .$$

Furthermore, it follows from the relation

$$\mathcal{T}_k^*(s) = 1 - \frac{s}{\theta_0}\mathcal{T}_k^*(s, \text{Ab}) - \frac{s}{\mu_0}\mathcal{T}_k^*(s, \text{Sr}) \quad k \geq 0$$

(or from $\mathcal{T}_k = \mathcal{W}_k + \mathcal{H}_k$) that

$$E[\mathcal{T}_k] = \frac{1}{\theta_0}\mathcal{P}_k\{\text{Ab}\} + \frac{1}{\mu_0}\mathcal{P}_k\{\text{Sr}\},$$

$$E[\mathcal{T}_k^\ell] = \frac{\ell}{\theta_0}E[\mathcal{T}_k^{\ell-1}, \text{Ab}] + \frac{\ell}{\mu_0}E[\mathcal{T}_k^{\ell-1}, \text{Sr}] \quad \ell = 2, 3, \dots .$$

5   Numerical Example

Numerical values are shown in Table 1, where we assume $m = 5$, $\mu = 1$, $\theta = 2$, and $\lambda = 10$ for a more patient customer ($\theta_0 = 1$) and for a less patient customer ($\theta_0 = 4$) with $\mu_0 = \mu$. The performance of an arriving unique customer can be found in the row of $k = 0$ in these tables.

From the numerical results for a more patient customer, we observe the following:

– The probability of service completion is higher.
– The time spent in the system is longer whether he abandons waiting or he gets served.
– The received service time is not much different from other customers.

This observation agrees with our feeling that we had better be more patient than other customers for secure service completion, though it takes us more time.

It remains us to investigate closely the trade-off between the probability of service completion and the time spent by a unique customer who gets served depending on the degree of his patience.

Table 1. Numerical example for a unique customer.

(a) More patient customer ($\theta_0 = 1 < \theta = 2$, $\mu_0 = \mu$)

k	$\mathcal{P}_k\{\mathrm{Ab}\}$	$\mathcal{P}_k\{\mathrm{Sr}\}$	$E[\mathcal{W}_k, \mathrm{Ab}]$	$E[\mathcal{H}_k, \mathrm{Ab}]$	$E[\mathcal{T}_k, \mathrm{Ab}]$	$E[\mathcal{W}_k, \mathrm{Sr}]$	$E[\mathcal{H}_k, \mathrm{Sr}]$	$E[\mathcal{T}_k, \mathrm{Sr}]$
0	0.49026	0.50974	0.45098	0.31260	0.76358	0.03928	0.19714	0.23642
1	0.53929	0.46071	0.49608	0.29483	0.79091	0.04320	0.16588	0.20909
2	0.59812	0.40188	0.55020	0.26861	0.81881	0.04792	0.13327	0.18119
3	0.66969	0.33031	0.61604	0.23042	0.84646	0.05365	0.09989	0.15354
4	0.75814	0.24186	0.69740	0.17503	0.87243	0.06074	0.06683	0.12757
5	0.86933	0.13067	0.79968	0.09496	0.89425	0.06964	0.03611	0.10575
6	0.91186	0.08814	0.84386	0.06379	0.90765	0.06800	0.02436	0.09235
7	0.93281	0.06719	0.86798	0.04862	0.91661	0.06483	0.01857	0.08339
8	0.94495	0.05505	0.88321	0.03984	0.92305	0.06174	0.01521	0.07965
9	0.95280	0.04720	0.89379	0.03415	0.92795	0.05901	0.01304	0.07205
10	0.95829	0.04171	0.90164	0.03018	0.93183	0.05665	0.01153	0.06817
15	0.97174	0.02826	0.92322	0.02045	0.94367	0.04852	0.00781	0.05633
20	0.97741	0.02259	0.93375	0.01635	0.95010	0.04366	0.00624	0.04990
30	0.98284	0.01716	0.94505	0.01242	0.95747	0.03779	0.00474	0.04253

(b) Equally patient customer ($\theta_0 = 2 = \theta$, $\mu_0 = \mu$)

k	$\mathcal{P}_k\{\mathrm{Ab}\}$	$\mathcal{P}_k\{\mathrm{Sr}\}$	$E[\mathcal{W}_k, \mathrm{Ab}]$	$E[\mathcal{H}_k, \mathrm{Ab}]$	$E[\mathcal{T}_k, \mathrm{Ab}]$	$E[\mathcal{W}_k, \mathrm{Sr}]$	$E[\mathcal{H}_k, \mathrm{Sr}]$	$E[\mathcal{T}_k, \mathrm{Sr}]$
0	0.51270	0.48730	0.24299	0.30992	0.55291	0.01336	0.17738	0.19074
1	0.56396	0.43604	0.26729	0.28965	0.55693	0.01470	0.14639	0.16108
2	0.62549	0.37451	0.29645	0.26019	0.55663	0.01630	0.11432	0.13062
3	0.70034	0.29966	0.33192	0.21777	0.54969	0.01825	0.08189	0.10014
4	0.79283	0.20717	0.37576	0.15678	0.53254	0.02066	0.05039	0.07105
5	0.90911	0.09089	0.43087	0.06878	0.49965	0.02369	0.02211	0.04579
6	0.94907	0.05093	0.45368	0.03854	0.49222	0.02085	0.01239	0.03324
7	0.96686	0.03314	0.46548	0.02508	0.49056	0.01795	0.00806	0.02601
8	0.97624	0.02376	0.47252	0.01798	0.49050	0.01560	0.00578	0.02138
9	0.98181	0.01819	0.47713	0.01377	0.49090	0.01378	0.00442	0.01820
10	0.98541	0.01459	0.48038	0.01104	0.49142	0.01233	0.00355	0.01588
15	0.99293	0.00707	0.48827	0.00535	0.49362	0.00819	0.00172	0.00991
20	0.99541	0.00459	0.49146	0.00347	0.49493	0.00624	0.00112	0.00736
30	0.99732	0.00268	0.49432	0.00203	0.49635	0.00434	0.00065	0.00499

(c) Less patient customer ($\theta_0 = 4 > \theta = 2$, $\mu_0 = \mu$)

k	$\mathcal{P}_k\{\mathrm{Ab}\}$	$\mathcal{P}_k\{\mathrm{Sr}\}$	$E[\mathcal{W}_k, \mathrm{Ab}]$	$E[\mathcal{H}_k, \mathrm{Ab}]$	$E[\mathcal{T}_k, \mathrm{Ab}]$	$E[\mathcal{W}_k, \mathrm{Sr}]$	$E[\mathcal{H}_k, \mathrm{Sr}]$	$E[\mathcal{T}_k, \mathrm{Sr}]$
0	0.52855	0.47145	0.12723	0.30714	0.43437	0.00490	0.16431	0.16921
1	0.58141	0.41859	0.13996	0.28500	0.42495	0.00540	0.13359	0.13899
2	0.64483	0.35517	0.15522	0.25314	0.40837	0.00598	0.10202	0.10801
3	0.72200	0.27800	0.17380	0.20760	0.38140	0.00670	0.07040	0.07710
4	0.81735	0.18265	0.19675	0.14250	0.33925	0.00758	0.04015	0.04773
5	0.93723	0.06277	0.22561	0.04897	0.27458	0.00870	0.01380	0.02249
6	0.97206	0.02794	0.23656	0.02180	0.25836	0.00646	0.00614	0.01260
7	0.98526	0.01474	0.24164	0.01150	0.25314	0.00467	0.00324	0.00791
8	0.99125	0.00875	0.24435	0.00682	0.25117	0.00347	0.00192	0.00539
9	0.99425	0.00565	0.24594	0.00441	0.25035	0.00265	0.00124	0.0389
10	0.99610	0.00390	0.24694	0.00304	0.24998	0.00209	0.00086	0.00294
15	0.99893	0.00107	0.24888	0.00084	0.24972	0.00085	0.00024	0.00108
20	0.99952	0.00048	0.24942	0.00037	0.24979	0.00046	0.00010	0.00057
30	0.99983	0.00017	0.24975	0.00013	0.24989	0.00020	0.00004	0.00024

(Continued)

Table 1. *(Continued)*

(d) More patient customer ($\theta_0 = 1 < \theta = 2$, $\mu_0 = \mu$)

k	$E[\mathcal{W}_k]$	$E[\mathcal{H}_k]$	$E[\mathcal{T}_k]$	$E[\mathcal{W}_k\mathcal{H}_k, \text{Ab}]$	$E[\mathcal{W}_k\mathcal{H}_k, \text{Sr}]$	$E[\mathcal{W}_k\mathcal{H}_k]$	$E[\mathcal{T}_k^2]$
0	0.49026	0.50974	1.00000	0.31598	0.03590	0.35187	2.0000
1	0.53929	0.46071	1.00000	0.30248	0.03556	0.22804	2.0000
2	0.59812	0.40188	1.00000	0.28117	0.03476	0.31563	2.0000
3	0.66969	0.33031	1.00000	0.25078	0.03328	0.28407	2.0000
4	0.75814	0.24186	1.00000	0.20496	0.03081	0.23577	2.0000
5	0.86933	0.13067	1.00000	0.13739	0.02682	0.16421	2.0000
6	0.91186	0.08814	1.00000	0.10788	0.02390	0.13178	2.0000
7	0.93281	0.06719	1.00000	0.09164	0.02181	0.11345	2.0000
8	0.94495	0.05505	1.00000	0.08132	0.02025	0.10157	2.0000
9	0.95280	0.04720	1.00000	0.07412	0.01904	0.09317	2.0000
10	0.95829	0.04171	1.00000	0.06876	0.01807	0.08673	2.0000
15	0.97174	0.02826	1.00000	0.05393	0.01505	0.06897	2.0000
20	0.97741	0.02259	1.00000	0.04663	0.01336	0.06001	2.0000
30	0.98284	0.01716	1.00000	0.03877	0.01144	0.05021	2.0000

(e) Equally patient customer ($\theta_0 = 2 = \theta$, $\mu_0 = \mu$)

k	$E[\mathcal{W}_k]$	$E[\mathcal{H}_k]$	$E[\mathcal{T}_k]$	$E[\mathcal{W}_k\mathcal{H}_k, \text{Ab}]$	$E[\mathcal{W}_k\mathcal{H}_k, \text{Sr}]$	$E[\mathcal{W}_k\mathcal{H}_k]$	$E[\mathcal{T}_k^2]$
0	0.25635	0.48730	0.74365	0.15700	0.01132	0.16832	0.93439
1	0.28198	0.43604	0.71802	0.14840	0.01112	0.15952	0.87910
2	0.31274	0.37451	0.68726	0.13565	0.01074	0.14639	0.81788
3	0.35017	0.29966	0.64983	0.11702	0.01011	0.12713	0.74997
4	0.39642	0.20717	0.60358	0.08994	0.00911	0.09905	0.67463
5	0.45456	0.09089	0.54544	0.05053	0.00755	0.05808	0.59124
6	0.47454	0.05093	0.52546	0.03405	0.00608	0.04012	0.55870
7	0.48343	0.03314	0.51567	0.02547	0.00502	0.03049	0.54257
8	0.48812	0.02376	0.51188	0.02033	0.00426	0.02459	0.53326
9	0.49090	0.01819	0.50910	0.01695	0.00371	0.02066	0.52729
10	0.49270	0.01459	0.50730	0.01456	0.00329	0.01785	0.52317
15	0.49647	0.00707	0.50353	0.00874	0.00213	0.01087	0.51345
20	0.49771	0.00459	0.50229	0.00637	0.00161	0.00798	0.50965
30	0.49866	0.00268	0.50134	0.00424	0.00111	0.00535	0.50633

(f) Less patient customer ($\theta_0 = 4 > \theta = 2$, $\mu_0 = \mu$)

k	$E[\mathcal{W}_k]$	$E[\mathcal{H}_k]$	$E[\mathcal{T}_k]$	$E[\mathcal{W}_k\mathcal{H}_k, \text{Ab}]$	$E[\mathcal{W}_k\mathcal{H}_k, \text{Sr}]$	$E[\mathcal{W}_k\mathcal{H}_k]$	$E[\mathcal{T}_k^2]$
0	0.13214	0.47145	0.60359	0.07776	0.00393	0.08169	0.55561
1	0.14535	0.41859	0.56394	0.07281	0.00383	0.07664	0.49046
2	0.16121	0.35517	0.51637	0.06560	0.00366	0.06927	0.42020
3	0.18050	0.27800	0.45850	0.05520	0.00340	0.05860	0.34489
4	0.20434	0.18265	0.38698	0.04022	0.00299	0.04321	0.26509
5	0.23431	0.06277	0.29708	0.01857	0.00237	0.02094	0.18228
6	0.24301	0.02794	0.27096	0.01028	0.00162	0.01190	0.15437
7	0.24632	0.01474	0.26105	0.00641	0.00113	0.00755	0.14240
8	0.24781	0.00875	0.25656	0.00435	0.00083	0.00517	0.13636
9	0.24859	0.00465	0.25424	0.00313	0.00062	0.00375	0.13296
10	0.24903	0.00390	0.25292	0.00236	0.00049	0.02850	0.13088
15	0.24973	0.00107	0.25080	0.00086	0.00019	0.00106	0.12703
20	0.24988	0.00048	0.25036	0.00045	0.00011	0.00056	0.12603
30	0.24996	0.00017	0.25013	0.00019	0.00005	0.00024	0.12543

References

1. Iravani, F., Balcıoğlu, B.: On priority queues with impatient customers. Queueing Syst. **58**(4), 239–260 (2008)
2. Takagi, H.: Times to service completion and abandonment in the M/M/m preemptive LCFS queue with impatient customers. In: QTNA 2016, Proceedings of the 11th International Conference on Queueing Theory and Network Applications, Article No. 2. ACM (2016). ISBN 978-1-4503-4842-3/16/12. https://dx.doi.org/10.1145/3016032.3016036
3. Takagi, H.: Distribution of the waiting and service time in an M/M/m preemptive-resume LCFS queue with impatient customers. Int. J. Pure Appl. Math. **116**(2), 501–545 (2017). https://dx.doi.org/10.12732/ijpam.v.116i2.21
4. Wolff, R.W.: Stochastic Modeling and the Theory of Queues. Prentice Hall, Englewood Cliffs (1989)

Multi-Server M/M/c Queue with N-Policy and d-Type Working Vacation

Xiuli Xu, Ming Li$^{(\boxtimes)}$, and Huining Wang

College of Science, Yanshan University, Qinhuangdao 066004, China
xxl-ysu@163.com, 752334915@qq.com, 1023548786@qq.com

Abstract. Introducing the N-policy and half exhaustive d-type working vacation into a multi-server M/M/c queue. If d or more than d servers are idle, we let d idled servers start a multiple synchronous working vacation, the servers on working vacation keep on serving new customers in a low rate, and the rest c-d servers work as usual. When a vacation is finished, once the customers in the system is greater or equal to N, terminating the vacation of these d servers, otherwise starting another working vacation. When a service is completed in a vacation period, what kind of service has been accepted by the customer who has left, depends on the customers in the system whether is more than c-d, if the number of customer is more than c-d, a customer who has left maybe served by normal rate or low rate service, if the number of customer isn't more than c-d, a customer who has left served by normal rate service. Using a quasi-birth and death process and matrix geometric method, we obtain the stationary distributions. Finally, the result is applied to hierarchical cellular system, and we make the numerical analysis.

Keywords: Half exhaustive · Partial working vacation · N-policy Hierarchical cellular system

1 Introduction

In the queueing system, vacation models are more flexible, and it is undoubtedly a hot topic in recent years. For instance, Zhang and Tian [1] analyzed queueing systems with synchronous vacation of partial servers. A multi-server retrial queue with vacations was studied by Yang [2]. Ammar [3] gave the transient solution of an M/M/1 vacation queue with a waiting server and impatient customers. Takhedmit and Abbas [4] analyzed a parametric uncertainty analysis method for queues with vacations. Ke et al. [5] researched an optimal (d, c) vacation policy, and obtained the stationary distribution of the number of customers in the system numerically. Xu and Tian [6,7] proposed M/M/c and M/M/c-e multiple vacation service system assembled model, and added N-policy, studied

---

The project was supported by the national natural science foundation item (11201408), China.

W. Yue et al. (Eds.): QTNA 2017, LNCS 10591, pp. 36–48, 2017.
https://doi.org/10.1007/978-3-319-68520-5_3

M/M/c queue with (e, d, N)-policy. Zhu and Xu [8] studied the N-policy and negative customer's partial working vacation queue. On the basis of previous researches on various strategies for vacation queueing, Wang et al. [9] proposed a hybrid vacation strategy, there are two vacation strategies in a queue model, studied the M/M/1 queue with successive two kinds of vacation, the steady state distribution and the conditional stochastic decomposition results are obtained.

The queueing model proposed in this paper is different from exhaustive service on the form of starting a vacation, and the vacation strategy is working vacation, the servers on working vacation still serve new customers in a low rate, and we introduce the N-policy into the model to avoid switching too frequently. The correlation model is established and the stationary distribution is obtained. The result we got can be applied to many queueing systems, In the section of numerical analysis, the result is applied to hierarchical cellular mobile communication system. Allocating channel resource reasonably, and making a numerical analysis for the relationship between system performance indicators and parameters.

The structure of this paper is organized as follows. A d-type working vacation with N-policy is presented in Sect. 2, and the process of quasi-birth and death is given. The necessary conditions for later calculation and proof are given in Sect. 3. The process of calculating and proving about the stationary distribution is given in Sect. 4. In Sect. 5, the application of the model in hierarchical cellular system is given and the numerical analysis is made. Finally, The work and achievements of this paper are summarized in Sect. 6.

2   System Model

In an M/M/c queueing system with c servers, introducing the N-policy and d-type working vacation into the model: once there are d $(0 < d < c)$ servers become idle, let the d idled servers start an any length of synchronized working vacation, The vacation time V is assumed to be exponentially distributed with mean of $1/\theta$. The servers under working vacation get into a low speed service status, and continue to serve new customers at the service rate μ_v. At the end of a vacation, if the number of customers in the system is greater or equal to N, the servers on vacation back to normal work, on the contrary, continue to another working vacation.

It is known that the coming process and service process are assumed to be exponential distribution, the arrival rate is λ, the normal service rate is μ_b. All processes are independent of each other, the service order is first come first service.

Let $L(t)$ denote the number of customers in the system at time t, defined that

$$J(t) = \begin{cases} 0, & d \text{ servers on vacation at time } t, \\ 1, & \text{no servers on vacation at time } t. \end{cases}$$

Then $\{L(t), J(t)\}$ is a quasi-birth and death process, the state space is

$$\Omega = \{(k, 0), 0 \le k \le c - d\} \cup \{(k, 1), k > c - d, j = 0, 1\}.$$

The infinitesimal generator can be written as

$$Q = \begin{bmatrix} A_0 & C_0 \\ B_1 & A_1 & C_1 \\ & B_2 & A_2 & C_2 \\ & & \ddots & \ddots & \ddots \\ & & & B_c & A_c & C_c \\ & & & & \ddots & \ddots & \ddots \\ & & & & & B_{N-1} & A_{N-1} & C_{N-1} \\ & & & & & & B & A & C \\ & & & & & & & B & A & C \\ & & & & & & & & \ddots & \ddots & \ddots \end{bmatrix}$$

where

$$A_k = \begin{cases} -(\lambda + k\mu_b), & 0 \leq k \leq c-d, \\ \begin{pmatrix} -[\lambda + (c-d)\mu_b + (k-c+d)\mu_v] & 0 \\ 0 & -(\lambda + k\mu_b) \end{pmatrix}, & c-d < k \leq c, \\ \begin{pmatrix} -[\lambda + (c-d)\mu_b + d\mu_v] & 0 \\ 0 & -(\lambda + c\mu_b) \end{pmatrix}, & c < k \leq N-1. \end{cases}$$

$$B_k = \begin{cases} k\mu_b, & 1 \leq k \leq c-d, \\ \begin{pmatrix} (c-d)\mu_b + \mu_v \\ (c-d+1)\mu_b \end{pmatrix}, & k = c-d+1, \\ \begin{pmatrix} (c-d)\mu_b + (k-c+d)\mu_v & 0 \\ 0 & k\mu_b \end{pmatrix}, & c-d+1 < k \leq c, \\ \begin{pmatrix} (c-d)\mu_b + d\mu_v & 0 \\ 0 & c\mu_b \end{pmatrix}, & c < k \leq N-1. \end{cases}$$

$$C_k = \begin{cases} \lambda, & 0 \leq k < c-d, \\ (\lambda \ 0), & k = c-d, \\ \begin{pmatrix} \lambda & 0 \\ 0 & \lambda \end{pmatrix}, & c-d < k \leq N-1. \end{cases} \qquad C = \begin{pmatrix} \lambda & 0 \\ 0 & \lambda \end{pmatrix}.$$

$$A = \begin{pmatrix} -[\lambda + \theta + (c-d)\mu_b + d\mu_v] & \theta \\ 0 & -(\lambda + c\mu_b) \end{pmatrix}, \quad B = \begin{pmatrix} (c-d)\mu_b + d\mu_v & 0 \\ 0 & c\mu_b \end{pmatrix}.$$

3   Analysis

This paper introduce the d-type working vacation with N-policy into M/M/c queue, When the number of customers in the system is not more than c-d during a vacation, a customer who has left was served by normal rate service, when the customers in the system is more than c-d, a customer who has left may be served by normal rate or low rate service. What's more, For avoiding the frequent switching between working and vacation, introducing N-policy. When a vacation is finished, if the customers in the system is greater or equal to N, terminating the vacation of these d servers, otherwise, starting another working vacation.

In order to analyze the quasi-birth and death process above, the minimum non-negative solution R of the matrix equation $R^2B + RA + C = 0$ is called the rate matrix. In order to express R, we introduce the following conclusions:

Lemma 1. *The quadratic equation* $[(c - d)\mu_b + d\mu_v]z^2 - [\lambda + \theta + (c - d)\mu_b + d\mu_v]z + \lambda = 0$ *has two different real roots* $r < r^*$, *and* $0 < r < 1$, $r^* > 1$. r *satisfies the relationship*

$$\lambda + \theta + [(c - d)\mu_b + d\mu_v](1 - r) = \frac{\theta}{1 - r} + [(c - d)\mu_b + d\mu_v] = \frac{\lambda}{r} \qquad (1)$$

where $r = \frac{\lambda + \theta + (c-d)\mu_b + d\mu_v - \sqrt{[\lambda + \theta + (c-d)\mu_b + d\mu_v]^2 - 4\lambda[(c-d)\mu_b + d\mu_v]}}{2[(c-d)\mu_b + d\mu_v]}$

Theorem 1. *When* $\rho = \frac{\lambda}{c\mu_b} < 1$, *The matrix equation* $R^2B + RA + C = 0$ *has the smallest non-negative solution*

$$R = \begin{pmatrix} r & \frac{\theta r}{c\mu_b(1-r)} \\ 0 & \rho \end{pmatrix} \qquad (2)$$

Proof. Due to the A, B, C are upper triangular matrix which are given above, The minimum non-negative solution R satisfies the matrix quadratic equation $R^2B + RA + C = 0$ must also be an upper triangular matrix, let

$$R = \begin{pmatrix} r_{11} & r_{12} \\ 0 & r_{22} \end{pmatrix}.$$

Substituting R into the matrix quadratic equation, we can get a set of equations:

$$\begin{cases} [(c - d)\mu_b + d\mu_v]r_{11}^2 - [\lambda + \theta + (c - d)\mu_b + d\mu_v]r_{11} + \lambda = 0, \\ c\mu_b r_{22}^2 - (\lambda + c\mu_b)r_{22} + \lambda = 0, \\ c\mu_b r_{12}(r_{11} + r_{12}) + \theta r_{11} - (\lambda + c\mu_b)r_{12} = 0. \end{cases}$$

By Lemma 1, let $r_{11} = r$, $r_{22} = \rho$. Substituting r and ρ into the last equation, we obtain $r_{12} = \frac{\theta r}{c\mu_b(1-r)}$.

4   Performance Measures

If $\rho < 1$, (L, J) represents the steady-state limit of process $\{L(t), J(t)\}$, let

$$\pi_{kj} = P\{L(t) = k, J(t) = j\} = \lim_{t \to \infty} P\{L(t) = k, J(t) = j\}, (k, j) \in \Omega.$$

$$\pi_k = \pi_{k0}, 0 \le k \le c - d; \pi_k = (\pi_{k0}, \pi_{k1}), k > c - d,$$
$$X = (\pi_0, \pi_1, \cdots, \pi_c, \pi_{c+1}, \cdots, \pi_N).$$

In order to show the distribution of (L, J), we introduce three sets of recursive sequences Φ_k, φ_k and γ_k

$$
\begin{cases}
\Phi_0 = 1, \\
\Phi_1 = \frac{1}{r}, \\
\Phi_k = \frac{1}{r}\left\{\left[\frac{(c-d)\mu_b + d\mu_v}{\lambda}\right]^{k-1} + \frac{\theta\, r}{\lambda(1-r)}\sum_{j=0}^{k-2}\left[\frac{(c-d)\mu_b + d\mu_v}{\lambda}\right]^j\right\}, 2 \le k \le N-c+1
\end{cases}
\tag{3}
$$

$$
\begin{cases}
\varphi_0 = 1,\ k = 0, \\
\varphi_k = 1 + \sum_{j=1}^{k}\dfrac{\prod\limits_{i=1}^{j}[(c-d)\mu_b + (d-c+j+i)\mu_v]}{\lambda^j},\ 1 \le k < c-1
\end{cases}
\tag{4}
$$

$$
\begin{cases}
\gamma_0 = 1, \\
\gamma_k = \left(\frac{\lambda}{\mu_b}\right)^k \frac{(c-d+1)!}{(k+c-d+1)!} + (c-d+1)\sum_{j=0}^{k-1}\left(\frac{\lambda}{\mu_b}\right)^{k-1-j}\frac{(j+c-d+1)!}{(k+c-d+1)!}, 1 \le k \le d
\end{cases}
\tag{5}
$$

Theorem 2. *When $\rho < 1$, the steady state probability distribution of (L, J) is*

$$
\pi_{j0} = \begin{cases}
K\frac{1}{j!}\left(\frac{\lambda}{\mu_b}\right)^j,\ 0 \le j \le c-d, \\
K\sigma\beta\left(\varphi_{c-1-j}\Phi_{N-c+1} - \frac{(c-d)\mu_b + d\mu_v}{\lambda}\varphi_{c-2-j}\Phi_{N-c}\right), \\
\qquad\qquad\qquad\qquad c-d+1 \le j \le c-2, \\
K\sigma\beta\,\Phi_{N-j},\ c-1 \le j \le N
\end{cases}
$$

$$
\pi_{j1} = \begin{cases}
K\sigma\beta\,\frac{\theta}{(c-d+1)\mu_b(1-r)}\gamma_{j-c+d-1},\ c-d+1 \le j \le c, \\
K\sigma\beta\left[\left(\frac{\lambda}{c\mu_b}\right)^{j-c}\gamma_{d-1}\frac{\theta}{(c-d+1)\mu_b(1-r)} + \sum_{v=0}^{j-c-1}\left(\frac{\lambda}{c\mu_b}\right)^v\right],\ c+1 \le j \le N
\end{cases}
$$

where

$$
\sigma = \left(\varphi_{d-1}\Phi_{N-c+1} - \frac{(c-d)\mu_b + d\mu_v}{\lambda}\varphi_{d-2}\Phi_{N-c}\right)^{-1},
$$

$$
\beta = \frac{1}{(c-d)!}\left(\frac{\lambda}{\mu_b}\right)^{c-d} = K^{-1}\pi_{c-d,0}.
$$

Proof. By the matrix geometric method, X satisfies the equation $XB[R] = 0$, where

$$
B[R] = \begin{bmatrix}
A_0 & C_0 & & & & \\
B_1 & A_1 & C_1 & & & \\
& \ddots & \ddots & \ddots & & \\
& & B_c & A_c & C_c & \\
& & & \ddots & \ddots & & \ddots \\
& & & & B_{N-1} & A_{N-1} & C_{N-1} \\
& & & & & B & RB+A
\end{bmatrix}
$$

Expanding the equation above we obtain the equations:

$$
\begin{cases}
-\lambda\pi_{00} + \mu_b\pi_{10} = 0, \\
\lambda\pi_{k-1,0} - (\lambda + k\mu_b)\pi_{k0} + (k+1)\mu_b\pi_{k+1,0} = 0, \quad 1 \le k < c-d, \\
\lambda\pi_{c-d-1,0} - [\lambda + (c-d)\mu_b]\pi_{c-d,0} + [(c-d)\mu_b + \mu_v]\pi_{c-d+1,0} \\
\qquad + (c-d+1)\mu_b\pi_{c-d+1,1} = 0, \quad k = c-d, \\
-[\lambda + (c-d+1)\mu_b]\pi_{c-d+1,1} + (c-d+2)\mu_b\pi_{c-d+2,1} = 0, \quad k = c-d+1, \\
\lambda\pi_{k-1,0} - [\lambda + (c-d)\mu_b + (k+d-c)\mu_v]\pi_{k0} \\
\qquad + [(c-d)\mu_b + (k+d-c+1)\mu_v]\pi_{k+1,0} = 0, \quad c-d+1 \le k < c, \\
\lambda\pi_{k-1,1} - (\lambda + k\mu_b)\pi_{k1} + (k+1)\mu_b\pi_{k+1,1} = 0, \quad c-d+1 < k < c, \\
\lambda\pi_{k-1,0} - [\lambda + (c-d)\mu_b + d\mu_v]\pi_{k0} \\
\qquad + [(c-d)\mu_b + d\mu_v]\pi_{k+1,0} = 0, \quad c \le k < N, \\
\lambda\pi_{k-1,1} - (\lambda + c\mu_b)\pi_{k1} + c\mu_b\pi_{k+1,1} = 0, \quad c \le k < N, \\
\lambda\pi_{N-1,0} - \frac{\lambda}{r}\pi_{N0} = 0, \quad k = N, \\
\lambda\pi_{N-1,1} + \frac{\theta}{1-r}\pi_{N0} - c\mu_b\pi_{N1} = 0, \quad k = N
\end{cases}
$$

$$(6)$$

For solving Eq. (6), the formulas (6) are denoted by (6-1) to (6-10), respectively. From Eqs. (6-1) and (6-2), let $\pi_{00} = K$, we obtain

$$
\pi_{j0} = K\frac{1}{j!}\left(\frac{\lambda}{\mu_b}\right)^j, \quad 0 \le j \le c-d.
$$

If $k = c-d$,

$$
\pi_{c-d,0} = K\frac{1}{(c-d)!}\left(\frac{\lambda}{\mu_b}\right)^{c-d}
$$

$$(7)$$

We can get $\pi_{N-1,0} = \frac{1}{r}\pi_{N0}$ from (6-9), substituting it into (6-7) and notice Eqs. (1) and (3) we can obtain that

$$
\pi_{j0} = \Phi_{N-j}\pi_{N0}, \quad c-1 \le j \le N
$$

In addition, from (6-5)

$$
-\lambda\pi_{c-1,0} + [(c-d)\mu_b + d\mu_v]\pi_{c0} + \lambda\pi_{j0} - [(c-d)\mu_b + (d-c+j+1)\mu_v]\pi_{j+1,0} = 0 \quad (8)
$$

Making the iteration of (8) and notice (4), we recursively have

$$
\pi_{j0} = \left(\varphi_{c-1-j}\Phi_{N-c+1} - \frac{(c-d)\mu_b + d\mu_v}{\lambda}\varphi_{c-2-j}\Phi_{N-c}\right)\pi_{N0}, \quad c-d \le j \le c-2
$$

When $j = c-d$,

$$
\pi_{c-d,0} = \left(\varphi_{d-1}\Phi_{N-c+1} - \frac{(c-d)\mu_b + d\mu_v}{\lambda}\varphi_{d-2}\Phi_{N-c}\right)\pi_{N0}
$$

Substituting the formula above into (7), there are

$$
\pi_{N0} = K\left(\varphi_{d-1}\Phi_{N-c+1} - \frac{(c-d)\mu_b + d\mu_v}{\lambda}\varphi_{d-2}\Phi_{N-c}\right)^{-1}\frac{1}{(c-d)!}\left(\frac{\lambda}{\mu_b}\right)^{c-d}
$$

$$
= K\sigma\beta
$$

Thus, the former equations in the Theorem 4.1 have been proved, and the following is the proof of the latter equations in Theorem 4.1.

By (6-4), there are

$$(c - d + 2)\mu_b\pi_{c-d+2,1} - \lambda\pi_{c-d+1,1} = (c - d + 1)\mu_b\pi_{c-d+1,1} \tag{9}$$

Substituting it into (6-6), obtaining

$$k\mu_b\pi_{k1} - \lambda\pi_{k-1,1} = (c - d + 1)\mu_b\pi_{c-d+1,1}, \quad c - d + 2 \le k \le c \tag{10}$$

Particularly,

$$c\mu_b\pi_{c1} - \lambda\pi_{c-1,1} = (c - d + 1)\mu_b\pi_{c-d+1,1} \tag{11}$$

(6-10) shows

$$c\mu_b\pi_{N1} - \lambda\pi_{N-1,1} = \frac{\theta}{1 - r}\pi_{N0}.$$

and from (6-8)

$$c\mu_b\pi_{k1} - \lambda\pi_{k-1,1} = c\mu_b\pi_{k+1,1} - \lambda\pi_{k1}, \quad c \le k \le N - 1.$$

From the two equations above recursively, we obtain

$$c\mu_b\pi_{k1} - \lambda\pi_{k-1,1} = \frac{\theta}{1 - r}\pi_{N0}, \quad c \le k \le N \tag{12}$$

Particularly,

$$c\mu_b\pi_{c1} - \lambda\pi_{c-1,1} = \frac{\theta}{1 - r}\pi_{N0}.$$

Substituting the equation above into (11), we have

$$\pi_{c-d+1,1} = \frac{\theta}{(c - d + 1)\mu_b(1 - r)}\pi_{N0} = K\sigma\beta\frac{\theta}{(c - d + 1)\mu_b(1 - r)}.$$

(10) shows,

$$\pi_{k1} = \frac{\lambda}{k\mu_b}\pi_{k-1,1} + \frac{c - d + 1}{k}\pi_{c-d+1,1}, \quad c - d + 2 \le k \le c.$$

Based on (5), recursively, we have

$$\pi_{k1} = \gamma_{k-c+d-1}\pi_{c-d+1,1} = K\sigma\beta\frac{\theta}{(c - d + 1)\mu_b(1 - r)}\gamma_{k-c+d-1}, c - d + 1 \le k \le c$$

If $k = c$, there is

$$\pi_{c1} = \gamma_{d-1}\pi_{c-d+1,1} = K\sigma\beta\frac{\theta}{(c - d + 1)\mu_b(1 - r)}\gamma_{d-1} \tag{13}$$

From (12)

$$\pi_{k1} = \frac{\lambda}{c\mu_b}\pi_{k-1,1} + \frac{\theta}{c\mu_b(1 - r)}\pi_{N0}, \quad c \le k \le N \tag{14}$$

Based on (13) and (14), recursively, we have

$$
\begin{aligned}
\pi_{k1} &= \left(\frac{\lambda}{c\mu_b}\right)^{k-c} \pi_{c1} + \sum_{v=0}^{k-c-1} \left(\frac{\lambda}{c\mu_b}\right)^v \pi_{N0} \\
&= \left(\frac{\lambda}{c\mu_b}\right)^{k-c} \gamma_{d-1}\pi_{c-d+1,1} + \sum_{v=0}^{k-c-1} \left(\frac{\lambda}{c\mu_b}\right)^v \pi_{N0} \\
&= K\sigma\beta \left[\left(\frac{\lambda}{c\mu_b}\right)^{k-c} \gamma_{d-1}\frac{\theta}{(c-d+1)\mu_b(1-r)} + \sum_{v=0}^{k-c-1} \left(\frac{\lambda}{c\mu_b}\right)^v \right], \; c \le k \le N
\end{aligned}
$$

Theorem 3. *If $\rho < 1$, and $0 \le j \le N$, the steady state distribution of (L, J) is given by Theorem 4.1, when $j > N$, we have*

$$
\begin{cases}
\pi_{j0} = K\sigma\beta r^{j-N}, & j > N \\
\pi_{j1} = \pi_{N1}\rho^{j-N} + \frac{\theta r}{c\mu_b(1-r)}\pi_{N0} \sum_{v=0}^{j-N-1} r^v \rho^{j-N-1-v}, & j > N
\end{cases} \tag{15}
$$

where

$$
\begin{aligned}
K = \Bigg\{ & 1 + \sum_{j=1}^{c-d} \frac{1}{j!} \left(\frac{\lambda}{\mu_b}\right)^j \\
& + \sigma\beta \sum_{j=c-d+1}^{c-2} \left(\varphi_{c-1-j}\Phi_{N-c+1} - \frac{(c-d)\mu_b+d\mu_v}{\lambda}\varphi_{c-2-j}\Phi_{N-c} \right) \\
& + \sigma\beta \sum_{j=c-1}^{N} \Phi_{N-j} + \sigma\beta\frac{\theta}{(c-d+1)\mu_b(1-r)} \sum_{j=c-d+1}^{c} \gamma_{j-c+d-1} \\
& + \sigma\beta \sum_{j=c+1}^{N} \left[\left(\frac{\lambda}{c\mu_b}\right)^{j-c} \gamma_{d-1}\frac{\theta}{(c-d+1)\mu_b(1-r)} + \sum_{v=0}^{j-c-1} \left(\frac{\lambda}{c\mu_b}\right)^v \right] \\
& + \sigma\beta\frac{\rho}{1-\rho} \left[\left(\frac{\lambda}{c\mu_b}\right)^{N-c} \gamma_{d-1}\frac{\theta}{(c-d+1)\mu_b(1-r)} + \sum_{v=0}^{N-c-1} \left(\frac{\lambda}{c\mu_b}\right)^v \right] \\
& + \sigma\beta\frac{r}{1-r} + \sigma\beta\,\Phi_0\frac{\theta r}{c\mu_b(1-r)} \sum_{j=N+1}^{\infty} \sum_{v=0}^{j-N-1} r^v \rho^{j-N-1-v} \Bigg\}^{-1}
\end{aligned}
$$

Proof. According to the matrix geometric solution method, when $k \ge N$, we have

$$
\pi_k = (\pi_{k0}, \pi_{k1}) = (\pi_{N0}, \pi_{N1})R^{k-N} \tag{16}
$$

From (2), we obtain

$$
R = \begin{pmatrix} r^k & \frac{\theta r}{c\mu_b(1-r)} \sum_{v=0}^{k-1} r^v \rho^{k-1-v} \\ 0 & \rho^k \end{pmatrix}, \; k \ge 1.
$$

Substituting it into (16), we have (15), K can be obtained by the equilibrium condition

$$
\sum_{k=0}^{\infty} \pi_{k0} + \sum_{k=c-d+1}^{\infty} \pi_{k1} = 1.
$$

The distribution of the number of customers L in the steady state is

$$
P\{L = k\} = \pi_{k0}, \; 0 \le k \le c - d; \; P\{L = k\} = \pi_{k0} + \pi_{k1}, \; k > c - d.
$$

Remark 1. Specially, If N = 1, $d = 0$, $c = 1$, the model can simply boil down to an M/M/1 queue with working vacation, and the Theorem 4.1 is equivalent to the result of article [10]. If N = 1, $d = 1$, $c = 1$, the model can simply boil down to an M/M/1 queue, just as the analysis in article [11].

5   Numerical Results

Zhang [12] studied the performance analysis of a queue with congestion-based staffing policy, maintaining the average queue length within a certain range is the primary objective, where the number of servers is adjusted according to the queue length during a planning period. On the basic of article [12], the model in this paper can be applied to allocating resources too, such as hierarchical cellular mobile communication system.

Considering a system with c channels, we suppose any calling occupies only one channel, and the arrival of calling is assumed to be exponentially distributed with arrival rate λ, if there are d channels become idle, we can make the d idled channels become adjacent higher or lower cellular level's overflow pool, providing channels both for the two cellular layers' incoming callings. The service time of the d channels for overflowing is assumed to be exponentially distributed with mean $1/\theta$. When a overflowing is completed, if the number of calling in the original cellular layer is not less than N, the d channels return to normal work, otherwise, start another time of overflowing. Where d and N can be seen as a threshold of overflowing, allowing the adjacent cellular layer to overflow while d channels of the original cellular layer are idle, when the calling in the original cellular layer reaches or exceeds N at the end of a overflowing, the overflowing is terminated. It avoid too frequent switching, allocate resources reasonable, and reduce the call blocking rate.

The transmission time of normal channels is assumed to be exponentially distributed with transmission rate μ_b. The transmission time of overflowing channels is assumed to be exponentially distributed with transmission rate μ_v. The arrival and transmission time are independent each other, the transmission order is FCFS.

The distribution of the number of calling L is

$$P\{L = k\} = \begin{cases} \pi_{k0}, & 0 \leq k \leq c - d, \\ \pi_{k0} + \pi_{k1}, & k \geq c - d + 1. \end{cases}$$

There are many indicators affect the performance of the system, according to the steady state distribution we have got in Theorems 4.2 and 4.3, we select blocking rate and using rate for numerical analysis. When the calling in the cellular layer is more than c, the calling is blocked, so the probability of call blocking is

$$P_b = P\{L > c\} = \sum_{k=c+1}^{\infty} (\pi_{k0} + \pi_{k1}) \tag{17}$$

Substituting π_{k0} π_{k1} that we have obtained from theorems above into (17)

$$P_b = K\sigma\beta \left[\sum_{j=c+1}^{N} \Phi_{N-k} + \frac{r}{1-r} + \gamma_{d-1} \frac{\theta}{(c-d+1)\mu_b(1-r)} \sum_{j=c+1}^{N} \rho^{j-c} \right.$$

$$\left. + \sum_{j=c+1}^{N} \sum_{v=0}^{j-c-1} \rho^v \right] + \frac{\rho}{1-\rho}\pi_{N1} + \frac{\theta r}{c\mu_b(1-r)}\pi_{N0} \sum_{j=N+1}^{\infty} \sum_{v=0}^{j-N-1} r^v \rho^{j-N-1-v}$$

$$(18)$$

Based on the steady state distribution, the average utilization rate of the channel U is

$$U = \frac{\sum_{j=1}^{c} j\pi_{j0} + \sum_{j=c+1}^{\infty} c\pi_{j0} + \sum_{j=c-d+1}^{c} j\pi_{j1} + \sum_{j=c+1}^{\infty} c\pi_{j1}}{c} \qquad (19)$$

Making numerical analysis on the indicators of (18) and (19) with matlab, setting the total number of channels $c = 24$; $\theta = 0.01$; $\mu_b = 0.05$; $\mu_v = 0.04$. Let N = 40; d are equal to 3, 4, 5, 7 respectively, when the arrival rate λ varies between 0.5 and 0.7. Figure 1 shows the curve of P_b changes with λ, and Fig. 2 shows the curve of U changes with λ. As it can be seen from Figs. 1 and 2, in the range of λ. Both P_b and U increased with the increasing of λ, all of P_b and U increased with the increasing of d. It is convenient for us to select the optimal P_b and U values based on the changing of λ.

For obvious observation, we take the arrival rate from 0.68 to 0.69. Take d = 10, N is equal to 39, 40, 41, 42 respectively. Figure 3 shows the curve of P_b changes with λ, Fig. 4 shows the curve of U changes with λ. As it can be seen from Figs. 3 and 4, both P_b and U increased with the increasing of λ, under the same conditions, the larger the value of N is, the larger the P_b and U are.

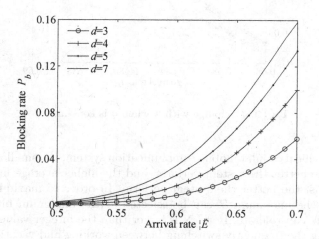

Fig. 1. P_b change with λ when N is constant.

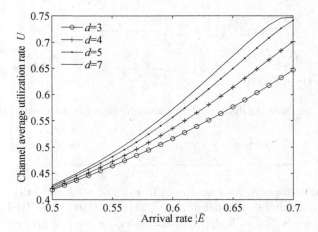

Fig. 2. U change with λ when N is constant.

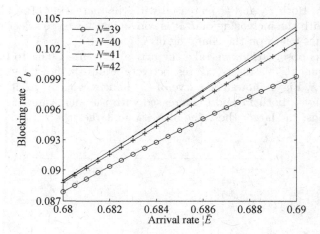

Fig. 3. P_b change with λ when d is constant.

In a hierarchical cellular mobile communication system, the smaller the blocking rate is, the better the system perform, and the higher average utilization of the channels is, the better the system perform. In order to maximize the system revenue, the maximum U can be selected at the maximum blocking rate that customers can tolerate. Here, N is larger than the integer value of c, effectively reducing the frequently switching between working and vacation. We can also set N to any integer value less than or equal to c, then get the stationary distribution, set the parameters to get the result of P_b and U.

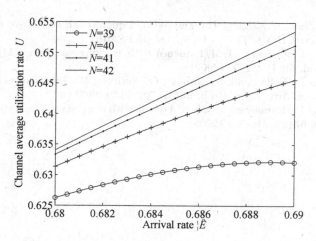

Fig. 4. U change with λ when d is constant.

6 Conclusions

In this paper, the N-policy and d-type working vacation queue model is proposed, the model is established and analysed, and the steady state distribution of queue length is got. Applying the result to the hierarchical cellular mobile communication system. Through the numerical analysis, the relationship between system performance and parameters is obtained, which provides a theoretical basis for the analysis of hierarchical cellular mobile communication system.

References

1. Zhang, Z.G., Tian, N.: Analysis on queueing systems with synchronous vacations of partial servers. Perform. Eval. **52**(4), 269–282 (2003)
2. Yang, W.S.: Algorithmic approach to Markovian multi-server retrial queues with vacations. Appl. Math. Comput. **250**(250), 287–297 (2015)
3. Ammar, S.I.: Transient solution of an M/M/1 vacation queue with a waiting server and impatient customers. J. Egypt. Math. Soc. 1–6 (2016)
4. Takhedmit, B., Abbas, K.: A parametric uncertainty analysis method for queues with vacations. J. Comput. Appl. Math. **312**, 143–155 (2017)
5. Ke, J.C., Lin, C.H., Yang, J.Y., Zhang, Z.G.: Optimal (d, c) vacation policy for a finite buffer M/M/c queue with unreliable servers and repairs. Appl. Math. Model. **33**(10), 3949–3962 (2009)
6. Xu, X., Tian, N.: The M/M/c queue with (e, d) setup time. J. Syst. Sci. Complex. **21**(3), 446–455 (2008)
7. Xu, X., Tian, N.: Analysis on M/M/c queue with multiple (e, d, N)-policy vacation. Chin. J. Eng. Math. **23**(6), 1095–1100 (2006). Gongcheng Shuxue Xuebao
8. Zhu, Y., Xu, H.: M/M/c queue of partial working vacation with N-policy and negative customer. J. Jiangsu Univ. **34**(3), 367–372 (2013). (Natural Science Edition)

9. Wang, H., Zhou, G., Zhang, H.: The M/M/1 queue with successive two kinds of vacation. Math. Pract. Theor. **46**(8), 188–192 (2016)
10. Servi, L.D., Finn, S.G.: M/M/1 queues with working vacations (M/M/1/WV). Perform Eval. **50**(1), 41–52 (2002)
11. Leguesdron, P., Pellaumail, J., Rubino, G., Sericola, B.: Transient analysis of the M/M/1 queue. Adv. Appl. Prob. **25**(3), 702–713 (1993)
12. Zhang, Z.G.: Performance analysis of a queue with congestion-based staffing policy. Manag. Sci. **55**(2), 240–251 (2009)

Group-Server Queues

Quan-Lin Li[1]([✉]), Jing-Yu Ma[1], Mingzhou Xie[2], and Li Xia[2]

[1] School of Economics and Management Sciences,
Yanshan University, Qinhuangdao 066004, China
`liquanlin@tsinghua.edu.cn`, `mjy0501@stumail.ysu.edu.cn`
[2] Department of Automation and TNList,
Center for Intelligent and Networked Systems,
Tsinghua University, Beijing 100084, China
`xmz16@mails.tsinghua.edu.cn`, `xial@tsinghua.edu.cn`

Abstract. By analyzing energy-efficient management of data centers, this paper proposes and develops a class of interesting *Group-Server Queues*, and establishes two representative group-server queues through loss networks and impatient customers, respectively. Furthermore, such two group-server queues are given model descriptions and necessary interpretation. Also, simple mathematical discussion is provided, and simulations are made to study the expected queue lengths, the expected sojourn times and the expected virtual service times. In addition, this paper also shows that this class of group-server queues are often encountered in many other practical areas including communication networks, manufacturing systems, transportation networks, financial networks and healthcare systems. Note that the group-server queues are always used to design effectively dynamic control mechanisms through regrouping and recombining such many servers in a large-scale service system by means of, for example, bilateral threshold control, and customers transfer to the buffer or server groups. This leads to the large-scale service system that is divided into several adaptive and self-organizing subsystems through scheduling of batch customers and regrouping of service resources, which make the middle layer of this service system more effectively managed and strengthened under a dynamic, real-time and even reward optimal framework. Based on this, performance of such a large-scale service system may be improved greatly in terms of introducing and analyzing such group-server queues. Therefore, not only analysis of group-server queues is regarded as a new interesting research direction, but there also exist many theoretical challenges, basic difficulties and open problems in the area of queueing networks.

Keywords: Group-server queue · Data center
Energy-efficient management · Loss network · Impatient customer

1 Introduction

In this paper, we propose and develop a class of interesting *Group-Server Queues* by analyzing energy-efficient management of data centers, and establish two

W. Yue et al. (Eds.): QTNA 2017, LNCS 10591, pp. 49–72, 2017.
https://doi.org/10.1007/978-3-319-68520-5_4

representative group-server queues by means of loss networks and impatient customers, respectively. Also, we show that this class of group-server queues are often encountered in many other practical areas including communication networks, manufacturing systems, transportation networks, financial networks and healthcare systems. Therefore, not only analysis of group-server queues is regarded as a new interesting research direction, but there also exist many theoretical challenges, basic difficulties and open problems in the area of queueing networks.

Data centers are main infrastructure platforms for various kinds of large-scale practical information systems, and always offer economies of scale for network, power, cooling, administration, security and surge capacity. So far it has been an interesting research direction to reduce the server farm energy requirements and to optimize the power efficiency which may be viewed as a ratio of performance improvement to power consumption reduction. For energy-efficient management of data centers, some authors have dealt with several key interesting issues, for example, data center network architecture by Al-Fares et al. [1], Guo et al. [20] and Pries et al. [35]; green networks and cloud by Kliazovich et al. [25], Mazzucco et al. [32], Gruber and Keller [18], Goiri et al. [13,14] with solar energy, Li et al. [27] with wind energy, Wu et al. [41] and Zhang et al. [43]; networks of data centers by Greenberg et al. [16,17], Gunaratne et al. [19], Shang et al. [38], Kliazovich et al. [26] and Wang et al. [39]; resiliency of data centers by Heller et al. [23] and Baldoni et al. [3]; revenues, cost and performance by Elnozahy et al. [8], Chen et al. [6], Benson et al. [4], Dyachuk and Mazzucco [7], Mazzucco et al. [31] and Aroca et al. [2]; analyzing key factors by Greenawalt [15] for hard disks, Chase et al. [5] for hosting centers (i.e., the previous one of data center), Guo et al. [21] for base station sleeping control, Guo et al. [22] for edge cloud systems, Horvath and Skadron [24] for multi-tier web server clusters, Lim et al. [30] for multi-tier data centers, Rivoire et al. [34] for a full-system power model, Sharma et al. [37] for QoS, Wierman et al. [40] for processor sharing, and Xu and Li [42] for part execution.

In analyzing energy-efficient management of data centers, queueing theory and Markov processes are two effective mathematical tools both from performance evaluation and from optimal control. Up till now, few papers have applied queueing theory, together with Markov processes, to performance analysis of data centers with energy-efficient management. Chen et al. [6] proposed a queueing model to control energy consumption of service provisioning systems subject to Service Level Agreements (SLAs). Nedevschi et al. [33] demonstrated that introducing a sleep state to power management in a data center can save much of the present energy expenditure, and even simple schemes for sleeping also offer substantial energy savings. Shang et al. [38] used network devices to routing service design, and made the idle network devices to shut down or to put into a sleep state. Gandhi et al. [9] modeled a server farms with setup cost by means of an $M/M/k$ queueing system, where each server can be in one of the following states: work, on, idle, off, or sleep, employed the popular metric of Energy-Response time Product (ERP) to capture the energy-performance tradeoff, and gave the first theoretical result on the optimality of server farm management policies. From a similar analysis to [9], Gandhi *et al.* [10] used an $M/M/k$ queueing system but each server has only three states: on, setup and off, and obtained the distributions of the response time and of the power consumption. Further, Gandhi and Harchol-Balter [11], Gandhi *et al.* [12] analyzed effectiveness of dynamic power management in data centers in terms of an $M/M/k$ model, and found that the dynamic power management policies are very effective when the setup time is small, the job size is large or the size of the data center is increasing; in contrast, the dynamic power management policies are ineffective for small data centers. Schwartz *et al.* [36] and Gunaratne *et al.* [19] provided the energy efficient-mechanism with dual thresholds. Schwartz *et al.* [36] presented a queueing model to evaluate the trade-off between the waiting time and the energy consumption, and also developed a queueing model with thresholds to turn-on reserve servers when needed. Gunaratne *et al.* [19] developed a single-server queue with state-dependent service rates and with dual thresholds for service rate transitions.

Contributions of this paper: The main contributions of this paper are threefold. The first one is to propose and develop a class of interesting *Group-Server Queues*, and establishes two representative group-server queues by means of loss networks and impatient customers, respectively, under a practical background for analyzing energy-efficient management of data centers. The second contribution is to set up a general framework for group-server queues, and to give a detailed discussion for basic issues, for instance, optimally structural division of server groups, transfer policies among server groups and/or buffers, dynamic control mechanism design, revenue management and cost control. The third contribution is to provide a simple mathematical analysis for a two-group-server loss queue, and also to design simulation experiments to evaluate the expected queue lengths, the expected sojourn times and the expected virtual service times for a three-group-server loss queue and a three-group-server queue with an infinite buffer. Note that this class of group-server queues are often encountered in many other practical areas including communication networks, manufacturing systems, transportation networks, financial networks and healthcare systems. Therefore, the methodology and results given in this paper provide highlights on a new class of queueing networks called group-server queues, and are applicable to the study of large-scale service networks in practice.

Organization of this paper: The structure of this paper is organized as follows. In Sect. 2, we propose and develop a class of new interesting queueing models: *Group-Server Queues*, and establish a basic framework of the group-server queues, such as, model structure, operations mechanism, necessary notation and key factors. In Sect. 3, we describe a group-server loss queue from analyzing energy-efficient management of data centers, where the loss mechanism makes a finite state space so that the group-server loss queue must be stable. In Sect. 4, we describe a group-server queue with impatient customers and with an infinite buffer, where some key factors of this system are discussed in detail. In Sect. 5, we provide a simple mathematical analysis for the two-group-server loss queue, and several open problems are listed. In Sect. 6, we design simulation experiments for performance evaluation of the three-group-server loss queues and of the three-group-server queues with an infinite buffer, and specifically, we analyze their expected queue lengths, the expected sojourn times and the expected virtual service times. Finally, some concluding remarks are given in Sect. 7.

2   A Basic Framework of Group-Server Queues

In this section, we propose a class of new interesting queueing models: *Group-Server Queues*, and establish a basic framework for the group-server queues, for instance, model structure, operations mechanism, necessary notation and key factors.

We propose such group-server queues by analyzing energy-efficient management of data centers. As analyzed in the next two sections, two representative group-server queues are established according to the need of energy saving. To realize energy saving in data center networks, a sleep (or off) state introduced to some servers is a class of key techniques. Using States on, sleep, off and others, a large-scale service system with more servers is divided into several subsystems, each of which contains some servers having certain common attributes. For example, a data center has a set of all different attributes: $E = \{$work, on, sleep, off$\}$, where 'work' denotes that a server is 'on' and is also serving a customer; 'on' denotes that a server is idle and ready to serve; 'sleep' denotes that a server is at the dormancy stage with lower power consumption; and 'off' denotes that a server is shut down, where a setup time may be needed if the server change its state from off to on (or sleep).

Now, we provide a concrete example how to establish the different groups of servers. Let the set of all the servers in the data center be $\Omega = \{$Server 1, Server 2, Server 3, ..., Server $N\}$. Then the service system of the data center can be divided into three subsystems whose server groups are given by

$\Omega_1 = \{$Server 1, Server 2, ..., Server $n\}$, each server is either at work or on;

$$\Omega_2 = \{\text{Server } n+1, \text{ Server } n+2, \ldots, \text{ Server } n+m\},$$
each server is either at work, on, or sleep;

$$\Omega_3 = \{\text{Server } n+m+1, \text{ Server } n+m+2, \ldots, \text{ Server } N\},$$
each server is either at work, on, sleep or off.

It is seen that the attributes of servers in Ω_2 are more than that in Ω_1, that is, sleep. While the sleep attribute makes Ω_2 practically different from Ω_1. See Fig. 1 in the next section for an intuitive understanding. At the same time, to further understand the attribute role played by the energy efficient-mechanism of data centers, readers may refer to the next two sections for establishing and analyzing two representative group-server queues.

In general, this class of group-server queues are often encountered in many other practical areas, for example, communication networks for green and energy saving, manufacturing systems for priority use of high-price devices, transportation networks with different crowded areas, and healthcare systems having different grade hospitals. Therefore, analyzing the group-server queues is an interesting research direction both in the queueing area and in many practical fields such as computer, communication, manufacturing, service, market, finance and so on. To our best knowledge, no previous work has looked at and summarized the group-server queues from a theoretical or practical framework yet.

Based on the above analysis, we provide a basic framework for the group-server queues, and discuss model structure, operations mechanism, necessary notation and key factors as follows:

Server Groups: We assume that a large-scale service system contains many servers whose set is given by

$$\Omega = \{\text{Server } 1, \text{ Server } 2, \text{ Server } 3, \ldots, \text{ Server } N\},$$

and they also have some different attributes whose set is given by

$$E = \{A_1, A_2, A_3, \ldots, A_r\},$$

where A_i is an attribute for $1 \leq i \leq s$. From practical need and physical behavior of a large-scale service system, the attribute set E is divided into some different subsets as follows:

$$E = E_1 \cup E_2 \cup \cdots \cup E_s, \ s \leq r.$$

Note that E_i and E_j may have common elements for $1 \leq i < j \leq s$. Applying the subsets E_i for $1 \leq i \leq s$ to system behavior, the server set Ω is divided into some different groups or subsets as follows:

$$\Omega = \Omega_1 \cup \Omega_2 \cup \cdots \cup \Omega_s,$$

where the server group Ω_i well corresponds to the attribute subset E_i for $1 \leq i \leq r$. Note that the server groups Ω_i for $1 \leq i \leq s$ are disjoint. In this case, the large-scale service system is divided into s subsystems, each of which contains the servers in the group Ω_i having certain common attributes in one of the subsets E_i for $1 \leq i \leq s$.

In practice, the attributes A_i for $1 \leq i \leq r$ have a wide range of meanings, for instance, states, properties, behaviors, control and mechanism. From such a setting, it is clear that the server grouping of a large-scale service system should not be unique.

Arrival processes: In the group-server queue, we assume that customer arrivals at this system are a renewal process with stationary arrival rate λ. For the customer arrivals, we shall encounter two issues: (a) Routing allocation mechanism, for example, joining the shortest queue length, and joining the shortest sojourn time. (b) Arrival rate control, for example, arrival rates depending on system states, arrival rates depending on prices, arrival rates depending on sojourn times, and arrival rates depending on threshold control.

Service processes: In the ith server group, the service times of customers are i.i.d. with general distribution function $F_i(x)$ of stationary service rate μ_i for $1 \leq i \leq N$. For the customer service processes, we shall encounter two issues: (a) service disciplines, for example, FCFS, LCFS, processor sharing, and matching service. (b) Service rate control, for example, service rates depending on system states, service rates depending on prices, service rates depending on sojourn times, and service rates depending on threshold control.

Customers transfer among server groups: The customers in heavy-load server groups are encouraged to transfer into light-load server groups; the customers in low-speed-service server groups are encouraged to transfer into high-speed-service server groups; the customers in high-cost-service server groups are encouraged to transfer into low-cost-service server groups; and so forth. Under customer transferring, the residual service times of non-exponential distributions always make model analysis more difficult and challenging.

Stability is a difficult issue: Since the group-server queue is always a large-scale complicated stochastic system, its stability and associated conditions are always very difficult to study. To easily deal with system stability, it is a simple method to introduce loss networks or impatient customers to the group-server queues, where the former is to use the finite state space, while the latter is to apply stability of the renewal processes. Therefore, this paper uses the loss networks and the impatient customers to set up some examples, which show how to simply guarantee stability of some group-server queues.

3   Group-Server Loss Queues

In this section, we describe a group-server loss queue from analyzing energy-efficient management of data centers. Note that the 'Loss Mechanism' is to set up a finite state space, whose purpose is to guarantee stability of the group-server loss queue.

In the energy-efficient management study of data center networks, it was an effective way to introduce two states: sleep and off for some servers. Based on this, we can make some different states: work, on, sleep, off and others, and the servers of the data center are grouped as $\Omega = \Omega_0 \cup \Omega_1 \cup \cdots \cup \Omega_N$. Concretely, a simple group-server loss queue is constructed under two states: work-on, and sleep. Here, work and on are assumed to have the same power consumption, hence work and on are regarded as a state: work-on. See Fig. 1, for understanding the $N+1$ server groups and even for intuitively understanding the group-server loss queue.

Now, we use the data center to describe a group-server loss queue, and set up energy efficient-mechanism, system parameters and model notation as follows:

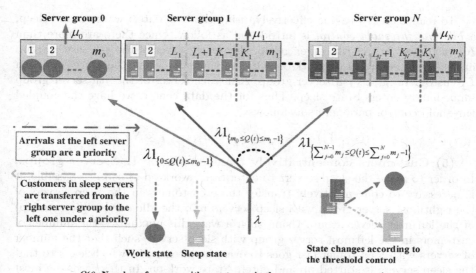

Fig. 1. Physical structure of the group-server loss queue

(1) Server groups: We assume that the data center contains $N + 1$ different server groups, each of which is a subsystem of the data center. For $j \in \{0, 1, 2, \ldots, N\}$, the jth server group contains m_j same servers. Thus the data center contains $\sum_{j=0}^{N} m_j$ servers in total. Note that the $N + 1$ different server groups can be divided into two basic categories: (a) Server group 0 is special, because its each server has only one state: work-on. Hence server group 0 with m_0 same servers is viewed as the base-line group in the data center. (b) Each server of the other N server groups has two states: work-on and sleep.

(2) Arrival processes: The arrivals of customers at the data center from outside are a renewal process with stationary arrival rate λ. An arriving customer preferentially enters one idle server of the leftmost server group with idle servers. We assume that each server and the data center all have no waiting room, while each server receives and serves only one customer at a time. Hence any new arrival is lost once all the servers contain their one customer, that is, the system is full when it has at most $\sum_{j=0}^{N} m_j$ customers.

(3) Service processes: In the jth server group, the service times of customers are i.i.d. with general distribution function $F_j(x)$ of stationary service rate μ_j for $0 \leq j \leq N$.

(4) Bilateral threshold control: Except server group 0, each server in the other N server groups have two states: work-on and sleep. To switch between work-on and sleep, it is necessary for the jth server group to introduce a positive integer pair (L_j, K_j) with $0 \leq L_j \leq K_j \leq m_j$, which leads to a class of bilateral threshold control applied to energy-efficient management of the data center networks.

To realize energy saving effectively, using the two states: work-on and sleep, a *bilateral threshold control* is introduced as follow: Once there are more than K_j customers in the jth server group, then each server of the jth server group immediately enters State work-on. On the contrary, when there are less than L_j customers in the jth server group, then each server of the jth server group immediately enters State sleep. Thus for the data center, we have the coupled threshold control parameters as follows:

$$\{(L_j, K_j) : 0 \le L_j \le K_j \le m_j, 1 \le j \le N\}.$$

(5) Customers concentratively transfer among the server groups: In order to make the sleep servers to enter State work-on as soon as possible, it is necessary to concentratively transfer those customers in the sleep servers of the rightmost server group with sleep servers into the idle servers (on or sleep) of the leftmost server group. Using such a way, this maximizes the number of customers in the leftmost server group with sleep servers such that the number of servers with a customer fast goes to over the integer K_j, which leads to that the sleep servers is started up and enters State work-on. In this case, the most sleep servers in the data center are setup at State work-on so that more and more customers are served as soon as possible.

(6) Energy consumption: We assume that the power consumption rates for the $1 + N$ server groups are given by $P_{W_0}, P_{W_1}, \ldots, P_{W_N}$ for State work-on; and $P_{S_0} = 0, P_{S_1}, P_{S_2}, \ldots, P_{S_N}$ for State sleep. To realize energy saving, let $0 < P_{S_j} < P_{W_j}$ for $j = 1, 2, \ldots, N$.

We assume that all the random variables in the system defined above are independent of each other.

A basic issue: Establishing some cost (or reward) functions is to evaluate a suitable trade-off between the sojourn time and the energy consumption. To concentratively reduce the sojourn time and to save energy, some effective methods are proposed and developed, such as, (a) the bilateral threshold control, (b) the customers at sleep servers concentratively transfer among the server groups, and (c) the residual service times are wasted or re-used. We need to analyze their performance and trade-off due to some mutual contradiction between reducing the sojourn time and saving the energy.

4   Group-Server Queues with Impatient Customers

In this section, we consider a group-server queue with impatient customers, which is refined and abstracted from energy-efficient management of data center networks, where the impatient customers are introduced to guarantee stability of this group-server queue with an infinite buffer.

In the data center, we still introduce two states: work-on and sleep for some servers. Based on the two different states, the servers of the data center are grouped as $\Omega = \Omega_0 \cup \Omega_1 \cup \Omega_2 \cup \cdots \cup \Omega_N$. See Fig. 2 both for the $N + 1$ server groups and for the group-server queue with impatient customers and with an infinite buffer.

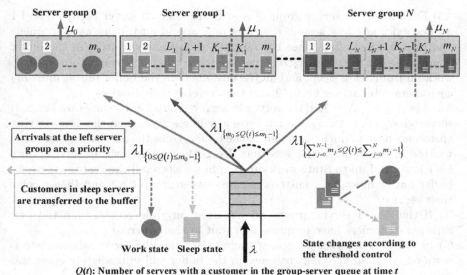

$Q(t)$: **Number of servers with a customer in the group-server queue at time t**

Fig. 2. Physical structure of the group-server queue with impatient customers

Now, we describe a group-server queue with impatient customers and with an infinite buffer, and also explain energy efficient mechanism, system parameters and model notation as follows:

(1) Server groups: We assume that the data center contains $N + 1$ different server groups, each of which is a subsystem of the data center. For a $j \in \{0, 1, 2, \ldots, N\}$, the jth server group contains m_j same servers. Thus the data center contains at most $\sum_{j=0}^{N} m_j$ servers. Note that the $N + 1$ different server groups can be divided into two basic categories: (a) Server group 0 is special, because its each server has only one state: work-on, hence server group 0 with m_0 same servers is viewed as the base-line group in the data center. (b) Each server of the other N server groups has two states: work-on and sleep.

(2) Arrival processes: The arrivals of customers at the data center from outside are a renewal process with stationary arrival rate λ. Each arrival customer must first enter the infinite buffer, then he is assigned to the $N + 1$ server groups according to the following allocation rules:

(2-A1) Each server in server group 0 is always at State work-on. If server group 0 have some idle servers, then the arriving customer can immediately enter one idle server in server group 0 and then receive his service.

(2-A2) Each server in server group 1 is at State sleep. If server group 0 does not exist any idle server, then the new arrival customers have to queue and wait in the buffer. Once the number of customers waiting in the buffer is not less than K_1, then each server in server group 1 is started up to State work-on from State sleep, and all the customers in the buffer but at most m_1 customers enter server group 1 and then receive their service.

(2-A3) Each server in server group 2 is at State sleep. If server groups 0 and 1 do not exist any idle server, then the new arrival customers have to queue and wait in the buffer. Once the number of customers waiting in the buffer is not less than K_2, then each server in server group 2 is started up to State work-on from State sleep, and all the customers in the buffer but at most m_2 customers enter server group 2 and then receive their service.

(2-A4) Let $2 \leq l \leq N - 1$. Each server in server group $l + 1$ is at State sleep. If server groups $0, 1, 2, \ldots, l$ do not have any idle server, then the new arriving customers have to queue and wait in the buffer. Once the number of customers waiting in the buffer is not less than K_{l+1}, then each server in server group $l + 1$ is started up to State work-on from State sleep, all the customers in the buffer but at most m_{l+1} customers enter server group $l + 1$ and then receive their service.

(2-B) If the $N + 1$ server groups do not exist any idle server, then the new arriving customers have to queue and wait in the buffer.

(2-C) In the $N + 1$ server groups, if there exists one idle server whose state is work-on, then an arriving customer in the buffer will immediately enter this server and then receive his service.

(3) **Service processes:** In the jth server group, the service times of customers are i.i.d. with general distribution function $F_j(x)$ of stationary service rate μ_j for $0 \leq j \leq N$.

(4) **Customer impatient processes:** Each customer in this system has an exponential patient time of impatient rate θ.

(5) **Bilateral threshold control:** Except server group 0, each server in the other N server groups have two states: work-on and sleep. To switch between work-on and sleep, it is necessary for the jth server group to introduce a positive integer pair (L_j, K_j) with $0 \leq L_j \leq K_j \leq m_j$, which leads to a bilateral threshold control by means of energy-efficient management of the data center networks. By using the two states: work-on and sleep, a *Bilateral Threshold Control* is introduced as follow:

(5-1) **From sleep to work-on:** We assume that each server in server group j is at State sleep, and an idle server does not exist in server groups $0, 1, 2, \ldots, j - 1$, for $1 \leq j \leq N$. If there are not less than K_j customers in the buffer, then each server in server group j is started up to State work-on from State sleep, all the customers in the buffer but at most m_j customers enter server group j and then receive their service.

(5-2) **From work-on to sleep:** We assume that each server in server group j is at State work-on. If there are less than L_j customers being served in server group j, then each server of server group j immediately enters State sleep from State work-on, and those customers being served at server group j are transferred to the head of the buffer. For such transferred customers, we assume that those service times obtained by the transferred customers, will become zero, and their service get start again (note that another useful case is that the received service times can be cumulative, once arriving at the total service time, the service is completed) immediately. Also, to allocate the customers in the buffer to some server groups, each customer transferred in the head of the buffer is the same as the new arriving customers. Thus we have the coupled threshold control parameters

$$\{(L_j, K_j) : 0 \leq L_j \leq K_j \leq m_j, 1 \leq j \leq N\}.$$

(6) Customers transfer to the buffer: In order to save energy as much as possible, if there are less than L_j customers being served in server group j, then it is necessary to transfer those customers in the server group j at State work-on into the head of the buffer, and each server of server group j immediately enters State sleep from State work-on.

(7) Energy consumption: We assume that the power consumption rates for the $1 + N$ server groups are given by $P_{W_0}, P_{W_1}, \ldots, P_{W_N}$ for State work-on; and $P_{S_0} = 0, P_{S_1}, P_{S_2}, \ldots, P_{S_N}$ for State sleep. For energy saving, let $0 < P_{S_j} < P_{W_j}$ for $j = 1, 2, \ldots, N$.

We assume that all the random variables in the system defined above are independent of each other.

A basic issue: Constructing some cost (or reward) functions is to evaluate a suitable trade-off between the sojourn time and the energy consumption. Note that some effective methods are proposed and developed, such as, (a) the bilateral threshold control, (b) customers in sleep servers transfer to the buffer, and (c) residual service times are wasted or re-used. We need to analyze their performance and the suitable trade-off due to some mutual contradiction between reducing the sojourn time and saving the energy.

Further discussion for stability: For the group-server queue with an infinite buffer drawn in Fig. 2, it is easy to give a sufficient condition of system stability: $\rho = \lambda/m_0\mu_0 < 1$, by means of a path coupling or comparison of a Markov process.

It is seen from Fig. 2 that server groups $1, 2, \ldots, N$ provide many service resources or ability in processing the customers, and this buffer also plays a key role in concentratively transferring the customers in sleep servers to improve service ability of the whole system. Therefore, it may be an interesting open problem to set up the necessary conditions under which the system stability is, how to be influenced by the key factors or parameters, as follows: (a) the bilateral threshold control, (b) customers in sleep servers transfer to the buffer, and (c) many residual service times are wasted or re-used.

5 Mathematical Analysis and Open Problems

In this section, we provide some simple mathematical analysis for a two-group-server loss queue with server group 0 and server group 1, a whole detailed investigation of which was given in Li *et al.* [29]. From such mathematical analysis, it is seen that analyzing more general group-server queues is interesting, challenging and difficult, and thus several open problems are listed for the future study of group-server queues.

5.1 Some Mathematical Analysis

A special model: $N = 1$. We consider a special two-group-server loss queue with server group 0 and server group 1. See model descriptions for a more general case given in Sect. 3 in more details.

For convenience of readers, it is still necessary to simply re-list some assumptions and descriptions for this special case as follows: **(a)** Server group 0 contains n servers, each of which has only one state: work-on; while server group 1 contains m servers, each of which has two different states: work-on and sleep. **(b)** No waiting room is available both at each server and in the group-server queue. Once there are $n + m$ customers in this systems (i.e., one server is serving a customer), any new customer arrival will be lost due to no waiting room. **(c)** The arrivals of customers at this two-group-server queue are a Poisson process with arrival rate λ. **(d)** The service times provided by server group 0 and by server group 1 are i.i.d. with two exponential distributions of service rate μ_0 and μ_1, respectively. **(e)** *Unilateral threshold control:* $L_1 = 0$ and $K_1 > 0$. In this case, if server group 0 are full with n serving customers, and if there are not less that K_1 customers waiting at server group 1 with m sleep servers, then each server of server group 1 immediately enters State work-on, and then provides service for its customer. **(f)** Because of $L_1 = 0$, once all the customers in server group 1 have completed their service, the m servers of server group 1 immediately switch to State sleep from State work-on; if there are some idle (on) servers in server group 0, and if there are customers waiting in server group 1 in which each server is at State sleep, then the customers waiting at the sleep servers of server group 1 can transfer to those idle (on) servers of server group 0. **(g)** For each server in the two server groups, the power consumption rates are listed as: P_{W_0} and P_{W_1} for State work-on, and P_{S_1} for State sleep ($P_{S_0} = 0$ because each server in server group 0 does not have State sleep). To save energy, we assume that $0 < P_{S_1} < P_{W_1}$.

A QBD process: For this two-group-server loss queue, the states of the corresponding Markov process are defined as the tuples: $(l_0, i; l_1, j)$ as shown in Fig. 3, where $l_0 = W$, $l_1 \in \{W, S\}$, W and S denote States work-on and sleep, respectively. Let i be the number of servers serving a customer in server group 0, and j the number of servers serving a customer in server group 1. It is seen from Fig. 3 that for $(W, i; l_1, j)$, we can set up four different sets as follows:

(1) In Block 1, $i \in \{0, 1, \ldots, n-1, n\}$; $l_1 = S$, $j = 0$;
(2) in Block 2, $i \in \{n\}$; $l_1 = S$, $j \in \{1, 2, \ldots, K-1\}$;
(3) in Block 3, $i \in \{0, 1, \ldots, n-1, n\}$; $l_1 = W$, $j \in \{1, 2, \ldots, K-1\}$;
(4) in Block 4, $i \in \{0, 1, \ldots, n-1, n\}$; $l_1 = W$, $j \in \{K, K+1, \ldots, m\}$.

We denote by $N_0(t)$ and $N_1(t)$ the numbers of servers serving a customer in server group 0 and server group 1, respectively; and $S_0(t)$ and $S_1(t)$ the states of servers in server groups 0 and 1, where $S_0(t) = W$ and $S_1(t) \in \{W, S\}$. Let $\mathbf{X}(t) = (S_0(t), N_0(t); S_1(t), N_1(t))$ with $S_0(t) = W$. Then $\{\mathbf{X}(t), t \geq 0\}$ is a QBD process with finitely many levels. From Fig. 3, it is seen that the QBD process $\{\mathbf{X}(t), t \geq 0\}$ has a state space $\boldsymbol{\Omega} = \boldsymbol{\Omega}_0 \cup \boldsymbol{\Omega}_1 \cup \boldsymbol{\Omega}_2 \cup \cdots \cup \boldsymbol{\Omega}_m$, where

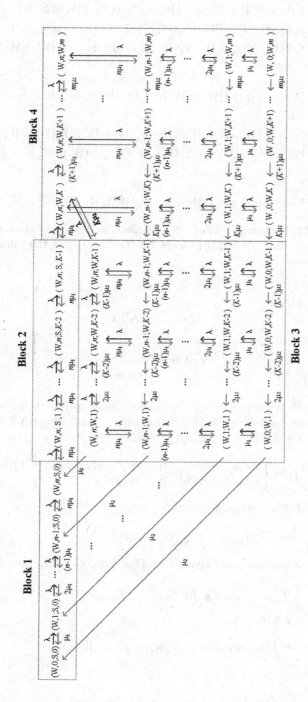

Fig. 3. State transition relations of Markov process in a two-group-server loss queue

$$\Omega_0 = \{(W, 0; S, 0), (W, 1; S, 0), \ldots, (W, n-1; S, 0), (W, n; S, 0)\},$$
$$\Omega_1 = \{(W, n; S, 1)\} \cup \{(W, n; W, 1), (W, n-1; W, 1), \ldots, (W, 0; W, 1)\},$$
$$\Omega_2 = \{(W, n; S, 2)\} \cup \{(W, n; W, 2), (W, n-1; W, 2), \ldots, (W, 0; W, 2)\},$$

$$\vdots \qquad\qquad \vdots$$

$$\Omega_{K-1} = \{(W, n; S, K-1)\} \cup \{(W, n; W, K-1), (W, n-1; W, K-1), \ldots,$$
$$(W, 0; W, K-1)\};$$
$$\Omega_K = \{(W, n; W, K), (W, n-1; W, K), \ldots, (W, 1; W, K), (W, 0; W, K)\};$$
$$\Omega_{K+1} = \{(W, n; W, K+1), (W, n-1; W, K+1), \ldots, (W, 0; W, K+1)\};$$

$$\vdots \qquad\qquad \vdots$$

$$\Omega_m = \{(W, n; W, m), (W, n-1; W, m), \ldots, (W, 1; W, m), (W, 0; W, m)\}.$$

Let the subset Ω_j be Level j. Then each element in Level j or the subset Ω_j is *a phase*. It is clear that the QBD process $\{\mathbf{X}(t), t \geq 0\}$ has the infinitesimal generator as follows:

$$Q = \begin{pmatrix} Q_{0,0} & Q_{0,1} & & & & \\ Q_{1,0} & Q_{1,1} & Q_{1,2} & & & \\ & Q_{2,1} & Q_{2,2} & Q_{2,3} & & \\ & & \ddots & \ddots & \ddots & \\ & & & Q_{m-1,m-2} & Q_{m-1,m-1} & Q_{m-1,m} \\ & & & & Q_{m,m-1} & Q_{m,m} \end{pmatrix}.$$

A Markov reward process: Using the power consumption rates, we define $f(x)$ as an instantaneous reward of the QBD process $\{\mathbf{X}(t), \ t \geq 0\}$ at the state $\mathbf{X}(t) = x$. It is clear that for $x = (W, i; l_1, j)$ with $i \in \{0, 1, \ldots, n-1, n\}$,

$$f(W, i; l_1, j) = \begin{cases} nP_{W_0} + mP_{S_1}, & l_1 = S, j \in \{0, 1, \ldots, K-1\}, \\ nP_{W_0} + mP_{W_1}, & l_1 = W, j \in \{1, 2, \ldots, m\}. \end{cases} \tag{1}$$

For simplification of description, we write

$$f_{W,i;l_1,j} = f(W, i; l_1, j).$$

It is seen from Fig. 3 and the state space $\Omega = \Omega_0 \cup \Omega_1 \cup \cdots \cup \Omega_m$ that for Level 0,

$$f_{S,0} = (f_{W,0;S,0}, f_{W,1;S,0}, \ldots, f_{W,n;S,0});$$

for Level j with $1 \leq j \leq K-1$,

$$f_{SW,j} = (f_{W,n;S,j}; f_{W,0;W,j}, f_{W,1;W,j}, \ldots, f_{W,n;W,j});$$

and for Level j with $K \leq j \leq m$,

$$f_{W,j} = (f_{W,0;W,j}, f_{W,1;W,j}, \ldots, f_{W,n;W,j}).$$

We write

$$f_j = \begin{cases} f_{S,0}, & j = 0, \\ f_{SW,j}, & 1 \le j \le K - 1, \\ f_{W,j}, & K \le j \le m, \end{cases}$$

and

$$f = (f_0, f_1, \ldots, f_{K-1}, f_K, \ldots, f_m)^T. \tag{2}$$

Now, for the two-group-server loss queue, we analyze some interesting performance measures: (a) The expected instantaneous power consumption rate $E[f(\mathbf{X}(t))]$; and (b) the cumulative power consumption $\Phi(t)$ during the time interval $[0, t)$. At the same time, the probability distribution and the first passage time involved in the energy saving are also computed in detail.

The expected instantaneous power consumption rate: We note that $f(\mathbf{X}(t))$ is the instantaneous power consumption rate of the two-group-server queue at time $t \ge 0$. If the QBD process Q is stable, then

$$\lim_{t \to +\infty} E[f(\mathbf{X}(t))] = \sum_{(W,i;l_1,j) \in \Omega} \pi_{W,i;l_1,j} f_{W,i;l_1,j} = \pi f,$$

where π is the stationary probability vector of the QBD process Q, and it can be obtained through solving the system of linear equations $\pi Q = 0$ and $\pi e = 1$ by means of the RG-factorizations given in Li [28], where e is a column vector of ones.

The Cumulative Power Consumption $\Phi(t)$ During the Time Interval $[0, t)$

Based on the instantaneous power consumption rate $f(\mathbf{X}(t))$, we define the cumulative power consumption during the time interval $[0, t)$ as

$$\Phi(t) = \int_0^t f(\mathbf{X}(u)) \, du.$$

(i) Computing the probability distribution: Now, we compute the probability distribution of the cumulative power consumption $\Phi(t)$ by means of a partial differential equation whose solution can explicitly be given in terms of the Laplace and Laplace-Stieltjes transforms. Let

$$\Theta(t, x) = P\{\Phi(t) \le x\}$$

and

$$H_{W,i;l_1,j}(t, x) = P\{\Phi(t) \le x, \ \mathbf{X}(t) = (W, i; l_1, j)\}.$$

We write that for Level 0,

$$H_{S,0}(t, x) = (H_{W,0;S,0}(t, x), H_{W,1;S,0}(t, x), \ldots, H_{W,n;S,0}(t, x));$$

for Level j with $1 \le j \le K - 1$,

$$H_{SW,j}(t, x) = (H_{W,n;S,j}(t, x); H_{W,0;W,j}(t, x), H_{W,1;W,j}(t, x), \ldots, H_{W,n;W,j}(t, x));$$

and for Level j with $K \leq j \leq m$,

$$H_{W,j}(t,x) = (H_{W,0;W,j}(t,x), H_{W,1;W,j}(t,x), \ldots, H_{W,n;W,j}(t,x)).$$

Based on this, we write

$$H_j(t,x) = \begin{cases} H_{S,0}(t,x), & j = 0, \\ H_{SW,j}(t,x), & 1 \leq j \leq K-1, \\ H_{W,j}(t,x), & K \leq j \leq m, \end{cases}$$

and

$$H(t,x) = (H_0(t,x), H_1(t,x), \ldots, H_{K-1}(t,x), H_K(t,x), \ldots, H_m(t,x)).$$

It is clear that

$$\Theta(t,x) = H(t,x)e.$$

For a column vector $a = (a_1, a_2, \ldots, a_r)$ of size r, we write

$$\Delta(a) = \text{diag}(a_1, a_2, \ldots, a_r).$$

Obviously for the column vector f, we have

$$\Delta = \text{diag}(\Delta(f_0), \Delta(f_1), \ldots, \Delta(f_{K-1}), \Delta(f_K), \ldots, \Delta(f_m)).$$

For the Markov reward process $\{\Phi(t), t \geq 0\}$, it follows from Sect. 10.2 of Chap. 10 in Li [28] that the vector function $H(t,x)$ is the solution to the Kolmogorov's forward equation

$$\frac{\partial H(t,x)}{\partial t} + \frac{\partial H(t,x)}{\partial x}\Delta = H(t,x)Q,$$

with the boundary condition

$$H(t,0) = \pi(0)\delta(t),$$

and the initial condition

$$H(0,x) = \pi(0)\delta(x),$$

$$\delta(x) = \begin{cases} 1, & x = 0, \\ 0, & x > 0. \end{cases}$$

(ii) **Computing the first passage time:** Let $\Gamma(x)$ be the first passage time of the cumulative power consumption $\Phi(t)$ arriving at a key power value x as follows:

$$\Gamma(x) = \min\{t : \Phi(t) = x\}.$$

We write

$$C(t,x) = P\{\Gamma(x) \leq t\}.$$

It is clear that the event $\{\Gamma(x) \leq t\}$ is equivalent to the event $\{\Phi(t) > x\}$. Hence we get

$$C(t,x) = 1 - P\{\Phi(t) \leq x\} = 1 - \Theta(t,x),$$

Hence, we have

$$P\{\Gamma(x) \leq t\} = 1 - P\{\Phi(t) \leq x\},$$

Let

$$M(r,k) = \frac{\partial^r}{\partial s^r}\left[(Q - sI)\Delta^{-1}\right]^k \big|_{s=0}, \ r, k \geq 0.$$

Then

$$\begin{aligned}
M(0,0) &= I, \\
M(r,0) &= 0, & r \geq 1, \\
M(0,k) &= \left(Q\Delta^{-1}\right)^k, & k \geq 1, \\
M(1,1) &= -\Delta^{-1}, \\
M(r,k) &= Q\Delta^{-1}M(r,k-1) - r\Delta^{-1}M(r-1,k-1), \ r, k \geq 1.
\end{aligned}$$

Therefore, we can provide expression for the rth moment $E[\Gamma(x)^r]$ as follows:

$$E[\Gamma(x)^r] = (-1)^{r+1}\pi(0)\sum_{k=0}^{\infty}\frac{x^k}{k!}M(r,k)e, \quad r \geq 1,$$

where $\pi(0)$ is any initial probability vector of the QBD process Q.

5.2 Open Problems

From the above analysis, it is seen that discussing more general group-server queues is interesting, challenging and difficult. Thus it may be valuable to list several open problems for the future study of group-server queues as follows:

- Setting up some suitable cost (or reward) functions, and provide and prove existence and structure of bilateral threshold control by means of Markov decision processes.
- Establishing fluid and diffusion approximations for more general group-server queues, and focus on how to deal with the residual service times of those concentratively transferred customers to the buffer or to the left-side server groups. In fact, the residual service times cause some substantial difficulties in model analysis.
- Constructing martingale problems or stochastic differential equations for more general group-server queues.
- Developing stochastic optimization and control, Markov decision processes and stochastic game theory in the study of group-server queues.

6 Simulation Experiments

In this section, we design some simulation experiments for performance evaluation of two different group-server queues: the group-server loss queues, and the group-server queues with infinite buffer. Specifically, we analyze the expected queue lengths, the expected sojourn times and the expected virtual service times for the two group-server queues.

In these following simulations, we use some common parameters for the two different group-server queues: the group-server loss queues and the group-server queues with infinite buffer, where there are three server groups in each queueing system. To that end, we take that $N = 2$, $m_0 = 4$, $m_1 = 4$, $m_2 = 3$, $\mu_0 = 5$, $\mu_1 = 4$, $\mu_2 = 3$, $K_1 = K_2 = 3$ and $L_1 = L_2 = 2$.

(1) The group-server loss queues. Figure 4 indicates how the expected customer number in the whole system depends on the arrival rate $\lambda \in (15, 45)$. It is seen from Fig. 4 that the expected customer number in the whole system strictly increase as λ increases.

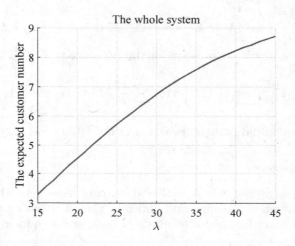

Fig. 4. The expected customer number in the whole system vs. λ

Figure 5 shows how the expected customer number in each of three server groups depends on the arrival rate $\lambda \in (15, 45)$. It is seen from Fig. 5 that the expected customer number in each of three server groups strictly increases as λ increases.

Note that each of the two group-server queues contains three different server groups with service rates: μ_0, μ_1, μ_2, it is easy to see that the service times of the three server groups are different from each other. In this case, for such a group-server queue, we need to introduce a *virtual service time* as follows: The virtual service time is defined as the average service time of any customer in the group-server queues. Thus the virtual service time describes the comprehensive service ability of a whole system through integrating the server groups with different service abilities.

Figure 6 demonstrates how the expected virtual service time and the expected sojourn time depend on the arrival rate $\lambda \in (15, 43)$, respectively. It is seen from Fig. 6 that both of them strictly increase as λ increases.

(2) The group-server queues with infinite buffer. Figure 7 indicates how the expected customer numbers in the whole system and in the buffer

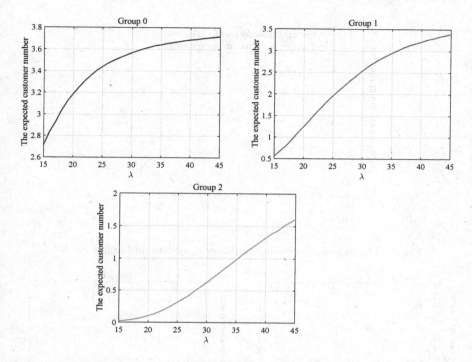

Fig. 5. The expected customer number in each of three server groups vs. λ

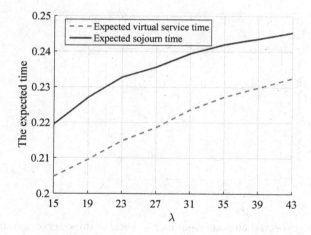

Fig. 6. The expected virtual service time and the expected sojourn time vs. λ

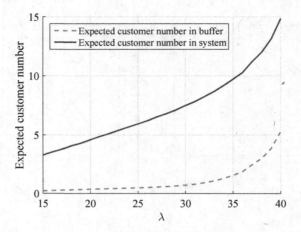

Fig. 7. The expected customer numbers in system and buffer vs. λ

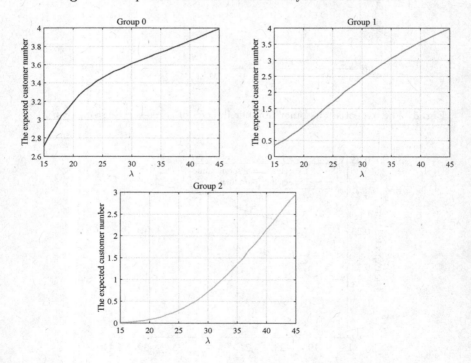

Fig. 8. The expected customer number in each of three server groups vs. λ

depend on the arrival rate $\lambda \in (15, 40)$, respectively. It is seen from Fig. 7 that both of them strictly increase as λ increases.

Figure 8 shows how the expected customer number in each of three server groups depends on the arrival rate $\lambda \in (15, 45)$. It is seen from Fig. 8 that the expected customer number in each of three server groups strictly increases as λ increases.

Figure 9 demonstrates how the expected virtual service time and the expected sojourn time depend on the arrival rate $\lambda \in (15, 40)$, respectively. It is seen from Fig. 9 that the expected virtual service time increases slowly as λ increases, while the expected sojourn time increases rapidly as λ increases.

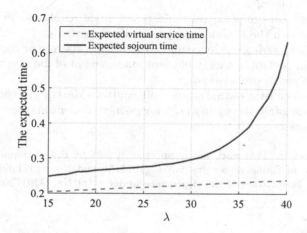

Fig. 9. The expected virtual service time and the expected sojourn time vs. λ

Here, we use some simulation experiments to give valuable observation and understanding with respect to system performance, and this is a valuable help for design, operations and optimization of energy-efficient management of data centers. Therefore, such a numerical analysis will also be useful and necessary in the energy-efficient management study of data center networks in practice.

7   Concluding Remarks

In this paper, we propose and develop a class of interesting *Group-Server Queues* by means of analyzing energy-efficient management of data centers, and establishes two representative group-server queues through loss networks and impatient customers, respectively. Some simple mathematical discussion is provided in the study of two-group-server loss queues, and simulations are made to study the expected queue lengths, the expected sojourn times and the expected virtual service times in the three-group-server queues with infinite buffer. Furthermore, we show that this class of group-server queues are often encountered in many other practical areas including communication networks, manufacturing systems, transportation networks, financial networks and healthcare systems. Therefore, not only analysis of group-server queues is regarded as a new interesting research direction, but there also exists many theoretic challenge and basic difficulties in the area of queueing networks. We hope the methodology and results given in this paper can be applicable to analyzing more general large-scale data center networks and service systems. Along these lines, we will continue our works in following directions in the future research:

- Establishing fluid and diffusion approximations for the group-server loss queues, and also for the group-server queues with impatient customers and with infinite buffers;
- making power consumption rate and power price regulation in data center networks through the Brownian approximation methods;
- setting up and proving existence and structure of (discrete or continue) bilateral threshold control in energy-efficient management of data centers in terms of Markov decision processes; and
- developing stochastic optimization and control, Markov decision processes and stochastic game theory in analyzing energy-efficient management of data center networks.

Acknowledgments. This work was supported in part by the National Natural Science Foundation of China under grant No. 71671158 and No. 71471160; and by the Natural Science Foundation of Hebei province under grant No. G2017203277.

References

1. Al-Fares, M., Loukissas, A., Vahdat, A.: A scalable, commodity data center network architecture. ACM SIGCOMM Comput. Commun. Rev. **38**, 63–74 (2008)
2. Aroca, J.A., Chatzipapas, A., Anta, A.N., Mancuso, V.: A measurement-based analysis of the energy consumption of data center servers. In: 5th International Conference on Future Energy Systems, pp. 63–74. ACM (2014)
3. Baldoni, R., Caruso, M., Cerocchi, A., Ciccotelli, C., Montanari, L., Nicoletti, L.: Correlating power consumption and network traffic for improving data centers resiliency, 1–6 (2014). arXiv:1405.2992 e-prints
4. Benson, T., Akella A., Maltz, D.: Network traffic characteristics of data centers in the wild. In: 10th ACM SIGCOMM Conference on Internet Measurement, pp. 267–280 (2010)
5. Chase, J.S., Anderson, D.C., Thakar, P.N., Vahdat, A.M., Doyle, R.P.: Managing energy and server resources in hosting centers. ACM SIGOPS Oper. Syst. Rev. **35**, 103–116 (2001)
6. Chen, Y., Das, A., Qin, W., Sivasubramaniam, A., Wang, Q., Gautam, N.: Managing server energy and operational costs in hosting centers. In: ACM SIGMETRICS International Conference on Measurement and Modeling of Computer Systems, pp. 303–314 (2005)
7. Dyachuk, D., Mazzucco, M.: On allocation policies for power and performance. In: 11th IEEE/ACM International Conference on Grid Computing, pp. 313–320 (2010)
8. Elnozahy, E.N.M., Kistler, M., Rajamony, R.: Energy-efficient server clusters. In: Falsafi, B., Vijaykumar, T.N. (eds.) PACS 2002. LNCS, vol. 2325, pp. 179–197. Springer, Heidelberg (2003). https://doi.org/10.1007/3-540-36612-1_12

9. Gandhi, A., Gupta, V., Harchol-Balter, M., Kozuch, M.: Optimality analysis of energy-performance trade-off for server farm management. Perform. Eval. **67**, 1155–1171 (2010)

10. Gandhi, A., Harchol-Balter, M., Adan, I.: Server farms with setup costs. Perform. Eval. **67**, 1123–1138 (2010)

11. Gandhi, A., Harchol-Balter, M.: How data center size impacts the effectiveness of dynamic power management. In: 49th Annual Allerton Conference on Communication, Control, and Computing, pp. 1164–1169. IEEE (2011)

12. Gandhi, A., Harchol-Balter, M., Kozuch, M.A.: Are sleep states effective in data centers? In: International Green Computing Conference, pp. 1–10. IEEE (2012)

13. Goiri, Í., Le, K., Haque, M.E., Beauchea, R.: GreenSlot: scheduling energy consumption in green data centers. In: International Conference for High Performance Computing, Networking, Storage and Analysis, pp. 1–11. ACM (2011)

14. Goiri, Í., Le, K., Nguyen, T.D., Guitart, J., Torres, J., Bianchini, R.: GreenHadoop: leveraging green energy in data-processing rameworks. In: 7th ACM European Conference on Computer Systems, pp. 57–70 (2012)

15. Greenawalt, P.M.: Modeling power management for hard disks. In: 2nd IEEE on Modeling, Analysis, and Simulation of Computer and Telecommunication Systems, pp. 62–66 (1994)

16. Greenberg, A., Hamilton, J., Maltz, D.A., Patel, P.: The cost of a cloud: research problems in data center networks. ACM SIGCOMM Comput. Commun. Rev. **39**, 68–73 (2009)

17. Greenberg, A., Hamilton, J.R., Jain, N., Kandula, S., Kim, C., Lahiri, P., Maltz, D.A., Patel, P., Sengupta, S.: VL2: a scalable and flexible data center network. ACM SIGCOMM Comput. Commun. Rev. **39**, 51–62 (2009)

18. Gruber, R., Keller, V.: HPC@Green IT: Green High Performance Computing Methods. Springer, Heidelberg (2010). https://doi.org/10.1007/978-3-642-01789-6

19. Gunaratne, C., Christensen, K., Nordman, B., Suen, S.: Reducing the energy consumption of ethernet with Adaptive Link Rate (ALR). IEEE Trans. Comput. **57**, 448–461 (2008)

20. Guo, C., Lu, G., Li, D., Wu, H., Zhang, X., Shi, Y., Tian, C., Zhang, Y., Lu, S.: BCube: a high performance, server-centric network architecture for modular data centers. ACM SIGCOMM Comput. Commun. Rev. **39**, 63–74 (2009)

21. Guo, X., Niu, Z., Zhou, S.: Delay-constrained energy-optimal base station sleeping control. IEEE J. Sel. Areas Commun. **34**, 1073–1085 (2016)

22. Guo, X., Singh, R., Zhao, T., Niu, Z.: An index based task assignment policy for achieving optimal power-delay tradeoff in edge cloud systems. In: International Conference on Communications, pp. 1–7. IEEE (2016)

23. Heller, B., Seetharaman, S., Mahadevan, P., Yiakoumis, Y., Sharma, P., Banerjee, S., McKeown, N.: Elastic tree: saving energy in data center networks. In: 7th USENIX Conference on Networked Systems Design and Implementation, pp. 249–264 (2010)

24. Horvath, T., Skadron, K.: Enhancing energy efficiency in multi-tier web server clusters via prioritization. In: IEEE International on Parallel and Distributed Processing Symposium, pp. 1–6 (2007)

25. Kliazovich, D., Bounvry, P., Audzevich, Y., Khan, S.U.: Greencloud: a packet-level simulator of energy-aware cloud computing data centers. J. Supercomput. **62**, 1263–1283 (2012)

26. Kliazovich, D., Bouvry, P., Khan, S.U.: DENS: data center energy-efficient network-aware scheduling. Clust. Comput. **16**, 65–75 (2013)

27. Li, C., Qouneh, A., Li, T.: Characterizing and analyzing renewable energy driven data centers. ACM SIGMETRICS Perform. Eval. Rev. **39**, 323–324 (2011)
28. Li, Q.L.: Constructive Computation in Stochastic Models with Applications: The RG-Factorizations. Springer, Heidelberg (2010). https://doi.org/10.1007/978-3-642-11492-2
29. Li, Q.L., Ma, J.Y., Xia, L., Xie, M.Z.: Energy management in data centers with threshold control analyzed by Markov reward processes. https://arxiv.org
30. Lim, S.H., Sharma, B., Tak, B.C., Das, C.R.: A dynamic energy management scheme for multi-tier data centers. In: IEEE International Symposium on Performance Analysis of Systems and Software, pp. 257–266 (2011)
31. Mazzucco, M., Dyachuk, D., Deters, R.: Maximizing cloud providers revenues via energy aware allocation policies. In: 3rd IEEE International Conference on Cloud Computing, pp. 131–138 (2010)
32. Mazzucco, M., Dyachuk, D., Dikaiakos, M.: Profit-aware server allocation for green internet services. In: IEEE the 18th International Symposium on Modeling, Analysis and Simulation of Computer and Telecommunication Systems, pp. 277–284 (2010)
33. Nedevschi, S., Popa, L., Iannaccone, G., Ratnasamy, S., Wetherall, D.: Reducing network energy consumption via rate-adaptation and sleeping. In: 5th USENIX Symposium on Networked Systems Design and Implementation, pp. 323–336 (2008)
34. Rivoire, S., Ranganathan, P., Kozyrakis, C.: A comparison of high-level full-system power models. In: HotPower 2008 Proceedings of the 2008 Conference on Power Aware Computing and System, p. 3 (2008)
35. Pries, R., Jarschel, M., Schlosser, D., Klopf, M., Tran-Gia, P.: Power consumption analysis of data center architectures. In: Rodrigues, J.J.P.C., Zhou, L., Chen, M., Kailas, A. (eds.) GreeNets 2011. LNICSSITE, vol. 51, pp. 114–124. Springer, Heidelberg (2012). https://doi.org/10.1007/978-3-642-33368-2_10
36. Schwartz, C., Pries, R., Tran-Gia, P.: A queuing analysis of an energy-saving mechanism in data centers. In: International Conference on Information Network, pp. 70–75. IEEE (2012)
37. Sharma, V., Thomas, A., Abdelzaher, T., Skadron, K., Lu, Z.: Poweraware QoS management in web servers. In: 24th IEEE International Real-Time Systems Symposium, pp. 63–72 (2003)
38. Shang, Y., Li, D., Xu, M.: Energy-aware routing in data center network. In: ACM SIGCOMM Workshop on Green Networking, pp. 1–8 (2010)
39. Wang, L., Anta, A.F., Zhang, F., Wu, J.: Multi-resource energy-efficient routing in cloud data centers with network-as-a-service. In: IEEE Symposium on Computers and Communication, pp. 694–699 (2015)
40. Wierman, A., Andrew, L.L.H., Tang, A.: Power-aware speed scaling in processor sharing systems. In: 28th IEEE Conference on Computer Communications, pp. 2007–2015 (2009)
41. Wu, J., Zhou, S., Niu, Z.: Traffic-aware base station sleeping control and power matching for energy-delay tradeoffs in green cellular networks. IEEE Trans. Wirel. Commun. **12**, 4196–4209 (2013)
42. Xu, H., Li, B.: Reducing electricity demand charge for data centers with partial execution. In: International Conference on Future Energy Systems, pp. 51–61. ACM (2014)
43. Zhang, S., Zhang, N., Zhou, S., Gong, J., Niu, Z., Shen, X.: Energy-aware traffic offloading for green heterogeneous networks. IEEE J. Sel. Areas Commun. **34**, 1116–1129 (2016)

Queueing Applications I

Transaction-Confirmation Time for Bitcoin: A Queueing Analytical Approach to Blockchain Mechanism

Yoshiaki Kawase[(✉)] and Shoji Kasahara

Graduate School of Information Science, Nara Institute of Science and Technology,
Takayama 8916-5, Ikoma, Nara 6300192, Japan
{kawase.yoshiaki.km3,kasahara}@is.naist.jp

Abstract. Bitcoin is a virtual currency based on a transaction-ledger database called blockchain. The blockchain is maintained and updated by mining process in which a number of nodes called miners compete for finding answers of very difficult puzzle-like problem. Transactions issued by users are grouped into a block, and the block is added to the blockchain when an algorithmic puzzle specialized for the block is solved. A recent study reveals that newly arriving transactions are not included in the block being under mining. In this paper, we model the mining process with a queueing system with batch service, analyzing the transaction-confirmation time. We consider an $M/G^B/1$ with batch service, in which a newly arriving transaction cannot enter the service facility even when the number of transactions in the service facility does not reach the maximum batch size, i.e., the block-size limit. In this model, the sojourn time of a transaction corresponds to its confirmation time. We consider the joint distribution of the number of transactions in system and the elapsed service time, deriving the mean transaction-confirmation time. In numerical examples, we show how the block-size limit affects the transaction-confirmation time.

Keywords: Bitcoin · Blockchain · Transaction-confirmation time

1 Introduction

Bitcoin is an autonomous decentralized virtual currency that does not have a central server or administrator, and it succeeds to prevent fraud such as multiple payment and impersonation by encryption and peer-to-peer network technologies [1]. Bitcoin can provide immediate and secure service of international money transfer, and it is expected to be used for micropayment such as small amount remittance and billing of a piece of Internet content, due to its low fee [2]. Bitcoin demand grows rapidly in recent years, and the average number of transactions per day in 2016 is 226,000, which is about twice as much as the previous year's 125,000.

Bitcoin virtual-currency system is based on two data types: transaction and block. A transaction includes information indicating a specific amount of money

© Springer International Publishing AG 2017
W. Yue et al. (Eds.): QTNA 2017, LNCS 10591, pp. 75–88, 2017.
https://doi.org/10.1007/978-3-319-68520-5_5

transferred from sender to recipient. On the other hand, a block consists of several transactions, and a newly generated block is confirmed by solving a puzzle-like problem. This confirmation process is called mining, and a number of nodes called miners compete for finding its answer. The difficulty of the puzzle-like problem is automatically adjusted so that mining finishes in 10 minutes on average [3]. A miner succeeds in mining, the miner receives reward called coinbase (currently 12.5 BTC in 2017) and transactions' remittance fees in the block. Note that miners keep mining work in order to get "coinbase" even when there are no transactions to be included into a block.

One of technical issues in Bitcoin is low transaction-processing speed due to the maximum block size [4]. In Bitcoin, newly arriving transactions are included in a block and confirmed by mining. Since the block size is limited to 1 MB and the time for mining is 10 min in average, the number of transactions processed per second is very small. In addition, Bitcoin system has transaction-priority mechanism, in which each transaction is prioritized according to its remittance amount, the elapsed time from previous approval, and the transaction-data size. A transaction is included in a block according to its priority value and the fee paid by user in advance [3]. Transactions with low remittance and/or low fee are likely not to be included in a block when the transaction arrival rate is high. It is reported in [5] that the recent block size is approaching the maximum block size, and therefore reducing the transaction size by separating part from transaction [6] have been proposed, however no agreement has been achieved for Bitcoin community.

In order to consider the scalability issue for Bitcoin, it is important to quantitatively characterize the transaction-confirmation process. Since transactions are processed in block basis, the transaction-confirmation process can be modeled as a single-server queueing system with batch service. In terms of the analysis for the single-server queue with batch service, the authors in [8] consider an $M/G^B/1$ queue and analyze the joint distribution of the number of customers in queue and the elapsed service time.

In [7], the authors consider an $M/G^B/1$ queue with priority mechanism, deriving the mean transaction-confirmation time for each transaction-priority class. In numerical experiments, they find quantitative difference between analysis and measured data. This results from the assumption based on default Bitcoin client mechanism, in which a newly arriving transaction is included in the block under mining if the number of transactions in the block is smaller than the block-size limit. They also find that the block-generation time follows an exponential distribution, conjecturing that a newly arriving transaction is not included in the block under mining.

In this paper, we consider a modified $M/G^B/1$ queueing model in which newly arriving transactions wait in queue even when the number of transactions is smaller than the maximum batch size. Because mining is done even if there is no transaction in system, we also assume that the system is always busy when there is no transaction in system. We analyze the joint distribution of the number of transactions in system and the elapsed service time, deriving the mean transaction-confirmation time. In numerical examples, we quantitatively evaluate the effects of the block size on the transaction-confirmation time.

The rest of this paper is organized as follows. We describe our queueing model in Sect. 2. The analysis of the queueing model is presented in Sect. 3 and some numerical examples are shown in Sect. 4. Finally, we conclude the paper in Sect. 5.

2 Queueing Model

Transactions arrive at the system according to a Poisson process with rate λ. The transactions are grouped into a block, and the block is confirmed when one of miners finds the answer of the puzzle-like problem. We define the block-generation time as the time interval between consecutive block-confirmation time points. Note that the block-generation time can be regarded as the service time for our queueing model. Let S_i $(i = 1, 2, \ldots)$ denote the ith block-generation time. We assume $\{S_i\}$'s are independent and identically distributed (i.i.d) and follow a distribution function $G(x)$. Let $g(x)$ denote the probability density function of $G(x)$. The mean block-generation time $E[S]$ is given by

$$E[S] = \int_0^\infty x \, dG(x) = \int_0^\infty x g(x) \, dx.$$

Transactions arriving to the system are served in a batch manner, and the maximum batch size is b. When a transaction arrives at the system, the transaction enters the queue. The transaction cannot enter the server at its arrival point even when the batch size under service is smaller than b or when the number of transactions in system is zero. In other words, the arriving transaction is served in the next block-generation time or later. This service is regarded as the gated service with multiple vacations [9], in which vacation periods are i.i.d and follow the same distribution of the service time.

3 Analysis

Let $N_s(t)$ denote the number of transactions in the server at time t, $N_q(t)$ the number of transactions in the queue at time t, and $X(t)$ the elapsed service time at t. We define $P_{m,n}(x, t)(m = 0, 1 \ldots, b, n = 0, 1, \ldots, x, t \geq 0)$ as

$$P_{m,n}(x, t) \, dx = Pr \{N_s(t) = m, N_q(t) = n, x < X(t) \leq x + dx\}.$$

Let $\xi(x)$ denote the hazard rate of the service time S, which is given by

$$\xi(x) = \frac{g(x)}{1 - G(x)}.$$

When $\lambda E[S] < b$ holds, the system is stable and limiting probabilities exist. Letting $P_{m,n}(x) = \lim_{t \to \infty} P_{m,n}(x, t)$, and $x(t)$ denote the differentiation of elapsed service time, we obtain

$$\frac{d}{dx} P_{m,n}(x) = -\{\lambda + \xi(x)\} P_{m,n}(x) + \lambda P_{m,n-1}(x), \quad 0 \leq m \leq b, \quad n \geq 1, \quad (1)$$

$$\frac{d}{dx} P_{m,0}(x) = -\{\lambda + \xi(x)\} P_{m,0}(x), \qquad 0 \leq m \leq b. \tag{2}$$

We have the following boundary conditions at $x = 0$

$$P_{b,n}(0) = \sum_{m=0}^{b} \int_0^{\infty} P_{m,n+b}(x)\xi(x)dx, \qquad n \geq 0,$$

$$P_{m,n}(0) = 0, \quad m = 0,1,\ldots,b-1, \quad n \geq 1,$$

$$P_{k,0}(0) = \sum_{m=0}^{b} \int_0^{\infty} P_{m,k}(x)\xi(x)dx, \qquad k = 0,1,\ldots,b.$$

The normalizing condition is given by

$$\sum_{n=0}^{\infty} \sum_{m=0}^{b} \int_0^{\infty} P_{m,n}(x)dx = 1.$$

We define the following probability generating functions (pgf's)

$$P(z_1, z_2; x) = \sum_{n=0}^{\infty} \sum_{m=0}^{b} P_{m,n}(x)z_1^m z_2^n, \tag{3}$$

$$P(z_1, z_2) = \int_0^{\infty} P(z_1, z_2; x)\, dx. \tag{4}$$

From (1) and (2), we obtain

$$\sum_{n=0}^{\infty} \sum_{m=0}^{b} \frac{d}{dx} P_{m,n}(x)\, z_1^m z_2^n = \sum_{n=0}^{\infty} \sum_{m=0}^{b} -\{\lambda + \xi(x)\} P_{m,n}(x)\, z_1^m z_2^n$$

$$+ \lambda z_2 \sum_{n=0}^{\infty} \sum_{m=0}^{b} P_{m,n}(x)\, z_1^m z_2^n.$$

From the above equation and (3), we obtain

$$\frac{d}{dx} P(z_1, z_2; x) = -\{\lambda + \xi(x)\} P(z_1, z_2; x) + \lambda z_2 P(z_1, z_2; x)$$

$$= -\{\lambda(1 - z_2) + \xi(x)\} P(z_1, z_2; x).$$

From this differential equation, $P(z_1, z_2; x)$ is given by

$$P(z_1, z_2; x) = P(z_1, z_2; 0)e^{-\lambda(1-z_2)x}\{1 - G(x)\}. \tag{5}$$

Multiplying (5) by $\xi(x)$, and integrating the equation, we obtain

$$\int_0^{\infty} P(z_1, z_2; x)\xi(x)\, dx = \int_0^{\infty} P(z_1, z_2; 0)e^{-\lambda(1-z_2)x}\{1 - G(x)\}\frac{dG(x)}{1 - G(x)}$$

$$= P(z_1, z_2; 0) \int_0^{\infty} e^{-\lambda(1-z_2)x}dG(x)$$

$$= P(z_1, z_2; 0)G^*(\lambda - \lambda z_2), \tag{6}$$

where $G^*(s)$ is the Laplace-Stieljes transform (LST) of $G(x)$ and given by

$$G^*(s) = \int_0^\infty e^{-s} dG(x).$$

From (3) and (6), we obtain

$$P(z_1, z_2; 0)G^*(\lambda - \lambda z_2) = \int_0^\infty \sum_{n=0}^\infty \sum_{m=0}^b P_{m,n}(x)\xi(x)\,dx\,z_1^m z_2^n. \tag{7}$$

Substituting $x = 0$ into (3) yields

$$P(z_1, z_2; 0) = \sum_{n=0}^\infty \sum_{m=0}^b P_{m,n}(0)\,z_1^m z_2^n$$

$$= \sum_{n=0}^\infty \sum_{m=0}^b \int_0^\infty P_{m,n+b}(x)\xi(x)\,dx\,z_1^b z_2^n$$

$$+ \sum_{n=0}^{b-1} \sum_{m=0}^b \int_0^\infty P_{m,n}(x)\xi(x)\,dx\,z_1^n. \tag{8}$$

Using (7) and (8), we obtain

$$P(z_1, z_2; 0) = \left(\frac{z_1}{z_2}\right)^b \left\{ P(1, z_2; 0)G^*(\lambda - \lambda z_2) - \sum_{n=0}^{b-1} \sum_{m=0}^b \int_0^\infty P_{m,n}(x)\xi(x)\,dx\,z_2^n \right\}$$

$$+ \sum_{n=0}^{b-1} \sum_{m=0}^b \int_0^\infty P_{m,n}(x)\xi(x)\,dx\,z_1^n. \tag{9}$$

Substituting $z_1 = 1$ into (9), we obtain

$$P(1, z_2; 0) = \left(\frac{1}{z_2}\right)^b \left\{ P(1, z_2; 0)G^*(\lambda - \lambda z_2) - \sum_{n=0}^{b-1} \sum_{m=0}^b \int_0^\infty P_{m,n}(x)\xi(x)\,dx\,z_2^n \right\}$$

$$+ \sum_{n=0}^{b-1} \sum_{m=0}^b \int_0^\infty P_{m,n}(x)\xi(x)\,dx.$$

Multiplying the above equation by z_2^b yields

$$\{z_2^b - G^*(\lambda - \lambda z_2)\}P(1, z_2; 0) = \sum_{n=0}^{b-1}(z_2^b - z_2^n)\sum_{m=0}^b \int_0^\infty P_{m,n}(x)\xi(x)\,dx.$$

From the above equation, we obtain

$$P(1, z_2; 0) = \frac{\sum_{n=0}^{b-1}(z_2^b - z_2^n)\alpha_n}{z_2^b - G^*(\lambda - \lambda z_2)}, \tag{10}$$

where α_n is given by

$$\alpha_n = \sum_{m=0}^{b} \int_0^{\infty} P_{m,n}(x)\xi(x)\,dx.$$

Applying Rouche's theorem [9] to (10), we can show that the equation

$$z_2^b - G^*(\lambda - \lambda z_2) = 0, \tag{11}$$

has b roots inside $|z_2| = 1 + \epsilon$ for a small real number $\epsilon > 0$. One of them is $z_2 = 1$. Let $z_{2,k}^*(k = 1, 2, ..., b - 1)$ denote the root of (11). From (10), we have the following $b - 1$ equations

$$\sum_{n=0}^{b-1} \left\{ \left(z_{2,k}^*\right)^b - \left(z_{2,k}^*\right)^n \right\} \alpha_n = 0, \qquad k = 1, 2, ..., b - 1. \tag{12}$$

From (9) and (10), we have

$$P(z_1, z_2; 0) = \left(\frac{z_1}{z_2}\right)^b \left\{ \frac{\sum_{n=0}^{b-1}\left(z_2^b - z_2^n\right)\alpha_n}{z_2^b - G^*(\lambda - \lambda z_2)} G^*(\lambda - \lambda z_2) - \sum_{n=0}^{b-1}\alpha_n z_2^n \right\}$$
$$+ \sum_{n=0}^{b-1}\alpha_n z_1^n. \tag{13}$$

From (4) and (5), we obtain

$$P(z_1, z_2) = P(z_1, z_2; 0)\int_0^{\infty} e^{-\lambda(1-z_2)x}\{1 - G(x)\}\,dx$$
$$= P(z_1, z_2; 0)\frac{1 - G^*(\lambda - \lambda z_2)}{\lambda(1 - z_2)}. \tag{14}$$

Multiplying (14) by $\lambda(1 - z_2)$ and partially differentiating it by z_2, we have

$$\frac{\partial P(z_1, z_2)}{\partial z_2}\lambda(1 - z_2) - P(z_1, z_2)\lambda = \frac{\partial P(z_1, z_2; 0)}{\partial z_2}\{1 - G^*(\lambda - \lambda z_2)\}$$
$$- P(z_1, z_2; 0)\frac{\partial G^*(\lambda - \lambda z_2)}{\partial z_2}.$$

Substituting $z_1 = z_2 = 1$ into the above equation, and noting that $P(1, 1) = 1$, we have

$$P(1, 1) = P(1, 1; 0)E[S] = 1.$$

Multiplying (14) by $z_2^b(z_2^b - G^*(\lambda - \lambda z_2))$ in order to calculate $P(1, 1; 0)$, we have

$$P(z_1, z_2; 0) z_2^b \left\{ z_2^b - G^*(\lambda - \lambda z_2) \right\} = z_1^b \sum_{n=0}^{b-1} \left(z_2^b - z_2^n \right) \alpha_n G^*(\lambda - \lambda z_2)$$

$$- z_1^b \left\{ z_2^b - G^*(\lambda - \lambda z_2) \right\} \sum_{n=0}^{b-1} \alpha_n z_2^n$$

$$+ z_2^b \left\{ z_2^b - G^*(\lambda - \lambda z_2) \right\} \left(\sum_{n=1}^{b-1} \alpha_n z_1^n + \alpha_0 z_1 \right).$$

Partially differentiating the above equation by z_2 and substituting $z_1 = z_2 = 1$, we obtain under the stability condition of $b > \lambda E[S]$

$$P(1, 1; 0) = \frac{\sum_{n=0}^{b-1} (b - n) \alpha_n}{b - \lambda E[S]}.$$

Hence, the normalizing condition is given by

$$\frac{\sum_{n=0}^{b-1} (b - n) \alpha_n}{b - \lambda E[S]} E[S] = 1. \tag{15}$$

From (12) and (15), α_n's are uniquely determined. From (13) and (14), we have

$$P(z_1, z_2) = \left\{ \left(\frac{z_1}{z_2} \right)^b \frac{\sum_{n=0}^{b-1} (z_2^b - z_2^n) \alpha_n}{z_2^b - G^*(\lambda - \lambda z_2)} G^*(\lambda - \lambda z_2) - \left(\frac{z_1}{z_2} \right)^b \sum_{n=0}^{b-1} \alpha_n z_2^n \right.$$

$$\left. + \sum_{n=0}^{b-1} \alpha_n z_1^n \right\} \frac{1 - G^*(\lambda - \lambda z_2)}{\lambda(1 - z_2)}. \tag{16}$$

Partially differentiating (16) by z_1 and substituting $z_1 = z_2 = 1$, we obtain the mean number of transactions in the server as

$$\left(\frac{\partial P(z_1, z_2)}{\partial z_1} \right)_{z_1=1, z_2=1} = \left\{ \frac{\sum_{n=0}^{b-1} (b - n) \alpha_n}{b - \lambda E[S]} E[S] \right\} \lambda E[S]$$

$$= \lambda E[S].$$

Similarly partially differentiating (16) by z_2, and substituting $z_1 = z_2 = 1$, we obtain the mean number of transactions in the queue as

$$\left(\frac{\partial}{\partial z_2} P(z_1, z_2) \right)_{z_1=1, z_2=1} = \frac{1}{2(b - \lambda E[S])} \left(\lambda^2 E[S^2] - 2b(b - \lambda E[S]) - b(b - 1) \right.$$

$$\left. + \sum_{n=0}^{b-1} \left\{ \lambda E[S^2](b - n) + E[S]\{b(b - 1) - n(n - 1)\} + 2bE[S](b - n) \right\} \alpha_n \right).$$

Hence, the mean number of transactions in the system $E[N]$ is given by

$$E[N] = \frac{1}{2(b - \lambda E[S])}\left(\lambda^2 E[S^2] - b(b-1) - 2(b - \lambda E[S])^2\right.$$

$$\left. + \sum_{n=0}^{b-1}\left\{\lambda E[S^2](b-n) + E[S]\{b(b-1) - n(n-1)\} + 2bE[S](b-n)\right\}\alpha_n\right).$$

Let T denote the transaction-confirmation time, the time interval from the arrival time point of a transaction to its departure one. From Little's theorem, the transaction-confirmation time is given by

$$E[T] = \frac{E[N]}{\lambda}$$

$$= \frac{1}{2\lambda(b - \lambda E[S])}\left(\lambda^2 E[S^2] - b(b-1) - 2(b - \lambda E[S])^2\right.$$

$$\left. + \sum_{n=0}^{b-1}\left\{\lambda E[S^2](b-n) + E[S]\{b(b-1) - n(n-1)\} + 2bE[S](b-n)\right\}\alpha_n\right). \tag{17}$$

4   Numerical Examples

4.1   Distribution of Block-Generation Time

It is reported in [7] that the distribution of the block-generation time $G(x)$ is the exponential one given by

$$G(x) = 1 - e^{-\mu x}, \qquad \text{where} \quad \mu = 0.0018378995.$$

Then, $E[S]$ and $E[S^2]$ are given by

$$E[S] = \frac{1}{\mu} = 544.0993884, \qquad E[S^2] = \frac{2}{\mu^2} = 592088.2889,$$

The Laplace-Stieltjes transform (LST) of $G(x)$ is given by

$$G^*(s) = \frac{\mu}{s + \mu}.$$

With these settings, we calculate the mean transaction-confirmation time $E[T]$.

4.2   Comparison of Analysis and Simulation

In order to confirm the validity of the results of analysis, we conduct the Monte-Carlo simulation of the same model as the analysis. Figure 1 shows the comparison of analysis and simulation model for the transaction-confirmation time. In this figure, the horizontal axis represents the transaction arrival rate λ and the vertical one is the mean transaction-confirmation time $E[T]$. The block size is fixed at $b = 1000$ in the numerical simulation. It is shown from Fig. 1 that the analytical result is the same as simulation, confirming the validity of analysis.

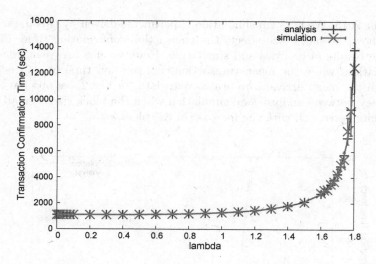

Fig. 1. Comparison of analysis and simulation model.

4.3   Comparison of Analysis and Measurement

In this subsection, we compare the analysis and measurement. In [7], the authors analyze two-year transaction data obtained from blockchain.info [5], reporting statistics such as the block-generation time, number of transactions in a block, and transaction-confirmation time. From the analysis of [7], the mean transaction size is 571.34 bytes, and hence the maximum block size b is set to 1750. Table 1 shows mean transaction-confirmation times of analysis and measurement. The analytical result is calculated with the mean transaction-arrival rate equal to 0.97091, which is obtained from measured data.

Table 1. Comparison of analysis and measurement.

Arrival rate	Measurement[s]	Analysis[s]
0.9709120529	1127.238651	1112.035745

Table 1 shows the results of measurement and analysis. We observe in this table that the analytical result is almost the same as the measurement value with relative error of 1.35%.

4.4   Comparison of Analysis and Trace-Driven Simulation

We conduct trace-driven simulation experiments for further validating our analytical model. We obtained two-year data of transaction-arrival time and block-generation time from [5], whose measurement period is from October 2013 to

September 2015. We perform simulation experiments driven by this data, investigating how the block size affects the transaction-confirmation time. Figure 2 shows the results of analysis and simulation. Here, we use two-year trace data for simulation, while the mean transaction-confirmation time of (17) is calculated with the mean arrival rate of two-year data. In Fig. 2, we observe a large discrepancy between analysis and simulation when the block size is small, while both results agree well with the increase in the block size.

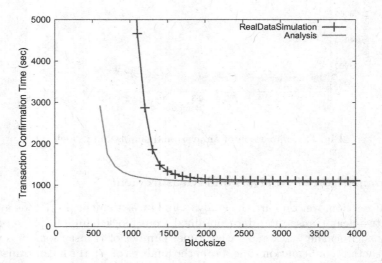

Fig. 2. Transaction-confirmation time vs. block size. The data measurement period is from October 2013 to September 2015, $\lambda = 0.9709120$, and $\mu = 0.0018378995$. The coefficient of variation of transaction inter-arrival time is 10.1789300.

Figures 3 and 4 represent the transaction-confirmation time against the block size. In Fig. 3, we use the trace data measured from October 2013 to September 2014, while the simulation result of Fig. 4 is based on the trace data measured from October 2014 to September 2015. Figure 3 shows a good agreement of analysis and simulation, however, we observe in Fig. 4 a discrepancy similar to Fig. 2.

In order to clarify the reason of these discrepancies, we investigate how the transaction arrival process evolves over time. Figure 5 shows the mean transaction-arrival rate per day. In this figure, we observe little variation during the first 12 months, while the mean transaction-arrival rate significantly varies for the last three months in the measured period.

Table 2 shows coefficients of variation for the three measurement periods: October 2013 to September 2014 (1st period), October 2014 to September 2015 (2nd period), and October 2013 to September 2015 (overall period). In this table, the coefficient of variation of the 2nd period is larger than that of 1st period. This large coefficient of variation of the 2nd period results in a large

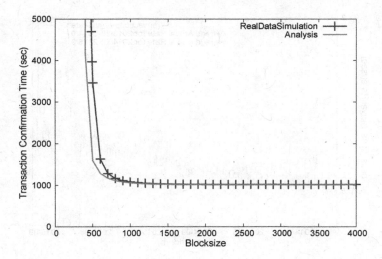

Fig. 3. Transaction-confirmation time vs. block size. The data measurement period is from October 2013 to September 2014, $\lambda = 0.7336929$, and $\mu = 0.0019748858$. The coefficient of variation of transaction inter-arrival time is 3.72401599.

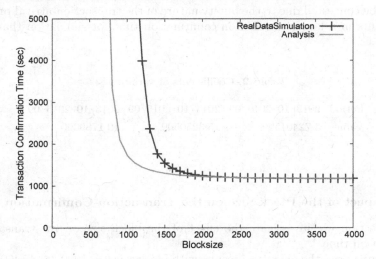

Fig. 4. Transaction-confirmation time vs. block size. The data measurement period is from October 2014 to September 2015, $\lambda = 1.2081311$, and $\mu = 0.0017009449$. The coefficient of variation of transaction inter-arrival time is 15.3250509.

coefficient of variation of the overall period, causing the discrepancy between analysis and simulation. When the block size is large, there is enough space to include transactions in the next block, and hence burst transaction arrivals are likely to be served in the next block. This causes little difference between analysis and simulation. On the other hand, when the block size is small, the system is

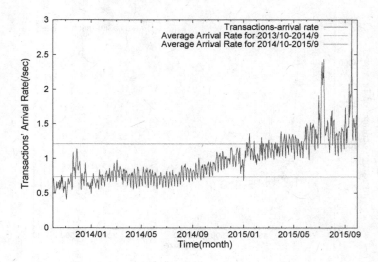

Fig. 5. The mean transaction-arrival rate.

likely to be congested due to the bursty nature of the transaction-arrival process. This results in a larger transaction-confirmation time for simulation than that for analysis.

Table 2. Coefficients of variation.

Period	2013/10–2014/09	2014/10–2015/09	2013/10–2015/09
Value	3.72401599	15.3250509	10.1789300

4.5 Impact of the Block Size on the Transaction-Confirmation Time

In this section, we investigate the effect of the block size on the transaction-confirmation time.

Figure 6 shows the analytical results with block size $b = 1000, 2000, 4000,$ and 8000. We observe that the transaction-confirmation time grows with the increase in the arrival rate. From [7], the maximum number of transactions included in the current maximum block size 1MB is approximately given by $b = 1750$. This value is close to $b = 2000$, diverging around $\lambda = 3.6$.

We also observe that enlarging the block size results in a small transaction-confirmation time. However, the transaction-confirmation time for $b = 8000$ rapidly increases when λ is greater than 13 transaction per second. This implies that enlarging the block size does not solve the scalability issue fundamentaly.

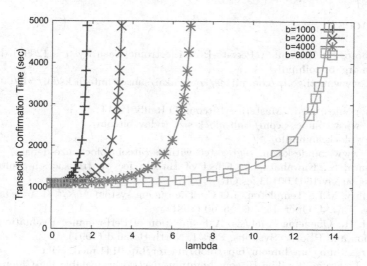

Fig. 6. The effects of the block size on the transaction-confirmation time.

5   Conclusion and Future Work

In this paper, we analyzed the transaction-confirmation time for Bitcoin using a single-server queue model with batch service $M/G^B/1$. In this queuing model, newly arriving transactions are temporarily stored in the queue first even when the number of transactions in the server is smaller than the batch size. We analyzed the mean transaction-confirmation time, and validated it by comparing simulation, and evaluated effects of the block size and transaction-arrival rate on the transaction-confirmation time. We found that the transaction-confirmation time can be decreased by changing the maximum block size. However, its improvement is not effective enough to increase the number of transactions processed per unit time.

In Bitcoin system, priority mechanism is implemented, in which the priority value of a transaction is determined according to transaction attributes such as remittance amount, data size, coin age and fee. It is important to analyze the transaction-confirmation time for the model in which the priority mechanism is taken into consideration. Other topic recently focused on in Bitcoin community is lightening network, which provides a channel dedicated to micropayment transactions [11]. The lightening network is expected to mitigate the overloaded block-generation process, however, it is not clear how the lightening network decreases the transaction-confirmation time. Developing analytical models for the above issues is also our future work.

Acknowledgments. This research was supported in part by SCAT Foundation, and Japan Society for the Promotion of Science under Grant-in-Aid for Scientific Research (B) No. 15H04008.

References

1. Nakamoto, S.: Bitcoin: A Peer-to-Peer Electronic Cash System (2008). https://bitcoin.org/bitcoin.pdf
2. http://www.meti.go.jp/committee/kenkyukai/sansei/fintech_kadai/pdf/003_02_00.pdf
3. Antonopoulos, A.M.: Mastering Bitcoin. O'Reilly (2014)
4. http://www.coindesk.com/1mb-block-size-today-bitcoin/
5. https://blockchain.info/
6. http://www.coindesk.com/segregated-witness-bitcoin-block-size-debate/
7. Kasahara, S., Kawahara, J.: Effect of Bitcoin fee on transaction-confirmation process. arXiv:1604.00103 [cs.CR]
8. Chaudhry, M.L., Templeton, J.G.C.: The queing system $M/G^B/1$ and its ramifications. Eur. J. Oper. Res. **6**, 56–60 (1981)
9. Takagi, H.: Queueing Analysis: A Foundation of Performance Evaluation vol. 1, Vacation and Priority Systems, Part 1, North-Holland (1991)
10. https://github.com/bitcoin/bips/blob/master/bip-0141.mediawiki
11. Poon, J., Dryja, T.: The Bitcoin Lightning Network: Scalable Off-Chain Instant Payments (2016). https://lightning.network/lightning-network-paper.pdf

Reliability Evaluation for a Portable Ethernet Data Acquisition Device Subjected to Competing Failure of Random Shocks

Min Lv[1], Qiang Xue[1], and Cong Lin[2(✉)]

[1] China Telecommunication Technology Labs,
China Academy of Information and Communications Technology,
Beijing 100191, China
{lvmin, xueqiang}@caict.ac.cn
[2] AVIC China Aero-Polytechnology Establishment, Beijing 100028, China
lincong19861209@gmail.com

Abstract. The growth of users and data throughput of mobile network requires regulation on the data service provider. Thus, the portable device is designed to be accessed to the core network and mirror data for further analysis about the quality of service. Since the portable device is easy to be exposed to shocks while delivered to the work place; and it is cost-dependent for preventive or corrective repair action, in this paper, we conduct a reliability evaluation for the portable Ethernet data acquisition device which is subject to two types of random shocks (extreme shock *vs.* consecutive-k minor shock). We define a two-tuple Markov chain to record the occurrence of the two competing failure models and we derive the equation for evaluating the reliability of the device. Furthermore, in order to help the crew to determine whether or not to check the disks after a long time delivery, we, according to the age-based replacement policy, derive the equation for optimizing the disk inspection window by making trade-off between the costs of preventive repair and corrective repair. Finally, we explain the proposed reliability evaluation method with a numerical example.

Keywords: Reliability · Extreme shock · Consecutive-k shock
Redundant array of independent disk (RAID)
Finite markov chain imbedding approach (FMCIA)

1 Introduction

The boom of mobile internet and related data services expand the capacity of the network rapidly. As the service provider evolves to 4G mobile network, the data throughput of the core network capacity increases dramatically. For example, the mobile subscriber has exceeded the number of fixed-line subscriptions in the European Union (EU) 27 countries since 2000 [1]. The data regulation is obligatory for billing and metering to ensure the customers is equity served [2]. Hence, a portable data acquisition device is designed for accessing the core network of the service provider and mirroring data, as described in Fig. 1, to evaluate the quality of the service, such as whether the customer is overcharged or whether the usage of data is correctly counted.

© Springer International Publishing AG 2017
W. Yue et al. (Eds.): QTNA 2017, LNCS 10591, pp. 89–100, 2017.
https://doi.org/10.1007/978-3-319-68520-5_6

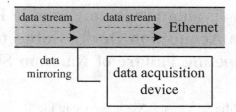

Fig. 1. Schematic diagram of Ethernet data mirroring.

As is depicted in Fig. 2, the device is accessed to the core network through an SFP+ optical module. The data transmitted in the Ethernet is in the form of packet. A packet is read into RAM through the FPGA. Then, the packet is buffered waiting for the FPAG to create a header for it and read out to the disk array [3]. The disk array of this device is composed of n disks. In order to achieve maximal acquisition speed, the disk array is applied with RAID-0. The packet is assigned by the RAID controller and store into disks. The header contains the traceable records of the packet, such as the identity information, packet type and sequence number [4], so that we can finally obtain sequential packets and complete data.

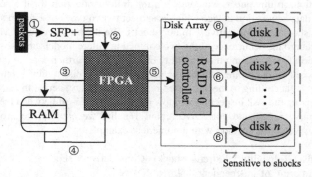

Fig. 2. Packet-flow schematic diagram of the data acquisition device.

However, because RAID-0 level series several disks [5], no redundancy is involved in the disk array, which means the failure of any single disk can lead the data stored in the disk array unavailable. For a portable device, it is very common to be exposed to shocks during the delivery, e.g., the vibration when it is carried to the workplace, the occasional drop of the device to the ground, or the accidental strike to the device caused by crews. These shocks are stochastic and may cause damage to the disk tracks, once the magnitude of a shock exceeds a certain level. Thus, before the device is setting-up to mirror data, the crew has an opportunity to decide whether a disk inspection is conducted or not. It is very straightforward that checking the disk array can ensure the device to perform data acquisition completely, but it is a cost-dependent action. If the device is transported to the work place in a relative long duration, the probability that the disks may be damaged becomes higher. If this probability is high enough, it would be better to conduct a preventive repair of the disk so that the crew can obtain completed data.

The reliability of the disk array not only depends on the magnitude of the shocks, but also it is related to the arrival of the shocks, i.e. the frequency of the shocks. Actually, for the portable device, those shocks parameters are in accordance with how the device is transported, e.g., delivered by airplane, consigned by train, or carried by person. We can obtain the probability distribution functions of the shock magnitude and shock arrival, by recording and analyzing the history data of shocks, for different transportation method, which provide the input for modeling the reliability of the device.

In this paper, we mainly consider two types of shocks which may cause failure to the device, the extreme shock and consecutive-k minor shocks. The former represents the shock whose magnitude exceeds a critical threshold; the latter refers to the situation when the device experiences multiple shocks with less magnitude. The occurrence of any one of shocks will lead the disk array unreliable. For the portable device, one can regard the extreme shock as the drop to the ground or a severe strike caused by crews; and one can regard the consecutive-k minor shocks as vibration during delivery. Thus, we establish a two-tuple Markov chain to record the two competing failure models and to obtain the reliability equation. Then, we employ age-based repair policy to determine the most economical disk inspection window, by considering the trade-off between the costs of preventive repair and corrective repair. The inspection window provides a reference to the crew to determine whether to check the disk array after delivery or not.

The following of this paper is organized as follows. We conduct the mathematical modeling for evaluating the portable Ethernet data acquisition device reliability in Sect. 2. Then, we obtain the equation for optimizing disk inspection window τ, which is helpful for determining whether or not to check the disk array before acquiring data, in Sect. 3. The proposed method is illustrated with a numerical example in Sect. 4 to demonstrate the applicability of our method. Finally, the conclusion is addressed in Sect. 5.

2   Reliability Modeling for Competing Failure of Random Shocks

The reliability is defined as the probability that the device can survive from the shocks by the time t. In order to present the reliability modeling for the portable device, we regard both of the occurrence and the magnitude of a shock as random variables, based on the following two primary assumptions:

- The shock occurs according to a homogeneous Poisson process

We assume that the shock arrival follows a homogeneous Poisson process $N_s(t)$ with rate of λ_s. Therefore the probability distribution of the number of shock by time t is obtained by Eq. (1), where the symbol ! is the factorial operator.

$$P\{N_s(t) = l\} = \frac{(\lambda_s t)^l e^{-\lambda_s t}}{l!}, \quad (l = 0, 1, \ldots). \tag{1}$$

- The magnitude of a shock on a disk is independent and identically distributed

The magnitude of the ith, $(i = 0, 1, \ldots)$ shock on the jth, $(j = 1, 2, \ldots, n)$ disk in the RAID-0 disk array, denoted as M_i^j, is independent of each other and is subject to an identical cumulative distribution function $F_M(x) = P\{M_i^j \leq x\}$.

Then, we can obtain the probability of that an extreme shock (which cause the device out of use immediately) occurs. The extreme shock is referred as the shock whose magnitude on the jth disk exceeds the critical level of threshold D_e [6], as depicted in Fig. 3. Let p_e represent the probability that the ith shock is an extreme shock. Therefore, we have

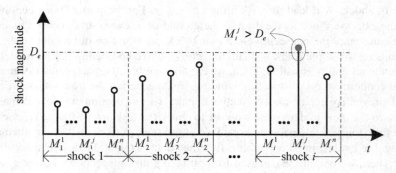

Fig. 3. Schematic diagram of extreme shock.

$$p_e = 1 - P\{\max_{1 \leq j \leq n} [M_i^j] \leq D_e\},$$

which can be calculated by Eq. (2), as we assume M_i^j is independent of each other.

$$p_e = 1 - P\{\bigcap_{j=1}^{n} M_i^j \leq D_e\} = 1 - \bigcap_{j=1}^{n} P\{M_i^j \leq D_e\}$$
$$= 1 - [F_M(D_e)]^n. \tag{2}$$

The extreme shock is easy to be identified, but the minor shocks, whose magnitude is greater than a level of D_m $(D_m < D_e)$, are not easy to be observed, as they do not generate the failure immediately. However, owing to the damage of the disk caused by the shocks is addictive, we cannot ignore them when perform the reliability analysis. Therefore, researchers employ consecutive-k model to evaluate the reliability associated with minor shocks [6–8].

In this paper, in terms of consecutive-k, the portable data acquisition device is considered as unreliable when it experiences consecutive-k minor shocks. As is illustrated in Fig. 4, for an arbitrary shock i, it can be regarded as a minor shock when there is at least one of the magnitude of the ith shock to the jth disk is greater than the threshold D_m. Let p_m represent the ith shock is a minor shock. We have

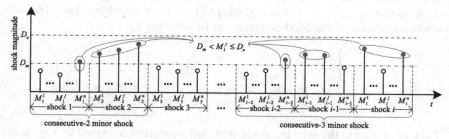

Fig. 4. Schematic diagram of consecutive-k minor shock.

$$p_m = P\left\{ \bigcup_{j=1}^{n} \left[M_i^j > D_m \right] \Big| \max_{1 \leq j \leq n} \left[M_i^j \right] \leq D_e \right\},$$

which can be calculated by Eq. (3), where $\dbinom{n}{j} = \dfrac{n!}{j!(n-j)!}$.

$$p_m = \frac{P\left\{ \bigcup_{j=1}^{n} \left[D_m < M_i^j \leq D_e \right] \right\}}{P\left\{ \max_{1 \leq j \leq n} \left[M_i^j \right] \leq D_e \right\}}$$

$$= \frac{\sum_{j=1}^{n} \dbinom{n}{j} [F_M(D_e) - F_M(D_m)]^j [F_M(D_m)]^{n-j}}{[F_M(D_e)]^n}. \tag{3}$$

We employ the finite Markov chain imbedding approach (FMCIA) to calculate the probability that the disk array survives from the shock by time t, given the number of shock $\{N_s(t) = l\}$. The FMCIA was proposed to study the distribution of runs in a sequence of independent or non-independent Bernoulli trials [9]. Then, it is widely applied by researchers to evaluate the system reliability with non-identical components, as the traditional combination method is too complicated to be applied, when the complex failure criterion is defined [10–15].

We define a two-tuple Markov chain $\{(C_e(t), C_m(t))\}$ to represent the count of the extreme shocks and the total count of the consecutive minor shocks that the device experiences within $\{N_s(t) = l\}$ shocks at time t. $C_e(t) \in \{0, 1\}$ is the count of the extreme shocks, where $C_e(t) = 1$ means there are at least one extreme shock within the l shocks at time t. $C_m(t) \in \{0, 1, \ldots, k\}$ is the total count of the consecutive minor shock, where $C_m(t) = j$ $(j = 0, 1, \ldots, k-1)$ means the device experiences consecutive-j minor shocks and $C_m(t) = k$ means the device suffers from at least consecutive-k minor shocks, within l shocks at time t. Therefore, the state space of the two-tuple Markov chain is presented as

$$S(t) = \{(0,0), (0,1), \ldots, (0, k-1), (0, k), (1, *)\}.$$

The set of the reliable state for the device is $\{(0,0), (0,1), \ldots, (0, k-1)\}$ and the set of unreliable state is $\{(0, k), (1, *)\}$, where $(1, *)$ is an absorbing state. The state transition probability of the Markov chain can be obtained by

$$P_{(0,u) \to (w,v)} = \begin{cases} p_e, & w = 1; \\ (1 - p_e)p_m, & w = 0 \quad \& \quad u = v - 1 \leq k - 1; \\ (1 - p_e)(1 - p_m), & w = 0 \quad \& \quad v = 0 \ \& \ u \leq k - 1; \\ 1 - p_e, & w = 0 \quad \& \quad u = v = k. \end{cases}$$

Thus, according to the state transition probability diagram described by Fig. 5, the transition probability matrix when the ith shock is imbedded into the Markov chain is presented as

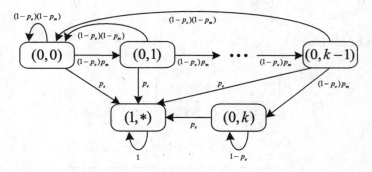

Fig. 5. State transition probability diagram of Markov chain $\{(C_e(t), C_m(t))\}$.

$$\Lambda_i(t) = \left[\begin{array}{cccc|cc} (1 - p_e)(1 - p_m) & (1 - p_e)p_m & & & & p_e \\ (1 - p_e)(1 - p_m) & & (1 - p_e)p_m & & & p_e \\ \vdots & & & \ddots & & \vdots \\ (1 - p_e)(1 - p_m) & & & & (1 - p_e)p_m & p_e \\ \hline & & & & (1 - p_e) & p_e \\ & & & & & 1 \end{array} \right]_{|k+2| \times |k+2|}$$

In terms of FMCIA, the probability that the device survives from the shocks, given the numbers of shocks at time t, $\{N_s(t) = l\}$, is calculated by Eq. (4), where the initial state probability vector is $\pi_0 = (1, 0, \ldots, 0)_{1 \times |k+2|}$; $\mathbf{U} = (1, \ldots, 1, 0, 0)_{1 \times |k+2|}$ is to sum up the probability values that the device is in working states; and T is the matrix transpose operator.

$$P_{\text{surv}|N_s(t)=l}(t) = \pi_0 \prod_{i=0}^{l} \Lambda_i(t)\mathbf{U}^T. \tag{4}$$

Therefore, the reliability of the device $R(t)$, can be presented as

$$R(t) = \sum_{l=0}^{\infty} P_{\text{surv}|N_s(t)=l}(t) \times P\{N_s(t) = l\}$$

$$= \sum_{l=0}^{\infty} \left[\pi_0 \prod_{i=0}^{l} \Lambda_i(t)\mathbf{U}^T \times \frac{(\lambda_s t)^l e^{-\lambda_s t}}{l!} \right].$$

Since the term $P\{N_s(t) = l\}$ decreases to zero as l grows to infinity, given the stop condition $l_s = \min\{l|P\{N_s(t) = l\} < \varepsilon, l = 0, 1, \ldots\}$, we can obtain the approximation of the reliability function with Eq. (5). The illustration of the equation is presented in Sect. 4.

$$R(t) \approx \sum_{l=0}^{l_s} \left[\pi_0 \prod_{i=0}^{l} \Lambda_i(t)\mathbf{U}^T \times \frac{(\lambda_s t)^l e^{-\lambda_s t}}{l!} \right]. \tag{5}$$

3 Optimizing Disk Array Inspection Window τ

For system that can record its operation condition, researchers applied the condition-based replacement policy to determine the replacement interval [16]. But, the portable device does not equip with the condition monitoring function. Owing to the reliability that the disk array survives from the shocks is a non-increase function, the longer the duration of the delivery phase is, the more opportunity there is bad track on disks. Whether the crew is going to conduct the disk inspection is related to the length of time that the device is delivered. Therefore, we apply age-based repair policy to obtained the optimal inspection window τ^*.

As is described in Fig. 6, for the disk array, the age-based repair policy assumes the inspection window to be τ. If the disk array is failed prior to the inspection window, the corrective repair is performed; otherwise, if the disk array survives by the inspection window, the preventive repair is conducted. The optimal inspection window can be determined by minimizing the long run expected repair cost rate which is the ratio of the expected cost in one inspection window to the expected length of one inspection window [17].

Let c_c be the cost of corrective repair; c_p be the cost of the preventive repair. Define $E[C(\tau)]$ to be the expected cost in one inspection window τ; $E[\tau]$ to be the expected length of one inspection window; and $Z(\tau)$ to be the expected repair cost rate.

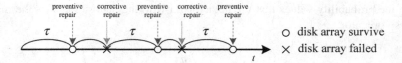

Fig. 6. Age-based repair policy.

If the disk array fails prior to τ, the expected cost is $c_c[1 - R(\tau)]$; if the system survives by τ, the expected cost is calculated by $c_p R(\tau)$. Therefore, the total expected cost in one inspection window is obtained by

$$E[C(\tau)] = c_c[1 - R(\tau)] + c_p R(\tau).$$

The expected length of time that the system fails prior to τ is calculated by $\int_0^\tau tf(t)\mathrm{d}t$, where $f(t) = -\frac{d}{dt}R(t)$. The expected length of time that the system survives by τ is calculated by $\tau\int_\tau^\infty f(t)\mathrm{d}t$. Thus, the total expected length of one inspection window is obtained by

$$E[\tau] = \int\limits_0^\tau tf(t)\mathrm{d}t + \tau R(\tau).$$

Thereafter, the long run expected repair cost rate is obtained by Eq. (6). The expected repair cost rate $Z(\tau)$, is a function of the inspection window τ. Therefore, we can obtain the optimal inspection window τ^*, by minimizing the expected repair cost rate.

$$Z(\tau) = \frac{E[C(\tau)]}{E[\tau]}. \tag{6}$$

4   Numerical Examples

In this section, we applied the proposed reliability evaluation method with some numerical examples. We set the threshold of extreme shock is $D_e = 1.55$ and the threshold of minor shock is $D_m = 1.20$. We assume that there are $n = 8$ disks in the RAID-0 disk array and we regard the device is unreliable when experiencing consecutive $k = 10$ minor shocks. The arrival rate and the magnitude of shocks depend on the way that the device is delivered, such as by personal carrying, by vehicle or by flight. The parameters of the shock arrival rate λ_s and the distribution of the shock magnitude $F_M(x)$ is listed in Table 1. Accordingly, the probability values, that an arbitrary shock belongs to extreme shock or minor shock, are also listed in Table 1 based on Eqs. (2) and (3).

Table 1. The parameter of shocks for different scenarios.

Scenario	λ_s	$F_M(x)$	p_e	p_m
1	0.015 / s	$N(1, 0.2^2)$	0.0236	0.7429
2	0.005 / s	$N(1.1, 0.2^2)$	0.0937	0.9423
3	0.003 / s	$N(1.2, 0.2^2)$	0.2790	0.9946

According to the state transition probability matrix for applying FMCIA, we can employ Eq. (5) to calculate the reliability function $R(t)$, i.e. the probability that the device survives from the random shocks by time t. The results for different scenarios are presented in Fig. 7.

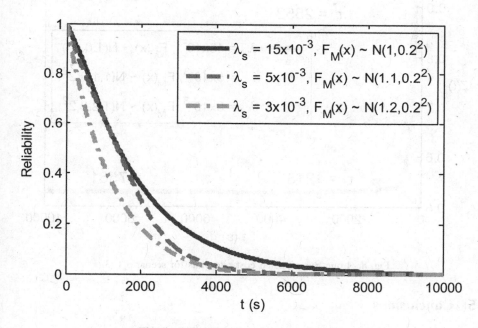

Fig. 7. Reliability plots for scenario 1–3.

As explained before, the optimal inspection window is provided to the crew for helping making decision whether the disk array is check or not. We assume the preventive repair cost is $c_p = 100$ and the corrective repair cost is $c_c = 1000$. By employing Eq. (6), we obtain the optimal inspection window, τ^*, for different scenarios which are listed in Table 2.

Table 2. The optimal inspection window for scenario 1–3.

Scenario	c_p	c_c	τ^* (sec)	$Z(\tau^*)$
1	100	1000	7561	0.5270
2	100	1000	1213	0.5488
3	100	1000	2552	0.8559

As is presented in Fig. 8, the long run expected repair cost rate various according to t. If the duration of the delivery phase, transported with scenario 1, is shorter than 7561 (s), the crew can start to acquire data directly. Otherwise, we suggest them to inspect the disks before starting data acquisition. The optimal disk inspection window for scenario 2 and 3 is 1213 (s) and 2552 (s), respectively; and the explanations for the optimal disk inspection window are similar to that of scenario 1, so we omit them here.

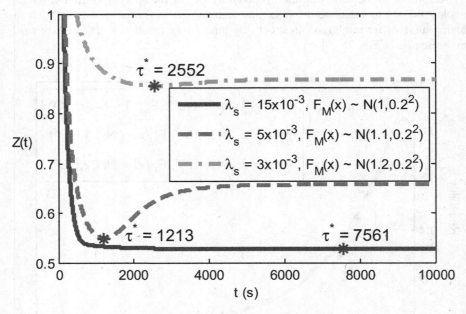

Fig. 8. Long run expected cost rate plots for scenario 1–3.

5   Conclusions

This paper conducts a reliability evaluation about a portable Ethernet data acquisition device subject to random shocks which may cause damage to the RAID-0 disk array. By considering two competing failure models (extreme shock *vs.* consecutive-k minor shock), we define a two-tuple Markov chain and employ FMCIA to derive the reliability equation. Involving the costs of preventive repair and corrective repair, we apply the age-based replacement policy to obtain the optimal disk inspection window τ^*, which is an index that helps the crew to decide whether to check the disks after the device is delivered to the work place. Finally, the proposed methods are illustrated and demonstrated by numerical examples.

The limitation of this paper is that we assume the shock occurs according to a homogeneous Poisson process, which is idealism in practice. In some situation, the homogeneous Poisson process assumption makes sense, such as the device is delivered by vehicle. While, if the device is transferred by flight, that assumption may be violated, because the device will experience extra fatal shocks when take-off and landing. In this case, one possible solution is that we can divide the transfer into three phases: take-off, cruise and landing; and regard each phase as a different Poisson process. The device is considered as reliable only if it can survive from all three phase. Then, we can apply the proposed methods, in this paper, to evaluate the reliability of the portable device.

Acknowledgements. This research is supported by National Natural Science Foundation of China under grant no. 71631001. The authors would express their appreciation to it. The authors also appreciate the valuable comments that enhance the quality of this paper, provided by the editors and reviewers.

References

1. Barth, A.K., Heimeshoff, U.: Does the growth of mobile markets cause the demise of fixed networks–evidence from the European Union. Telecommun. Policy **38**(11), 945–960 (2014)
2. Hasbi, M.: Universal service obligations and public payphone use: is regulation still necessary in the era of mobile telephony. Telecommunications Policy **39**(5), 421–435 (2015)
3. Qu, Y.R., Prasanna, V.K.: High-performance and dynamically updatable packet classification engine on FPGA. IEEE Trans. Parallel Distrib. Syst. **27**(1), 197–209 (2016)
4. Lane, R.G., Daniels, S., Yuan, X.: An empirical study of reliable multicast protocols over Ethernet-connected networks. Perform. Eval. **64**(3), 210–228 (2007)
5. Shooman, A.M., Shooman, M.L.: A comparison of RAID storage schemes: reliability and efficiency. In: Proceedings of Annual Reliability and Maintainability Symposium (RAMS), pp. 255–258 (2012)
6. Rafiee, K., Feng, Q., Coit, D.W.: Condition-based maintenance for repairable deteriorating systems subject to a generalized mixed shock model. IEEE Trans. Reliab. **64**(4), 1164–1174 (2015)
7. Chen, Y.L.: A bivariate optimal imperfect preventive maintenance policy for a used system with two-type shocks. Comput. Ind. Eng. **63**(4), 1227–1234 (2012)
8. Yun, W.Y., Endharta, A.J.: A preventive replacement policy based on system critical condition. Proceedings of the Institution of Mechanical Engineers Part O-Journal of Risk and Reliability **230**(1), 93–100 (2016)
9. Fu, J.C., Koutras, M.V.: Distribution-theory of runs-a Markov-chain approach. J. Am. Stat. Assoc. **89**(427), 1050–1058 (1994)
10. Cui, L.R., Lin, C., Du, S.J.: "*m*-Consecutive-*k*, *l*-Out-of-*n* Systems. IEEE Trans. Reliab. **64** (1), 386–393 (2015)
11. Fu, J.C.: On the distribution of the number of occurrences of an order-preserving pattern of length three in a random permutation. Methodol. Comput. Appl. Probab. **14**(3), 831–842 (2012)
12. Fu, J.C., Hsieh, Y.F.: On the distribution of the length of the longest increasing subsequence in a random permutation. Methodol. Comput. Appl. Probab. **17**(2), 489–496 (2015)
13. Lee, W.C.: Power of discrete scan statistics: a Finite Markov Chain Imbedding Approach. Methodol. Comput. Appl. Probab. **17**(3), 833–841 (2015)
14. Zhao, X., Cui, L.R.: Reliability evaluation of generalized multi-state *k*-out-of-*n* systems based on FMCI approach. Int. J. Syst. Sci. **41**(12), 1437–1443 (2010)
15. Zhao, X., Sun, G., Xie, W.J., Lin, C.: On generalized multi-state start-up demonstration tests. Appl. Stoch. Models Bus. Ind. **31**(3), 325–338 (2015)

16. Castro, I.T., Caballe, N.C., Perez, C.J.: A condition-based maintenance for a system subject to multiple degradation processes and external shocks. Int. J. Syst. Sci. **46**(9), 1692–1704 (2015)
17. Coit, D.W., Chatwattanasiri, N., Wattanapongsakorn, N., Konak, A.: Dynamic *k*-out-of-*n* system reliability with component partnership. Reliab. Eng. Syst. Safety **138**, 82–92 (2015)

Delay Analysis of Three-State Markov Channels

Jian Zhang[1($\boxtimes$)], Zhiqiang Zhou[1], Tony T. Lee[2], and Tong Ye[1]

[1] Shanghai Jiao Tong University, Shanghai 200240, China
{201laad, zhouzhiqiang, yetong}@sjtu.edu.cn
[2] The Chinese University of Hong Kong, Shenzhen, Shenzhen 518000, China
tonylee@cuhk.edu.cn

Abstract. The wireless fading channel can be modeled as an M/MMSP/1 queue in which packets service rate varies with time. The existing analysis only provides a closed-form result for 2-state case. In this paper, we focus on a 3-state Markov channel and one of which has service rate 0. We use hybrid embedded Markov chain to describe queueing process of the packets and transform this queueing problem into a linear system. We provide a closed-form formula for mean waiting time of 3-state M/MMSP/1 queue and show that the state transition rate significantly influences the mean waiting time. Our method is innovative and can be easily generalize to any finite-state Markov channel.

Keywords: Markov channel · Hybrid embedded Markov chain
Start service probability · Mean waiting time

1 Introduction

Nowadays, with the rapid development of wireless technology, the range of communication area has been extended beyond the confines of wired technology. The wireless communication offers great flexibility and convenience to mobile users. However, due to multipath effect of communication environment, the channel fading will happen, and thus quality of wireless channel is not stable but varying with time [1]. This characteristic affects the channel transmission delay, which is vital in designing delay-sensitive applications such as video conferencing, voice over Internet protocol (VoIP) and utility computing. Thus, a detailed analysis is needed to characterize the channel transmission delay under the influence of fading.

In existing literatures, a wireless communication fading channel is commonly modeled as a Markov channel. The study of Markov channel emerges from the early work of Gilbert [2] and Elliott [3], in which they study a two-state Markov channel known as the Gilbert-Elliott channel. The quality of the channel is either totally noisy or noiseless. However, the two-state model is not adequate to describe the features of wireless fading channel when the quality of fading channel varies dramatically.

As the fading channel is described as a Markov chain, the service of the fading channel is called Markov modulated service process (MMSP). There have already been some studies on the analysis of queuing models with Markov modulated service processes. In earlier times, via generating function, Eisen and Tainiter [4], Yechiali and

© Springer International Publishing AG 2017
W. Yue et al. (Eds.): QTNA 2017, LNCS 10591, pp. 101–117, 2017.
https://doi.org/10.1007/978-3-319-68520-5_7

Naor [5] considered a queue where the arrival process as well as the service process is affected by a two-state Markov chain. Later the generalized n-state case was analyzed by Yechiali [6]. However, the method of generating functions has failed to give an explicit solution in general case. To cope with this issue, Mahabhashyam and Gautam introduced the concept of start service probability in [7]. Huang and Lee adopted this concept in [8] and derived the generalized P-K formula for two-state Markov channels. However, their method is difficult to generalize to the case with more than 2 states.

In this paper, we'll focus on the three-state Markov model, which consists of a good state, a bad state and a failure state, that may happen when the system crashes. With clear physical interpretations, the hybrid embedded Markov chain is used to analyze the system. The analytical expressions of start service probability and mean queue length are given, which are helpful in figuring out which parameter will affect the transmission delay of a three-state Markov channel. At last, we'll generalize our method to any finite-state Markov channel and provide a systematical way to acquire the mean service time and mean waiting time. The main contribution of this paper is the method we adopted to transform the M/MMSP/1 queueing problem into a set of linear dynamical system.

The rest of the paper is organized as follows: The three-state Markov channel model is introduced in Sect. 2. In Sect. 3, we define the embedded Markov chain and the embedded points and discuss the relationship between the embedded points. We also provide the closed-form formula for start service probability in Sect. 3. In Sect. 4, we set up equations of conditional expected delay and derive the P-K formula for our M/MMSP/1 queueing model. In Sect. 5, we generalize our method to any finite-state Markov channel and finally a conclusion is provided in Sect. 6.

2   M/MMSP/1 Model of Three-State Markov Channel

2.1   Three-State Markov Channel Model

The communication environment of Markov channel is changing with time due to multi-path effect. In accordance with the received SNR, the Markov channel can be divided into three states: a crash state 0, a bad state 1 and a good state 2. In this paper, a three-state Markov chain is used to govern the transition of channel state, which is completely independent of the actions of packets. The transition of the three-state Markov chain is represented in Fig. 1.

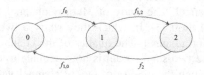

Fig. 1. Transition of three-state Markov chain.

By definition, the infinitesimal generator Q of the three-state Markov chain is given by:

$$Q = \begin{pmatrix} -f_0 & f_0 & 0 \\ f_{1,0} & -f_1 & f_{1,2} \\ 0 & f_2 & -f_2 \end{pmatrix}. \tag{1}$$

The parameter $f_1 = f_{1,0} + f_{1,2}$ corresponds to the transition rate out of the state 1. Let π_j denote the steady-state probability that the channel is in state j, we have

$$\pi Q = 0. \tag{2}$$

where $\pi = (\pi_0, \pi_1, \pi_2)$ is defined as steady-state probability vector. Together with the following relationship that the sum of all steady-state probabilities equals to 1:

$$\sum_{j=0}^{2} \pi_j = 1, \tag{3}$$

the steady-state probability π_j can be obtained and given as follows:

$$\pi_0 = \frac{f_{1,0}f_2}{f_0f_2 + f_2f_{1,0} + f_0f_{1,2}}, \tag{4}$$

$$\pi_1 = \frac{f_0f_2}{f_0f_2 + f_2f_{1,0} + f_0f_{1,2}}, \tag{5}$$

$$\pi_1 = \frac{f_0f_2}{f_0f_2 + f_2f_{1,0} + f_0f_{1,2}}. \tag{6}$$

Suppose the service rate of the channel is μ_j in state $j(j = 0, 1, 2)$. As state 0 means that the channel crashes, the service rate μ_0 is zero. And naturally, the service rate μ_1 in bad state is smaller than the service rate μ_2 in good state.

2.2 M/MMSP/1 Queuing Model

In the last subsection, we focus on the property of Markov channels which is independent of packets. Taking the packet arrival process into consideration, the wireless fading channel can be modeled as an M/MMSP/1 queueing system.

Implied by the notations of queuing theories, the M/MMSP/1 queue consists of a Poisson arrival process at rate λ, a service process that is Markov modulated, a single server queue with infinite buffer. We assume that the service of packets follows the first in first out (FIFO) policy. The evolution process of system is shown in Fig. 2.

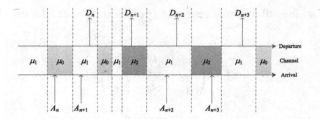

Fig. 2. Evolution process of system.

For the M/MMSP/1 model, the state of the system is determined by channel state and the number of packets in the buffer. At time t define $X(t)$ as the total number of packets in the buffer and $Y(t)$ as the channel state. Thus, the stochastic process $\{(X(t), Y(t)), t \geq 0\}$ is a two-dimensional continuous time Markov chain with state space $\{(n,j), n = 0, 1, 2, \ldots, j = 0, 1, 2\}$. The state transition diagram is depicted as follows:

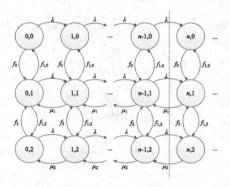

Fig. 3. State transition diagram for two-dimensional Markov chain.

When the system is stable, the joint steady-state probability that n packets are in the buffer and the channel is in state j is

$$p_{n,j} = \lim_{t \to \infty} P\{X(t) = n, Y(t) = j; n = 0, 1, 2, \ldots, j = 0, 1, 2\}. \tag{7}$$

We obtain the following relationship directly from the definition of π_j and $p_{n,j}$:

$$\pi_j = \lim_{t \to \infty} P\{Y(t) = j\} = \sum_{n=0}^{\infty} p_{n,j}. \tag{8}$$

The marginal steady-state probability that there are n packets in the buffer is given by:

$$p_n = \lim_{t \to \infty} P\{X(t) = n\} = \sum_{j=0}^{2} p_{n,j}. \tag{9}$$

From the above steady-state probabilities, conditioning on the channel state, the partial generating functions are defined below:

$$G_j(z) = \sum_{n=0}^{\infty} p_{n,j} z^n, |z| \le 1, j = 0, 1, 2, \tag{10}$$

and the overall generating function is given by:

$$G(z) = \sum_{n=0}^{\infty} p_n z^n = \sum_{j=0}^{2} G_j(z). \tag{11}$$

From the state transition diagram depicted in Fig. 3, we can obtain the balance equations for the continuous time Markov chain as follows:

(1) When $n = 0$,

$$(\lambda + f_0)p_{0,0} = f_{1,0}p_{0,1}, \tag{12}$$

$$(\lambda + f_1)p_{0,1} = \mu_1 p_{1,1} + f_0 p_{0,0} + f_2 p_{0,2}, \tag{13}$$

$$(\lambda + f_2)p_{0,2} = \mu_2 p_{1,2} + f_{1,2}p_{0,1}. \tag{14}$$

(2) When $n \ge 1$,

$$(\lambda + f_0)p_{n,0} = \lambda p_{n-1,0} + f_{1,0}p_{n,1}, \tag{15}$$

$$(\lambda + f_1 + \mu_1)p_{n,1} = \lambda p_{n-1,1} + \mu_1 p_{n+1,1} + f_0 p_{n,0} + f_2 p_{n,2}, \tag{16}$$

$$(\lambda + f_2 + \mu_2)p_{n,2} = \lambda p_{n-1,2} + \mu_2 p_{n+1,2} + f_{1,2}p_{n,1}. \tag{17}$$

From Eqs. (12)–(17), the following equations on $G_j(z)$ can be acquired:

$$(\lambda(1-z) + f_0)G_0(z) - f_{1,0}G_1(z) = 0, \tag{18}$$

$$(\lambda z(1-z) + f_1 z + \mu_1(z-1))G_1(z) - z f_0 G_0(z) - z f_2 G_2(z) = \mu_1 p_{0,1}(z-1), \tag{19}$$

$$(\lambda z(1-z) + f_2 z + \mu_2(z-1))G_2(z) - z f_{1,2}G_1(z) = \mu_2 p_{0,2}(z-1). \tag{20}$$

The Rouche's Theorem can be used to calculate $p_{0,1}$ and $p_{0,2}$, and $G_j(z)$ can be derived by solving the set of linear equations.

The typical way to derive the average queue length is to take the first order derivative of the generating function $G(z)$ with respect to z and substitute $z = 1$. Then by Little's Law, the mean delay can be obtained. The deficiency of this method is that it provides little insight about the physical service process of the packets. Thus, little information can be obtained from the numerical results. In this paper, we derive the well-organized results with full physical insight by using the method of hybrid embedded Markov chain.

From the balance equations of the dashed line in Fig. 3, we have

$$\lambda \sum\nolimits_{j=0}^{2} p_{n-1,j} = \lambda p_{n-1} = \sum\nolimits_{j=1}^{2} \mu_j p_{n,j}. \tag{21}$$

Summing (21) over all n, we have

$$\lambda = \sum\nolimits_{j=1}^{2} \mu_j \left(\pi_j - p_{0,j} \right). \tag{22}$$

Define the capacity of the channel $\hat{\mu} = \sum_{j=1}^{2} \pi_j \mu_j$, which is the maximum rate of the system to serve the packets in theory. The sufficient condition for the system to be stable is that the arrival rate λ is less than the channel capacity, otherwise the system will be unstable and the queue length will approach infinite.

3   Hybrid Embedded Markov Chain

In this section, the hybrid embedded Markov chain is introduced, which offers us an alternative way to analyze the M/MMSP/1 queue. The system behavior is modeled by selecting a set of embedded points, and we set up equations to describe the transitions between these embedded points.

3.1   Selection of Embedded Points

In queuing theory, the epochs when packets end (or start) their services are usually selected as the embedded points. However, due to channel transitions, the services between packets in M/MMSP/1 queue are dependent, thus memory exists between end service epochs. Hence, a new set of embedded points will be added to cope with the memory of channel states.

As the dependency of services is caused by channel state transitions during the service of packets, the epochs when channel state transits are needed in our study. Since the time interval between the epochs when service starts or state transits is exponentially distributed, a hybrid embedded Markov chain can be constructed to model the M/MMSP/1 queue. The embedded points can be categorized into two types: service-start points and state-transition points. In respect to channel state $j(j = 0, 1, 2)$, we are interested in the following embedded points:

(1) Φ_j: Epoch when channel state transits to state j,
(2) S_j: Epoch when service starts with channel state j.

When the system is busy, as Fig. 4(a) shows, the channel is in state j immediately after an embedded point. The next event may be a state transition to other states with time interval T_1, which is exponentially with parameter f_j or a service start with time interval T_2, which is exponentially with parameter μ_j. Thus, the time until the occurrence of the next event is $T = min\{T_1, T_2\}$, which is exponentially with parameter $\mu_j + f_j$.

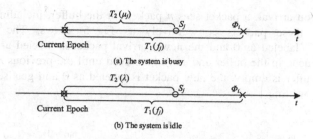

Fig. 4. Relationship between embedded points.

According to the properties of exponentially distributed random variable, the probability that the time interval T_1 is less than the interval T_2 is $f_j/(\mu_j+f_j)$. Thus, we obtain the following properties:

(1) The probability that the next embedded point is a state transition point, after which the channel state is $i (i \neq j)$, is $f_{j,i}/(\mu_j+f_j)$;

(2) The probability that the next embedded point is a state transition point is $\sum_{i \neq j} f_{j,i}/(\mu_j+f_j) = f_j/(\mu_j+f_j)$;

(3) The probability that the next embedded point is a service start point is $\mu_j/(\mu_j+f_j)$;

(4) The holding time, which is defined as the time elapsed from the current embedded point to the next embedded point, follows an exponential distribution with parameter μ_j+f_j.

Similarly, as Fig. 4(b) shows, when the system is idle, the epoch when a new packet arrives in channel state j is a service start point S_j. Using similar arguments, we obtain the following properties:

(1) The probability that the next embedded point is a state transition point, after which the channel state is $i (i \neq j)$, is $f_{j,i}/(\lambda+f_j)$;

(2) The probability that the next embedded point is a state transition point is $\sum_{i \neq j} f_{j,i}/(\lambda+f_j) = f_j/(\lambda+f_j)$;

(3) The probability that the next embedded point is a service start point is $\lambda/(\lambda+f_j)$;

(4) The holding time, which is defined as the time elapsed from the current embedded point to the next embedded point, follows an exponential distribution with parameter $\lambda+f_j$.

3.2 Start Service Probability

The start service probability, denoted as $\hat{\pi}_j$, is defined as the probability that a packet starts its service in channel state j. The difference between $\hat{\pi}_j$ and π_j is that the former is averaged over all packets start service epochs while the latter is averaged over all time. In some limiting cases, the value of $\hat{\pi}_j$ will approach to π_j.

Suppose upon arrival, a packet sees n packets in the buffer, including the one in service. Label these packets consecutively as Fig. 5 shows. The head-of-line (HOL) packet is labeled as 0 and the newly-arrival packet is labeled as n. The new packet has to queue in the buffer and can't be served until the previous n packets get served. If the buffer is empty, the new packet is labeled as 0 and gets service immediately without waiting.

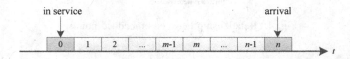

Fig. 5. New packet arrives seeing n packets in the buffer.

From the epoch when the new packet arrives to the epoch when it becomes the HOL packet, there exist n service starts and quite a few channel state transitions, which are those embedded points defined above. For the process, we need to define two classes of conditional probabilities:

(1) $\hat{\pi}_{n,j}(m) = P\{$the mth packet starts service in state $j\,|$ an arrival sees n packets in the system$\}$,
(2) $\hat{\varphi}_{n,j}(m) = P\{$during the service of the mth packet, the channel transits into state $j\,|$ an arrival sees n packets in the system$\}$.

The conditional start service probability $\hat{\pi}_{n,j}(m) =$ means the mth packet starts service in channel state j, which stands for an embedded point S_j. By definition, the channel state will remain unchanged before and after the service start point, thus the last embedded point will lie in the service of $(m-1)^{th}$ packet with channel state j. The probability that the last embedded point transits to S_j is $\mu_j/(\mu_j+f_j)$. Due to the reason that the service rate of state 0 is zero, a packet can't finish its service in state 0, which implies that $\hat{\pi}_{n,0}(m)$ equals to zero for packets numbered from 1 to n. Therefore, for $1 \leq m \leq n$,

$$\hat{\pi}_{n,0}(m) = 0, \tag{23}$$

$$\hat{\pi}_{n,1}(m) = \frac{\mu_1}{\mu_1+f_1}(\hat{\pi}_{n,1}(m-1) + \hat{\varphi}_{n,1}(m-1)), \tag{24}$$

$$\hat{\pi}_{n,2}(m) = \frac{\mu_2}{\mu_2+f_2}(\hat{\pi}_{n,2}(m-1) + \hat{\varphi}_{n,2}(m-1)). \tag{25}$$

On the other hand, the conditional probability $\hat{\varphi}_{n,j}(m)$ means the channel state transits to j during the service of mth packet, which stands for an embedded point Φ_j. By definition, the channel state will change before and after the state transition point, thus the last embedded point will lie in the service of mth packet with channel state

$i(i \neq j)$. The probability that the last embedded point transits to S_j is $f_{i,j}/(\mu_i + f_i)$. Therefore, for $0 \leq m \leq n$,

$$\hat{\varphi}_{n,0}(m) = \frac{f_{1,0}}{\mu_1 + f_1} \left(\hat{\pi}_{n,1}(m) + \hat{\varphi}_{n,1}(m) \right), \tag{26}$$

$$\hat{\varphi}_{n,1}(m) = \frac{f_0}{\mu_0 + f_0} \left(\hat{\pi}_{n,0}(m) + \hat{\varphi}_{n,0}(m) \right) + \frac{f_2}{\mu_2 + f_2} \left(\hat{\pi}_{n,2}(m) + \hat{\varphi}_{n,2}(m) \right), \tag{27}$$

$$\hat{\varphi}_{n,2}(m) = \frac{f_{1,2}}{\mu_1 + f_1} \left(\hat{\pi}_{n,1}(m) + \hat{\varphi}_{n,1}(m) \right). \tag{28}$$

By solving Eqs. (23)–(28) the following relationship between $\hat{\pi}_{n,j}(m)$ and $\hat{\pi}_{n,j}(m-1)$ can be established as follows:

$$\begin{pmatrix} \hat{\pi}_{n,0}(m) \\ \hat{\pi}_{n,1}(m) \\ \hat{\pi}_{n,2}(m) \end{pmatrix} = \hat{Q} \begin{pmatrix} \hat{\pi}_{n,0}(m-1) \\ \hat{\pi}_{n,1}(m-1) \\ \hat{\pi}_{n,2}(m-1) \end{pmatrix} = \hat{Q}^m \begin{pmatrix} \hat{\pi}_{n,0}(0) \\ \hat{\pi}_{n,1}(0) \\ \hat{\pi}_{n,2}(0) \end{pmatrix}, \tag{29}$$

where

$$\hat{Q} = \begin{pmatrix} 0 & 0 & 0 \\ \beta\left(1 + \frac{f_2}{\mu_2}\right) & \beta\left(1 + \frac{f_2}{\mu_2}\right) & \beta\frac{f_2}{\mu_2} \\ \beta\frac{f_{1,2}}{\mu_1} & \beta\frac{f_{1,2}}{\mu_1} & \beta\left(1 + \frac{f_{1,2}}{\mu_1}\right) \end{pmatrix}. \tag{30}$$

and 0, 1 and $\beta = \mu_1\mu_2/(\mu_1\mu_2 + \mu_1 f_2 + \mu_2 f_{1,2})$ are three eigenvalues of matrix $\hat{Q}$.

Solving Eq. (29), for all $1 \leq m \leq n$,

$$\hat{\pi}_{n,0}(m) = 0, \tag{31}$$

$$\hat{\pi}_{n,1}(m) = \eta_1 + \eta_2 \beta^m \hat{\pi}_{n,0}(0) + \eta_2 \beta^m \hat{\pi}_{n,1}(0) - \eta_1 \beta^m \hat{\pi}_{n,2}(0), \tag{32}$$

$$\hat{\pi}_{n,2}(m) = \eta_2 - \eta_2 \beta^m \hat{\pi}_{n,0}(0) - \eta_2 \beta^m \hat{\pi}_{n,1}(0) + \eta_1 \beta^m \hat{\pi}_{n,2}(0), \tag{33}$$

where $\eta_1 = f_2\mu_1/(f_2\mu_1 + f_{1,2}\mu_2)$ and $\eta_2 = f_{1,2}\mu_2/(f_2\mu_1 + f_{1,2}\mu_2)$.

As far as $m = 0$, by definition, probability $\hat{\pi}_{n,j}(0)$ is the conditional start service probability of the HOL packet. Due to the memoryless property of Markov chain, the start service state of the HOL packet can be regarded as the channel state when the new packet arrives. Using the property of PASTA [10], upon arrival, the probability that a packet sees n packets in the buffer is p_n and the probability that a packet sees n packets while the channel state is j is just $p_{n,j}$. Consequently, for $j = 0, 1, 2$, the initial conditional probability is given by:

$$\hat{\pi}_{n,j}(0) = p_{n,j}/p_n. \tag{34}$$

As probability $\hat{\pi}_{n,j}(n)$ is the conditional start service probability of the newly-arrived packet, conditioning on the number of packets in the buffer upon arrival, the start service probabilities can be derived from the conditional service probabilities

$$\hat{\pi}_j = \sum_{n=0}^{\infty} p_n \hat{\pi}_{n,j}(n). \tag{35}$$

Thus, combining Eqs. (31)–(35), the analytical expressions of the start service probabilities is given in the following theorem.

Theorem 1: For three-state Markov channel, the start service probability of a packet in an M/MMSP/1 queue is given as follows:

$$\hat{\pi}_0 = p_{0,0}, \tag{36}$$

$$\hat{\pi}_1 = \eta_1 + \eta_2 G_0(\beta) + \eta_2 G_1(\beta) - \eta_1 G_2(\beta) - p_{0,0}, \tag{37}$$

$$\hat{\pi}_2 = \eta_2 - \eta_2 G_0(\beta) - \eta_2 G_1(\beta) + \eta_1 G_2(\beta). \tag{38}$$

The service time of each packet is associated with the start service state. The concept of conditional service time T_j is brought up in [7], which is the time needed to serve a packet when the service begins with channel state j. From theorem 1 of [7], we have the following equations:

$$E[T_0] = \frac{1}{f_0} + E[T_1], \tag{39}$$

$$E[T_1] = \frac{\mu_1}{\mu_1 + f_1} \frac{1}{\mu_1 + f_1} + \frac{f_{1,0}}{\mu_1 + f_1} \left(\frac{1}{\mu_1 + f_1} + E[T_0] \right) + \frac{f_{1,2}}{\mu_1 + f_1} \left(\frac{1}{\mu_1 + f_1} + E[T_2] \right), \tag{40}$$

$$E[T_2] = \frac{\mu_2}{\mu_2 + f_2} \frac{1}{\mu_2 + f_2} + \frac{f_2}{\mu_2 + f_2} \left(\frac{1}{\mu_2 + f_2} + E[T_1] \right). \tag{41}$$

From Eqs. (39)–(41), the conditional service time $E[T_j]\,(j = 0, 1, 2)$ can be easily solved. Hence, conditioning on the start service state, the average service time of the three-state Markov model is

$$E[T] = \sum_{j=0}^{2} \hat{\pi}_j E[T_j]. \tag{42}$$

Two limiting cases are of interest: the offered load is relatively low or the offered load is rather high. In the case when the offered load is low, suppose that the arrival rate of packets λ approaches to zero, there are nearly no packets in the buffer and a new packet gets service as soon as it arrives. The channel state of start service is just the channel state when it arrives. Hence, the start service probability $\hat{\pi}_j$ equals to π_j,

$$\lim_{\lambda \to 0} E[T] = \sum_{j=0}^{2} \pi_j E[T_j].$$ (43)

When the offered load is rather high, the arrival rate λ approaches to the channel capacity $\hat{\mu}$. In this case, a large number of packets are backlogged in the buffer. The probability that there are no packets in the buffer approaches to zero. Thus, by definition, the start service probability $\hat{\pi}_0$ is zero. When a new packet arrives, the number of packets in the buffer n is very large. Thus, the factor $\beta^n (0 < \beta < 1)$ in expression $\hat{\pi}_{n,j}(n)$ approach to zero and the value of $\hat{\pi}_j$ is η_j,

$$\lim_{\lambda \to \hat{\mu}} E[T] = \eta_1 E[T_1] + \eta_2 E[T_2] = 1/\hat{\mu},$$ (44)

and the service rate of the M/MMSP/1 queue reaches its maximum service rate in theory.

4 P-K Formula of Waiting Time

In literature, based on the concept of the residual service time, the waiting time can be calculated as the sum of the service time of previous packets and residual service time of the HOL packet [9, 10]. However, in generalizing to any finite-state Markov channel, this method becomes very complicated due to its tedious algebraic calculations. In this paper, similar to the process of obtaining start service probability, we'll provide an alternative approach to calculate the mean waiting time using the idea of embedded Markov chain.

Similar to the process described in Fig. 5, a new packet sees n packets in the buffer upon arrival. Label these packets from 0 to n. As time goes by, the packets finish services and depart the buffer consecutively. The new packet moves forward in the buffer and finally becomes the HOL packet. The position of the new packet changes from n to $n-1$ and eventually 0. The waiting time of this packet is the time elapsed from the epoch when it arrives in the system to the epoch when it becomes the HOL packet.

Distinct to the classical queuing system, in M/MMSP/1 queue, the waiting time of the newly arrived packet correlates not only to the number of existing packets in the buffer, but also to the channel state when it arrives. Therefore, we need to define two classes of conditional waiting time that are associated with the position of the packet in the buffer.

(1) $W_{n,j}(k)$: A packet sees n packets in the buffer upon arrival, the conditional expected delay from the epoch when the newly arrived packet becomes the $k^{th} (0 \le k \le n)$ packet in the buffer while the channel is in state j to the epoch when it becomes an HOL packet.

(2) $V_{n,j}(k)$: A packet sees n packets in the buffer upon arrival, the conditional expected delay from the epoch when the channel transits to state j while the newly arrived packet is the $k^{th} (0 \le k \le n)$ packet in the buffer to the epoch when it becomes an HOL packet.

By definition, $W_{n,j}(n)$ is the waiting time of the newly arrived packet with channel state j, and the initial condition $W_{n,j}(0)$ equals zero (Fig. 6).

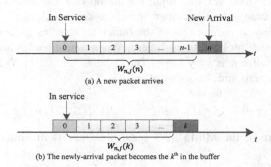

Fig. 6. Conditional waiting time $W_{n,j}(k)$.

The conditional expected delay $W_{n,j}(k)$ means the newly arrived packet turns the kth packet in the buffer while in channel state j, which stands for an embedded point S_j. The probability that next embedded point is S_j is $\mu_j/(\mu_j+f_j)$, which means the HOL packet has finished its service and the position of the new packet moves one step to $k-1$ with conditional delay $W_{n,j}(k-1)$.

On the other hand, the probability that next embedded point is Φ_i is $f_{j,i}/(\mu_j+f_j)$, which means the channel state transits to i with the position of the new packet unchanged. The conditional delay for the next embedded point is $V_{n,i}(k+1)$. Together with the property that the time interval between two successive embedded points is exponentially distributed with parameter μ_j+f_j, for $1 \le k \le n$,

$$W_{n,0}(k) = V_{n,1}(k) + \frac{1}{f_0}, \tag{45}$$

$$\begin{aligned} W_{n,1}(k) = {}& \frac{f_{1,0}}{\mu_1+f_1}\left(V_{n,0}(k) + \frac{1}{\mu_1+f_1}\right) + \frac{f_{1,2}}{\mu_1+f_1}\left(V_{n,2}(k) + \frac{1}{\mu_1+f_1}\right) \\ & + \frac{\mu_1}{\mu_1+f_1}\left(W_{n,1}(k-1) + \frac{1}{\mu_1+f_1}\right), \end{aligned} \tag{46}$$

$$W_{n,2}(k) = \frac{f_2}{\mu_2+f_2}\left(V_{n,1}(k+1) + \frac{1}{\mu_1+f_1}\right) + \frac{\mu_2}{\mu_2+f_2}\left(W_{n,2}(k-1) + \frac{1}{\mu_2+f_2}\right). \tag{47}$$

We can use the same logic to derive $V_{n,j}(k)$, however, it is easier to use memoryless property of the Markov chain to find that:

$$V_{n,j}(k) = W_{n,j}(k), \tag{48}$$

and this result is consistent with directly derived result.

From Eqs. (45)–(48) the following relationship can be established:

$$\begin{pmatrix} W_{n,0}(k) \\ W_{n,1}(k) \\ W_{n,2}(k) \end{pmatrix} = \hat{Q}^T \begin{pmatrix} W_{n,0}(k-1) \\ W_{n,1}(k-1) \\ W_{n,2}(k-1) \end{pmatrix} + \begin{pmatrix} E[T_0] \\ E[T_1] \\ E[T_2] \end{pmatrix}, \tag{49}$$

and we can obtain $W_{n,j}(k)$ as follows:

$$W_{n,0}(k) = \frac{1}{f_0} + \frac{1}{\hat{\mu}}k + \frac{1-\beta^k}{1-\beta}\left(E[T_1] - \frac{1}{\hat{\mu}}\right), \tag{50}$$

$$W_{n,1}(k) = \frac{1}{\hat{\mu}}k + \frac{1-\beta^k}{1-\beta}\left(E[T_1] - \frac{1}{\hat{\mu}}\right), \tag{51}$$

$$W_{n,2}(k) = \frac{1}{\hat{\mu}}k + \frac{1-\beta^k}{1-\beta}\left(E[T_2] - \frac{1}{\hat{\mu}}\right). \tag{52}$$

Consequently, by conditioning on the number of packets in the buffer and channel state upon arrival, the mean waiting time can be derived:

$$W = \sum_{j=0}^{2} \sum_{n=0}^{\infty} W_{n,j}(n)p_{n,j}. \tag{53}$$

Thus, combining with (34), (50)–(53), the analytical expression of P-K formula for the mean waiting time of M/MMSP/1 queue is offered in the following theorem.

Theorem 2: For three-state Markov channel, the P-K formula of mean waiting time for a packet in an M/MMSP/1 queue is given as follows:

$$W = \frac{\frac{1}{\hat{\mu}}\lambda E[T] + \frac{1}{f_0}(\pi_0 - p_{0,0}) - \frac{1}{f_0}\frac{1}{1-\beta}(\pi_0 - G_0(\beta)) + \frac{1}{1-\beta}\sum_{j=0}^{2}\left(E[T_j] - \frac{1}{\hat{\mu}}\right)(\pi_j - G_j(\beta))}{1 - \frac{\lambda}{\hat{\mu}}}. \tag{54}$$

Replacing the $G_j(\beta)$ in Theorem 2 by the start service probabilities, Theorem 2 can be rewritten in another form as follows:

$$W = \frac{\frac{1}{\hat{\mu}}\lambda E[T] + \frac{1}{1-\beta}\sum_{j=0}^{2} E[T_j]\left(\pi_j - \hat{\pi}_j\right) - \frac{\beta}{1-\beta}\frac{1}{f_0}(\pi_0 - \hat{\pi}_0)}{1 - \frac{\lambda}{\hat{\mu}}}, \tag{55}$$

which shows the connection between P-K formula and start service probabilities.

From Eq. (55), we see that the parameter β, which indicates the speed of state transition, has a significant influence on mean waiting time.

When β approaches to 0, $\hat{\pi}_j$ approaches to π_j, thus Eq. (55) can be simplified as:

$$\lim_{\beta \to 0} W = \frac{\frac{1}{\hat{\mu}}\lambda E[T]}{1 - \frac{\lambda}{\hat{\mu}}}. \tag{56}$$

Equation (56) is consistent with the expression of the waiting time of an M/M/1 queue with service rate $\hat{\mu}$. As β approaches to 0 means transition rates approach to infinity, the channel state transits between all states infinite times and the dependency between the states is eliminated. The service rate may be regarded as uniform and be the average of service rates in each state in time.

On the other hand, when β approaches to 1, Eq. (55) becomes much larger than Eq. (56). In this scenario, since state transition rates are small, the wireless channel may keep in one state for a long time. If the channel stays in failure state or bad state with service rate smaller than packet arrival rate, the queue length may grow to infinity. And this portion of time greatly influence the mean waiting time.

The mean waiting times with different transition rates are shown in Fig. 7, the curve of M/M/1 queue in black line is the lower bound of the waiting time. We can see that the mean waiting time will go up with the decreasing of transition rate.

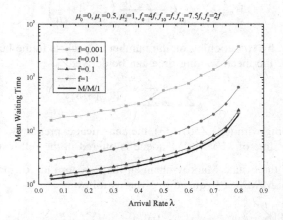

Fig. 7. The mean waiting time when changing all transition rate.

5 A Generalization to Finite-State Markov Channel

The deficiency of the work in [8] is that they can't generalize the method of calculating mean waiting time to any finite-state Markov channel, even to the three-state Markov channel due to tedious algebraic calculations. To show the generality of the method of hybrid Markov chain, in this section, we'll generalize our method to any finite-state Markov channel.

Suppose the channel is governed by an underlying Markov chain with N states from state 0 to state $N - 1$. The channel state can transit between each other freely. The infinitesimal generator Q for this homogenous continuous time Markov chain is given by:

$$\boldsymbol{Q} = \begin{pmatrix} -f_0 & f_{0,1} & f_{0,2} & \cdots & f_{0,N-1} \\ f_{1,0} & -f_1 & f_{1,2} & \cdots & f_{1,N-1} \\ f_{2,0} & f_{2,1} & -f_2 & \cdots & f_{2,N-1} \\ \vdots & \vdots & \vdots & \ddots & \vdots \\ f_{N-1,0} & f_{N-1,1} & f_{N-1,2} & \cdots & -f_{N-1} \end{pmatrix}. \tag{57}$$

The service rate in state j is defined as $\mu_j (0 \le j \le N-1)$ and thus the diagonal matrix $\boldsymbol{D} = diag(\mu_0, \mu_1, \ldots, \mu_{N-1})$. The capacity of the channel is $\hat{\mu} = \sum_{j=0}^{N-1} \pi_j \mu_j$.

Use the same definition of $\hat{\pi}_{n,j}(m)$ and $\hat{\varphi}_{n,j}(m)$ and follow the procedure of establishing the state equations of conditional start service probabilities in three-state Markov channel. Define $\hat{\pi}_n(\boldsymbol{m}) = (\hat{\pi}_{n,0}(m), \hat{\pi}_{n,1}(m), \ldots, \hat{\pi}_{n,N-1}(m))^T$. We immediately obtain the following theorem.

Theorem 3: For finite-state Markov model, the state equations of conditional start service probabilities are given as follows:

$$\hat{\pi}_{n,j}(m) = \frac{\mu_j}{\mu_j + f_j} \left(\hat{\pi}_{n,j}(m-1) + \hat{\varphi}_{n,j}(m-1) \right), \tag{58}$$

$$\hat{\varphi}_{n,j}(m) = \sum_{i \neq j} \frac{f_{i,j}}{\mu_i + f_i} \left(\hat{\pi}_{n,i}(m) + \hat{\varphi}_{n,i}(m) \right), \tag{59}$$

for $j = 0, 1, \ldots, N-1$, and $1 \le m \le n$. From Eqs. (53) and (54), we can derive the following differential equation for $\hat{\pi}_n(\boldsymbol{m})$:

$$\hat{\pi}_n(\boldsymbol{m}) = \hat{\boldsymbol{Q}} \hat{\pi}_n(\boldsymbol{m}-1), \tag{60}$$

where $\hat{\boldsymbol{Q}} = \left((\boldsymbol{D} - \boldsymbol{Q})^{-1} \boldsymbol{D} \right)^T$.

By solving Eq. (60), the conditional start service probability $\hat{\pi}_{n,j}(n)$ is obtained and the start service probability can be derived from Eq. (35).

Similarly, the derivation of P-K formula of finite-state Markov channel can be obtained as follows. Define $\boldsymbol{W}_n(k) = (W_{n,0}(k), W_{n,1}(k), \ldots, W_{n,N-1}(k))^T$, and Theorem 4 gives equations of conditional expected delay.

Theorem 4: For $j = 0, 1, \ldots, N-1$ and $1 \le k \le n$, the state equations of conditional waiting time for finite-state Markov model are:

$$W_{n,j}(k) = \frac{\mu_j}{\mu_j + f_j} \left(\frac{1}{\mu_j + f_j} + W_{n,j}(k-1) \right) + \sum_{i \neq j} \frac{f_{j,i}}{\mu_j + f_j} \left(\frac{1}{\mu_j + f_j} + V_{n,i}(k) \right), \tag{61}$$

$$V_{n,j}(k) = \frac{\mu_j}{\mu_j + f_j} \left(\frac{1}{\mu_j + f_j} + W_{n,j}(k-1) \right) + \sum_{i \neq j} \frac{f_{j,i}}{\mu_j + f_j} \left(\frac{1}{\mu_j + f_j} + V_{n,i}(k) \right). \tag{62}$$

From Eqs. (61) and (62), we can derive the following differential equation for $W_n(m)$:

$$W_n(k) = \hat{Q}^T W_n(k-1) + (D-Q)^{-1}1. \tag{63}$$

Note that $W_n(1) = (D-Q)^{-1}1$. By the definition of $W_{n,j}(k)$, $W_{n,j}(1)$ just equal the conditional service time with state j. Thus,

$$(D-Q)^{-1}1 = (E[T_0], E[T_1], \ldots, E[T_{N-1}])^T. \tag{64}$$

By solving Eq. (63), the conditional expected delay $W_{n,j}(n)$ can be acquired. The P-K formula can be obtained from Eq. (53) in a similar manner, by summing over the state j from 0 to $N-1$.

6 Conclusions

In this paper, we model the wireless channel as a three-state Markov channel and discuss the delay of M/MMSP/1 queue. Different from other related works, we use hybrid embedded Markov chain to describe the queueing process. From the result, we find that the mean waiting time is significantly influenced by the parameter β which represents the state transition rate of the system. When $\beta \to 0$, the state transits much faster than packet service, and the queue performs like an M/M/1 queueing system. On the other hand, when $\beta \to 1$, the state transits slower than packet service, and mean waiting time will get larger. From the derivation of three-state case, we generalize our method to any finite-state M/MMSP/1 queueing system.

References

1. Rappaport, T.S.: Wireless Communications: Principles and Practice, vol. 2. Prentice Hall PTR, New Jersey (1996)
2. Gilbert, E.N.: Capacity of a burst-noise channel. Bell Syst. Techn. J. **39**(5), 1253–1265 (1960)
3. Elliott, E.O.: Estimates of error rates for codes on burst-noise channels. Bell Syst. Tech. J. **42** (5), 1977–1997 (1963)
4. Eisen, M., Tainiter, M.: Stochastic variations in queuing processes. Oper. Res. **11**(6), 922–927 (1963)
5. Yechiali, U., Pinhas, N.: Queuing problems with heterogeneous arrivals and service. Oper. Res. **19**(3), 722–734 (1971)
6. Yechiali, U.: A queuing-type birth-and-death process defined on a continuous-time Markov chain. Oper. Res. **21**(2), 604–609 (1973)
7. Mahabhashyam, S.R., Natarajan, G.: On queues with Markov modulated service rates. Queueing Syst. **51**(1), 89–113 (2005)

8. Huang, L., Lee, T.: Generalized pollaczek-khinchin formula for markov channels. IEEE Trans. Commun. **61**(8), 3530–3540 (2013)
9. Bertsekas, D.P., Gallager, R.G., Humblet, P.: Data Networks, vol. 2. Prentice-Hall International, New Jersey (1992)
10. Wolff, R.W.: Poisson arrivals see time averages. Oper. Res. **30**(2), 223–231 (1982)

A Nonlinear Solution to Closed Queueing Networks for Bike Sharing Systems with Markovian Arrival Processes and Under an Irreducible Path Graph

Quan-Lin Li, Rui-Na Fan$^{(\boxtimes)}$, and Zhi-Yong Qian

School of Economics and Management Sciences, Yanshan University,
Qinhuangdao 066004, China
liquanlin@stinghua.edu.cn, {fanruina,qianzhiyong0926}@stumail.ysu.edu.cn

Abstract. As a favorite urban public transport mode, the bike sharing system is a large-scale and complicated system, and there exists a key requirement that a user and a bike should be matched sufficiently in time. Such matched behavior makes analysis of the bike sharing systems more difficult and challenging. To design a better bike sharing system, it is a key to analyze and compute the probabilities of the problematic (i.e., full or empty) stations. In fact, such a computation is established for some fairly complex stochastic systems. To do this, this paper considers a more general large-scale bike sharing system from two important views: (a) Bikes move in an irreducible path graph, which is related to geographical structure of the bike sharing system; and (b) Markovian arrival processes (MAPs) are applied to describe the non-Poisson and burst behavior of bike-user (abbreviated as user) arrivals, while the burstiness demonstrates that the user arrivals are time-inhomogeneous and space-heterogeneous in practice. For such a complicated bike sharing system, this paper establishes a multiclass closed queueing network by means of some virtual ideas, for example, bikes are abstracted as virtual customers; stations and roads are regarded as virtual nodes. Thus user arrivals are related to service times at station nodes; and users riding bikes on roads are viewed as service times at road nodes. Further, to deal with this multiclass closed queueing network, we provide a detailed observation practically on physical behavior of the bike sharing system in order to establish the routing matrix, which gives a nonlinear solution to compute the relative arrival rates in terms of the product-form solution to the steady-state probabilities of joint queue lengths at the virtual nodes. Based on this, we can compute the steady-state probability of problematic stations, and also deal with other interesting performance measures of the bike sharing system. We hope that the methodology and results of this paper can be applicable in the study of more general bike sharing systems through multiclass closed queueing networks.

Keywords: Bike sharing system · Closed queueing network · Product-form solution · Irreducible path graph · Problematic station · Markovian arrival process

© Springer International Publishing AG 2017
W. Yue et al. (Eds.): QTNA 2017, LNCS 10591, pp. 118–140, 2017.
https://doi.org/10.1007/978-3-319-68520-5_8

1   Introduction

In this paper, we propose a more general bike sharing system with Markovian arrival processes and under an irreducible path graph. Note that the bike sharing system always has some practically important factors, for example, time-inhomogeneity, geographical heterogeneity, and arrival burstiness. To analyze such a bike sharing system, we establish a multiclass closed queueing network by means of virtual customers, virtual nodes and virtual service times. Further, when studying this multiclass closed queueing network, we set up a routing matrix which gives a nonlinear solution to compute the relative arrival rates, and provide the product-form solution to the steady-state probabilities of joint queue lengths at the virtual nodes. Based on this, we can compute the steady-state probability of problematic stations, and also deal with other interesting performance measures of the bike sharing system.

During the last decades bike sharing systems have emerged as a public transport mode devoted to short trip in more than 600 major cities around the world. Bike sharing systems are regarded as a promising way to jointly reduce, such as, traffic and parking congestion, traffic noise, air pollution and greenhouse effect. Several excellent overviews and useful remarks were given by DeMaio [6], Meddin and DeMaio [23], Shu et al. [35], Labadi et al. [16] and Fishman et al. [7].

Few papers applied queueing theory and Markov processes to the study of bike sharing systems. On this research line, it is a key to compute the probability of problematic stations. However, so far there still exist some basic difficulties and challenges for computing the probability of problematic stations because computation of the steady-state probability, in the bike sharing system, needs to apply the theory of complicated or high-dimensional Markov processes. For this, readers may refer to recent literatures which are classified and listed as follows. **(a) Simple queues:** Leurent [17] used the M/M/1/C queue to study a vehicle-sharing system, and also analyzed performance measures of this system. Schuijbroek et al. [32] evaluated the service level by means of the transient distribution of the M/M/1/C queue, and the service level was used to establish some optimal models to discuss vehicle routing. Raviv et al. [29] and Raviv and Kolka [28] employed the transient distribution of the time-inhomogeneous $M(t)/M(t)/1/C$ queue to compute the expected number of bike shortages at each station. **(b) Closed queueing networks:** Adelman [1] applied a closed queueing network to propose an internal pricing mechanism for managing a fleet of service units, and also used a nonlinear flow model to discuss the price-based policy for establishing the vehicle redistribution. George and Xia [11] used the closed queueing networks to study the vehicle rental systems, and determined the optimal number of parking spaces for each rental location. Li et al. [20] proposed a unified framework for analyzing the closed queueing networks in the study of bike sharing systems. **(c) Mean-field method.** Fricker et al. [8] considered a space-inhomogeneous bike-sharing system with multiple clusters, and expressed the minimal proportion of problematic stations. Fricker and Gast [9] provided a detailed analysis for a space-homogeneous bike-sharing system in terms of the M/M/1/K queue as well as some simple mean-field models, and crucially, they derived the closed-form solution to find the minimal proportion of problematic stations. Fricker and Tibi [10] studied the central limit and local limit theorems for the independent (non-identically distributed) random variables, which provide support on analysis of a generalized Jackson network with product-form solution. Further, they used the limit theorems to give an outline of stationary asymptotic analysis for the locally space-homogeneous bike-sharing systems. Li *et al.* [21] provided a complete picture on how to jointly use the mean-field theory, the time-inhomogeneous queues and the nonlinear birth-death processes to analyze performance measures of the bike-sharing systems. Li and Fan [19] discussed the bike sharing system under an Markovian environment by means of the mean-field computation, the time-inhomogeneous queues and the nonlinear Markov processes. **(d) Markov decision processes.** To discuss the bike-sharing systems, Waserhole and Jost [36,37,39] and Waserhole *et al.* [38] used the simplified closed queuing networks to establish the Markov decision models, and computed the optimal policy by means of the fluid approximation which overcame the state space explosion of multi-dimensional Markov decision processes.

There has been much key research on closed queueing networks. Readers may refer to, such as, three excellent books by Kelly [13,14] and Serfozo [34]; multiclass customers by Baskett *et al.* [2], multiple closed chains by Reiser and Kobayashi [30], computational algorithms by Bruell and Balbo [4], mean-value computation by Reiser [31], sojourn time by Kelly and Pollett [15], survey for blocks by Onvural [26], and batch service by Henderson *et al.* [12].

Markovian arrival process (MAP) is a useful mathematical model for describing bursty traffic in, for example, communication networks, manufacturing systems, transportation networks and so forth. Readers may refer to recent publications for more details, among which are Ramaswami [27], Chap. 5 in Neuts [24], Lucantoni [22], Neuts [25], Chakravarthy [5] and Li [18].

Contributions of this paper: The main contributions of this paper are twofold: The first contribution is to propose a more general bike sharing system with Markovian arrival processes and under an irreducible path graph. Note that Markovian arrival processes, as well as the irreducible path graph indicate that burst arrival behavior and geographical structure of the bike sharing system are more general and practical. Specifically, the burstiness is to well express that the user arrivals are time-inhomogeneous and space-heterogeneous in practice. For such a bike sharing system, this paper establishes a multiclass closed queueing network by means of virtual customers, virtual nodes and virtual service times. The second contribution is to deal with such a multiclass closed queueing network with virtual customers, virtual nodes and virtual service times, and to establish a routing matrix which gives a nonlinear solution to compute the relative arrival rates in terms of the product-form solution to the steady-state probabilities of joint queue lengths at the virtual nodes. By using the product-form solution, this paper computes the steady-state probability of problematic stations, and also deals with other interesting performance measures of the bike sharing system. Therefore, the methodology and results of this paper can be applicable in the study of more general bike sharing systems by means of multiclass closed queueing networks.

Organization of this paper: The remainder of this paper is organized as follows. In Sect. 2, we describe a large-scale bike sharing system with Markovian arrival processes and under an irreducible path graph. In Sect. 3, we abstract the bike sharing system as a multiclass closed queueing network with virtual customers, virtual nodes and virtual service times. Further, we establish the routing matrix, and compute the relative arrival rate in each node, where three examples are given to express and compute the routing matrix and the relative arrival rate. In Sect. 4, we give a product-form solution to the steady-state probabilities of joint queue lengths at the virtual nodes, and provide a nonlinear solution to determine the N undetermined constants which are related to the probability of problematic stations. Moreover, we compute the steady-state probability of problematic stations, and also analyze other performance measures of the bike sharing system. Finally, some concluding remarks are given in Sect. 5.

2   Model Description

In this section, we describe a more general large-scale bike sharing system, where arrivals of bike users are non-Poisson and are characterized as Markovian arrival processes (MAPs), and users riding bikes travel in an irreducible path graph which is constituted by N different stations and some different directed roads.

In a large-scale bike sharing system, a user arrives at a station, rents a bike, and uses it for a while; then he returns the bike to another station, and immediately leaves this system. Based on this, we describe a more general large-scale space-heterogeneous bike sharing system, and introduce operational mechanism, system parameters and basic notation as follows:

(1) Stations: We assume that there are N different stations in the bike sharing system. The N stations may be different due to their geographical location and surrounding environment. We assume that every station has C bikes and K parking positions at the initial time $t = 0$, where $1 \leq C < K < \infty$, and $NC \geq K$. Note that such a condition $NC \geq K$ is to make at least a full station.

(2) Roads: Let Road $i \rightarrow j$ be a road relating Station i to Station j. Note that Road $i \rightarrow j$ and Road $j \rightarrow i$ may be different. To express all the roads beginning from Station i for $1 \leq i \leq N$, we write

$$R(i) = \{\text{Road } i \rightarrow j : j \neq i, 1 \leq j \leq N\}.$$

Similarly, to express all roads be over at Station j for $1 \leq j \leq N$, we write

$$\overline{R}(j) = \{\text{Road } i \rightarrow j : i \neq j, 1 \leq i \leq N\}.$$

It is easy to see that there are at most $N - 1$ different directed roads in the set $R(i)$ or $\overline{R}(j)$. We denote by $|R(i)|$ the number of elements or roads in the set $R(i)$. Thus $|R(i)| \leq N - 1$ for $1 \leq i \leq N$ and $\sum_{i=1}^{N} |R(i)| \leq N(N-1)$.

To express all the stations in the near downlink of Station i, we write

$$\Theta_i = \{\text{Station } j : \text{Road } i \to j \in R(i)\}.$$

Similarly, the set of all stations in the near uplink of Station i is written as

$$\Delta_i = \{\text{Station } j : \text{Road } j \to i \in \overline{R}(i)\}.$$

(3) An irreducible path graph: To express the bike moving paths, it is easy to observe that the bikes dynamically move either among the stations or among the roads. To record the bike dynamic positions, it is better to introduce two classes of virtual nodes: (a) station nodes; and (b) road nodes. The set of all the virtual nodes of the bike sharing system is given by

$$\Theta = \{\text{Station } i : 1 \le i \le N\} \cup \left\{ \underset{i=1}{\overset{N}{\cup}} R(i) \right\}.$$

In this bike sharing system, it is easy to calculate that there are $N + \cup_{i=1}^{N} |R(i)|$ virtual nodes.

If Station i has a near downstream Road $i \to j$, then we call that Node i (i.e. Station i) can be accessible to Node $i \to j$ (i.e. Road $i \to j$), denoted as Node $i \Longrightarrow$ Node $i \to j$; otherwise Node i can not be accessible to Node $i \to j$. If Station j has a near upstream Road $i \to j$, then we call that Node $i \to j$ can be accessible to Node j, denoted as Node $i \to j \Longrightarrow$ Node j; otherwise Node $i \to j$ can not be accessible to Node j.

If there exist some virtual nodes $n_1, n_2, \ldots, n_r$ in the set Θ such that

$$\text{Node } n_1 \Longrightarrow \text{Node } n_2 \Longrightarrow \cdots \Longrightarrow \text{Node } n_r,$$

then we call that there is an accessible path formed by the virtual nodes $n_1, n_2, \ldots, n_r$.

If for any two virtual nodes m_a and m_b in the set Θ, there always exist some virtual nodes $n_1, n_2, \ldots, n_r$ in the set Θ such that

$$\text{Node } m_a \Longrightarrow \text{Node } n_1 \Longrightarrow \text{Node } n_2 \Longrightarrow \cdots \Longrightarrow \text{Node } n_r \Longrightarrow \text{Node } m_b,$$

then we call that the path graph of the bike sharing system is irreducible.

In this paper, we assume that the bike sharing system exists an irreducible path graph. In this case, we call that the bike sharing system is path irreducible. Note that this irreducibility is guaranteed through setting up an appropriate road construction with $R(i)$ for $1 \le i \le N$. In general, such a road construction is not unique in order to guarantee the irreducible path graph.

(4) Markovian arrival processes: Arrivals of outside bike users at Station i are a Markovian arrival process (MAP) of irreducible matrix descriptor $(\mathbf{C}_i, \mathbf{D}_i)$ of size m, denoted as $\text{MAP}(\mathbf{C}_i, \mathbf{D}_i)$, where

$$\mathbf{C}_i = \begin{pmatrix} c_{1,1}^{(i)} & c_{1,2}^{(i)} & \cdots & c_{1,m}^{(i)} \\ c_{2,1}^{(i)} & c_{2,2}^{(i)} & \cdots & c_{2,m}^{(i)} \\ \vdots & \vdots & \ddots & \vdots \\ c_{m,1}^{(i)} & c_{m,2}^{(i)} & \cdots & c_{m,m}^{(i)} \end{pmatrix}$$

and

$$\mathbf{D}_i = \begin{pmatrix} d_{1,1}^{(i)} & d_{1,2}^{(i)} & \cdots & d_{1,m}^{(i)} \\ d_{2,1}^{(i)} & d_{2,2}^{(i)} & \cdots & d_{2,m}^{(i)} \\ \vdots & \vdots & \ddots & \vdots \\ d_{m,1}^{(i)} & d_{m,2}^{(i)} & \cdots & d_{m,m}^{(i)} \end{pmatrix}.$$

Let $c_{k,l}^{(i)} \geq 0$ with $l \neq k$, $d_{r,s}^{(i)} \geq 0$, $c_{k,k}^{(i)} = -\left(\sum_{l \neq k}^{m} c_{k,l}^{(i)} + \sum_{r=1}^{m} d_{k,r}^{(i)} \right)$, and hence $(\mathbf{C}_i + \mathbf{D}_i) e = 0$. We assume that Markov chain $\mathbf{C}_i + \mathbf{D}_i$ is irreducible, finite-state and aperiodic, hence it is positive-recurrent due to the finite state space. Further, in the Markov chain $\mathbf{C}_i + \mathbf{D}_i$ there exists the unique stationary probability vector $\widetilde{\theta}^{(i)} = \left(\theta_1^{(i)}, \theta_2^{(i)}, \cdots, \theta_m^{(i)} \right)$ for $1 \leq i \leq N$, that is, the vector $\theta^{(i)}$ is the unique solution to the system of linear equations $\widetilde{\theta}^{(i)} (\mathbf{C}_i + \mathbf{D}_i) = 0$ and $\widetilde{\theta}^{(i)} e = 1$. In this case, the stationary average arrival rate of the MAP$(\mathbf{C}_i + \mathbf{D}_i)$ is $\lambda_i = \widetilde{\theta}^{(i)} D^{(i)} e$. Specifically, we write that $\overrightarrow{\lambda}_i = \left(\lambda_i^{(1)}, \lambda_i^{(2)}, \cdots, \lambda_i^{(m)} \right) = \widetilde{\theta}^{(i)} \mathbf{D}_i$ for $1 \leq i \leq N$.

(5) The first riding-bike time: An outside bike user arrives at the ith station to rent a bike. If there is no bike in the ith station (i.e., the ith station is empty), then the user immediately leaves this bike sharing system. If there is at least one available bike at the ith station, then the user rents a bike and goes to Road $i \rightarrow j$ for $j \neq i$ with probability $p_{i,j}$ for $\sum_{j \in \Theta_i} p_{i,j} = 1$, and his riding-bike time on Road $i \rightarrow j$ is an exponential random variable with riding-bike rate $\mu_{i,j} > 0$.

(6) The bike return times: Notice that for any user, his first bike return process may be different from those retrial processes with successively returning the bike to one station for at least twice due to his pasted arrivals at the full stations. In this situation, his road selection as well as his riding-bike time in the first process may be different from those in any retrial return process.

The first return – When the user completes his short trip on Road $i \rightarrow j$, he needs to return his bike to the jth station. If there is at least one available parking position (i.e., a vacant docker), then the user directly returns the bike to the jth station, and immediately leaves this bike sharing systems.

The second return – If no parking position is available at the jth station, then the user has to ride the bike to the l_1th station with probability α_{j,l_1} for $l_1 \neq j$ and $\sum_{l_1 \in \Theta_j} \alpha_{j,l_1} = 1$; and his future riding-bike time on Road $j \rightarrow l_1$ is also an exponential random variable with riding-bike rate $\xi_{j,l_1} > 0$. If there is at least one available parking position, then the user directly returns his bike to the l_1th station, and immediately leaves this bike sharing system.

The $(k+1)$st return for $k \geq 2$ – We assume that this bike has not been returned at any station yet through k consecutive returns. In this case, the user has to try his $(k+1)$st lucky return. Notice that the user goes to the l_kth station from the l_{k-1}th full station with probability α_{l_{k-1},l_k} for $l_k \neq l_{k-1}$ and

$\sum_{l_k \in \Theta_{l_{k-1}}} \alpha_{l_{k-1},l_k} = 1$; and his riding-bike time on Road $l_{k-1} \to l_k$ is an exponential random variable with riding-bike rate $\xi_{l_{k-1},l_k} > 0$. If there is at least one available parking position, then the user directly returns his bike to the l_kth station, and immediately leaves this bike sharing system; otherwise he has to continuously ride his bike in order to try to return the bike to another station again.

We further assume that the returning-bike process is persistent in the sense that the user must find a station with an empty position to return his bike because the bike is a public property.

It is seen from the above description that the parameters: $p_{i,j}$ and $\mu_{i,j}$, for $j \neq i$ and $1 \leq i,j \leq N$, of the first return, may be different from the parameters: $\alpha_{i,j}$ and $\xi_{i,j}$, for $j \neq i$ and $1 \leq i,j \leq N$, of the kth return for $k \geq 2$. This is due to a simple observation that the user possibly deal with more things (for example, tourism, shopping, visiting friends and so on) in the first return process, but he becomes only one return task for returning his bike to one station during the k successive return processes for $k \geq 2$.

(7) The departure discipline: The user departure process has two different cases: (a) An outside user directly leaves the bike sharing system if he arrives at an empty station; and (b) if one user rents and uses a bike, and he finally returns the bike to a station, then the user completes his trip, and immediately leaves the bike sharing system.

We assume that all the above random variables are independent of each other. For such a bike sharing system, Fig. 1 provides some intuitive physical interpretation for the bike sharing system.

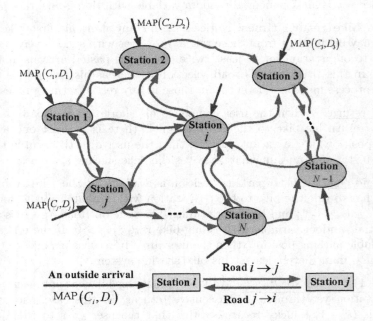

Fig. 1. The physical structure of the bike sharing system

3   A Closed Queueing Network

In this section, we describe the bike sharing system as a closed queueing network according to the fact that the number of bikes in this system is fixed. To study such a closed queueing network, we need to determine the service rates, the routing matrix and the relative arrival rates in all the virtual nodes.

For the bike sharing system, we need to abstract it as a closed queueing network as follows:

(1) Virtual nodes: Although the stations and the roads have different physical attributes, such as, different functions, different geographical topologies and so forth, it is seen that here the stations and the roads are all regarded as the same abstracted nodes in a closed queueing network.

(2) Virtual customers: The bikes either at the stations or on the roads are viewed as virtual customers as follows:

A closed queueing network under virtual idea: The   virtual   customers   are abstracted by the bikes from either the stations or the roads. In this case, the service processes are taken either from user arrivals at the station nodes or from users riding bikes on the road nodes. Since the total number of bikes in the bike sharing system is fixed as the positive integer NC, thus the bike sharing system can be regarded as a closed queueing network with such virtual customers, virtual nodes and virtual service times.

Two classes of virtual customers: From Assumptions (2), (5) and (6) in Sect. 2, it is seen that there are two different classes of virtual customers in the road nodes, where the first class of virtual customers are the bikes ridden on the roads for the first time; while the second class of virtual customers are the bikes which are successively ridden on the roads at least twice due to his arrivals at full stations.

We abstract the virtual nodes both from the stations and from the roads, and also find the virtual customers corresponding to the NC bikes. This sets up a multiclass closed queueing network. To compute the steady-state probabilities of joint queue lengths in the bike sharing system, it is seen from Chap. 7 in Bolch *et al.* [3] that we need to determine the service rate and the relative arrival rate for each virtual node in the multiclass closed queueing network.

(a) The service rates at nodes

We discuss the service processes of the closed queueing network from two different cases: One for the station nodes, and the other for the road nodes. Figure 2 shows how the two classes of service times are given from the multiclass closed queueing network.

Case one: A road node in the set $\cup_{i=1}^{N} R(i)$

The first class of virtual customers: We denote the number of virtual customers of the first class on Road $i \to l$ by $m_{i,l}^{(1)}$. The return process of bikes of the first class from Road $i \to l$ to Station l for the first time is Poisson with service rate

$$a_{i,l}^{(1)} = m_{i,l}^{(1)} \mu_{i,l}.$$

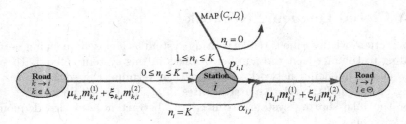

Fig. 2. The queueing processes in the multiclass closed queueing network

The second class of virtual customers: We denote the number of virtual customers of the second class on Road $i \rightarrow l$ by $m_{i,l}^{(2)}$. The retrial return process of customers of the second class from Road $i \rightarrow l$ to Station l is Poisson with service rate

$$a_{i,l}^{(2)} = m_{i,l}^{(2)} \xi_{i,l}.$$

Case two: The N station nodes

Let n_i be the number of bikes packed in Station i. The departure process of bikes from the ith station is due to those customers who rent the bikes at the ith station and then immediately enter one road in $R(i)$. Thus if the ith station is not empty, then the service process (i.e. renting bikes) is a MAP with a stationary service rate of phase v

$$a_i^{(v)} = \lambda_i^{(v)} \mathbf{1}_{\{1 \leq n_i \leq K\}} \sum_{l \neq i}^{N} p_{i,l} = \lambda_i^{(v)} \mathbf{1}_{\{1 \leq n_i \leq K\}}, 1 \leq v \leq m, \tag{1}$$

where $\sum_{l \neq i}^{N} p_{i,l} = 1$, and $\overrightarrow{\lambda}_i = \left(\lambda_i^{(1)}, \lambda_i^{(2)}, \ldots, \lambda_i^{(m)} \right)$ is given by the MAP (C_i, D_i) through $\overrightarrow{\lambda}_i = \widetilde{\theta}^{(i)} D_i$ for $1 \leq i \leq N$.

(b) The relative arrival rates

For the multiclass closed queueing network, to determine the steady-state probability distribution of joint queue lengths at any virtual node, it is necessary to firstly give the relative arrival rates at the virtual nodes. To this end, we must establish the routing matrix in the first step.

Based on Chap. 7 in Bolch *et al.* [3], we denote by e_i and $e_{R_{i \rightarrow j}}^{(r)}$ the relative arrival rates of the ith station, and of Road $i \rightarrow l$ with bikes of class r, respectively. We write

$$\overrightarrow{e} = \{\overrightarrow{e}_i : 1 \leq i \leq N\},$$

where

$$\overrightarrow{e}_i = \left\{ \mathbf{e}_i, \mathbf{e}_{R_{i \rightarrow j}}^{(r)}, j \in \Theta_i, r = 1, 2 \right\}.$$

Note that this bike sharing system is large-scale, thus the routing matrix of the closed queueing network corresponding to the bike sharing system will be very complicated. To understand how to set up such a routing matrix, in what follows we first give three simple examples for the purpose of writing the routing

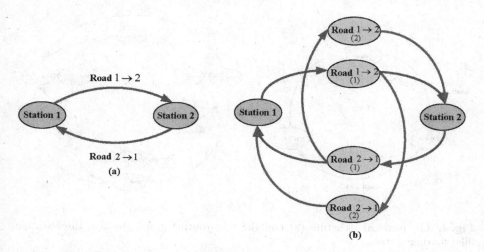

Fig. 3. The physical structure (a) and the bike routing graph (b) for a two-station bike sharing system

matrix, using the physical structure and the routing graph of the bike sharing system. See Figs. 3, 4 and 5 for more details.

Let $Q_i(t)$ be the number of bikes parked at Station i at time $t \geq 0$. From the exponential and MAP assumptions, it is seen that an irreducible finite state Markov chain is used to express and analyze the bike sharing system, while the Markov chain is aperiodic and positive recurrent. In this case, there exists stationary probability vector in the Marokov chain, and thus we give the limit

$$\pi_{i,K} = \lim_{t \to +\infty} P\{Q_i(t) = K\}.$$

Example One: We consider a simple bike sharing system with two stations, and the physical structure of the stations and roads is depicted in (a) of Fig. 3. Note that there exist two classes of virtual customers in the road nodes, and the bike routing graph of the bike sharing system is depicted in (b) of Fig. 3. Since there are only two stations in this bike sharing system, we have $p_{i,j} = \alpha_{i,j} = 1$. Based on this, we obtain the routing matrix of order 6 as follow:

$$P = \begin{pmatrix} & 1 & & & & \\ & & 1-\pi_{2,K} & \pi_{2,K} & & \\ & & 1-\pi_{2,K} & \pi_{2,K} & & \\ & & & & 1 & \\ 1-\pi_{1,K} & \pi_{1,K} & & & & \\ 1-\pi_{1,K} & \pi_{1,K} & & & & \end{pmatrix},$$

where all those elements that are not expressed are viewed as zeros, and $\pi_{i,K}$ is a undetermined constant, and it is also the stationary probability of the ith full station for $i = 1, 2$.

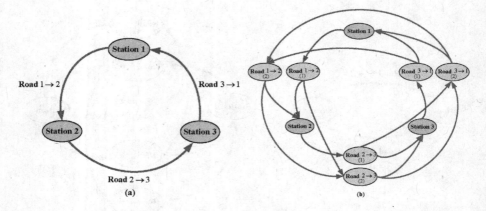

Fig. 4. The physical structure (a) and the bike routing graph (b) of a three-station bike sharing system

To determine the relative arrival rate at each virtual node, using the system of linear equations $\vec{e}\,P = \vec{e}$ and $e_1 = 1$, we obtain

$$\begin{cases} e_1 = \left(e^{(1)}_{R_{2\to1}} + e^{(2)}_{R_{2\to1}} \right)(1 - \pi_{1,K}), \\ e^{(1)}_{R_{1\to2}} = e_1, \\ e^{(2)}_{R_{1\to2}} = \left(e^{(1)}_{R_{2\to1}} + e^{(2)}_{R_{2\to1}} \right)\pi_{1,K}, \\ e_2 = \left(e^{(1)}_{R_{1\to2}} + e^{(2)}_{R_{1\to2}} \right)(1 - \pi_{2,K}), \\ e^{(1)}_{R_{2\to1}} = e_2, \\ e^{(2)}_{R_{2\to1}} = \left(e^{(1)}_{R_{1\to2}} + e^{(2)}_{R_{1\to2}} \right)\pi_{2,K}. \end{cases}$$

Using $e_1 = 1$, we get

$$\begin{cases} e_1 = e^{(1)}_{R_{1\to2}} = 1, \\ e^{(2)}_{R_{1\to2}} = \frac{\pi_{1,K}}{1-\pi_{1,K}}, \\ e_2 = e^{(1)}_{R_{2\to1}} = \frac{1-\pi_{2,K}}{1-\pi_{1,K}}, \\ e^{(2)}_{R_{2\to1}} = \frac{\pi_{2,K}}{1-\pi_{1,K}}, \end{cases} \tag{2}$$

where the two undetermined positive constants $\pi_{1,K}$ and $\pi_{2,K}$ will be given in the next section, and they determine the relative arrival rates at the six virtual nodes.

Example Two: We consider a bike sharing system with three stations, and the physical structure of the stations and roads can be seen in (a) of Fig. 4. There exist two classes of virtual customers in the road nodes, and the bike routing graph of the bike sharing system is depicted in (b) of Fig. 4. It is seen from (a) of Fig. 4 that $p_{1,2} = p_{2,3} = p_{3,1} = \alpha_{1,2} = \alpha_{2,3} = \alpha_{3,1} = 1$. Based on this, the routing matrix of order 9 is given by

$$
\begin{pmatrix}
 & 1 & & & & & & & \\
 & & 1-\pi_{2,K} & \pi_{2,K} & & & & & \\
 & & 1-\pi_{2,K} & \pi_{2,K} & & & & & \\
 & & & & 1 & & & & \\
 & & & & & 1-\pi_{3,K} & \pi_{3,K} & & \\
 & & & & & 1-\pi_{3,K} & \pi_{3,K} & & \\
 & & & & & & & 1 & \\
1-\pi_{1,K} & \pi_{1,K} & & & & & & & \\
1-\pi_{1,K} & \pi_{1,K} & & & & & & &
\end{pmatrix} .
$$

To determine the relative arrival rate at each virtual node, using the system of linear equations $\overrightarrow{e}P = \overrightarrow{e}$ and $e_1 = 1$, we obtain

$$
\begin{cases}
e_1 = \left(e^{(1)}_{R_{3\to1}} + e^{(2)}_{R_{3\to1}} \right)(1-\pi_{1,K}), \\[4pt]
e^{(1)}_{R_{1\to2}} = e_1, \\[4pt]
e^{(2)}_{R_{1\to2}} = \left(e^{(1)}_{R_{3\to1}} + e^{(2)}_{R_{3\to1}} \right)\pi_{1,K}, \\[4pt]
e_2 = \left(e^{(1)}_{R_{1\to2}} + e^{(2)}_{R_{1\to2}} \right)(1-\pi_{2,K}), \\[4pt]
e^{(1)}_{R_{2\to3}} = e_2, \\[4pt]
e^{(2)}_{R_{2\to3}} = \left(e^{(1)}_{R_{1\to2}} + e^{(2)}_{R_{1\to2}} \right)\pi_{2,K}, \\[4pt]
e_3 = \left(e^{(1)}_{R_{2\to3}} + e^{(2)}_{R_{2\to3}} \right)(1-\pi_{3,K}), \\[4pt]
e^{(1)}_{R_{3\to1}} = e_3, \\[4pt]
e^{(2)}_{R_{3\to1}} = \left(e^{(1)}_{R_{2\to3}} + e^{(2)}_{R_{2\to3}} \right)\pi_{3,K}.
\end{cases}
$$

Using $e_1 = 1$, we get

$$
\begin{cases}
e_1 = e^{(1)}_{R_{1\to2}} = 1, \\[4pt]
e^{(2)}_{R_{1\to2}} = \frac{\pi_{1,K}}{1-\pi_{1,K}}, \\[4pt]
e_2 = e^{(1)}_{R_{2\to3}} = \frac{1-\pi_{2,K}}{1-\pi_{1,K}}, \\[4pt]
e^{(2)}_{R_{2\to3}} = \frac{\pi_{2,K}}{1-\pi_{1,K}}, \\[4pt]
e_3 = e^{(1)}_{R_{3\to1}} = \frac{1-\pi_{3,K}}{1-\pi_{1,K}}, \\[4pt]
e^{(2)}_{R_{3\to1}} = \frac{\pi_{3,K}}{1-\pi_{1,K}},
\end{cases}
\tag{3}
$$

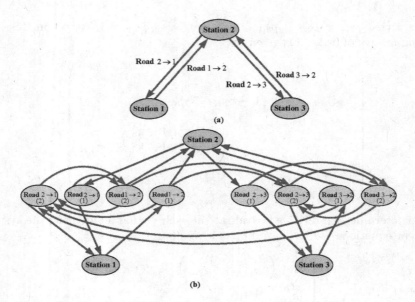

Fig. 5. The physical structure (a) and the bike routing graph (b) of a three-station bike sharing system

where the three undetermined positive constants $\pi_{1,K}$, $\pi_{2,K}$ and $\pi_{3,K}$ will be given in the next section, and they determine the relative arrival rates for the nine virtual nodes.

Example Three: We consider a bike sharing system with three stations, and the physical structure of the stations and roads can be seen in (a) of Fig. 5. There exist two classes of virtual customers in the road nodes, and the bike routing graph of the bike sharing system is depicted in (b) of Fig. 5. Based on this, we obtain the routing matrix of order 11 as follow:

$$
\begin{pmatrix}
 & 1 & & & & & & & \\
 & & 1-\pi_{2,K} & \alpha_{2,1}\pi_{2,K} & \alpha_{2,3}\pi_{2,K} & & & & \\
 & & 1-\pi_{2,K} & \alpha_{2,1}\pi_{2,K} & \alpha_{2,3}\pi_{2,K} & & & & \\
 & & & p_{2,1} & & p_{2,3} & & & \\
1-\pi_{1,K} & \pi_{1,K} & & & & & & & \\
1-\pi_{1,K} & \pi_{1,K} & & & & & & & \\
 & & & & & & 1-\pi_{3,K} & \pi_{3,K} & \\
 & & & & & & 1-\pi_{3,K} & \pi_{3,K} & \\
 & & & & & & & 1 & \\
 & & 1-\pi_{2,K} & \alpha_{2,1}\pi_{2,K} & \alpha_{2,3}\pi_{2,K} & & & & \\
 & & 1-\pi_{2,K} & \alpha_{2,1}\pi_{2,K} & \alpha_{2,3}\pi_{2,K} & & & &
\end{pmatrix}.
$$

To determine the relative arrival rate at each virtual node, using the system of linear equations $\vec{e}P = \vec{e}$ and $e_1 = 1$, we obtain

$$
\begin{cases}
e_1 = \left(e^{(1)}_{R_2 \to 1} + e^{(2)}_{R_2 \to 1} \right) (1 - \pi_{1,K}), \\[4pt]
e^{(1)}_{R_1 \to 2} = e_1, \\[4pt]
e^{(2)}_{R_1 \to 2} = \left(e^{(1)}_{R_2 \to 1} + e^{(2)}_{R_2 \to 1} \right) \pi_{1,K} \\[4pt]
e_2 = \left(e^{(1)}_{R_1 \to 2} + e^{(2)}_{R_1 \to 2} + e^{(1)}_{R_3 \to 2} + e^{(2)}_{R_3 \to 2} \right) (1 - \pi_{2,K}), \\[4pt]
e^{(1)}_{R_2 \to 1} = p_{2,1} e_2, \\[4pt]
e^{(2)}_{R_2 \to 1} = \left(e^{(1)}_{R_1 \to 2} + e^{(2)}_{R_1 \to 2} + e^{(1)}_{R_3 \to 2} + e^{(2)}_{R_3 \to 2} \right) \alpha_{2,1} \pi_{2,K}, \\[4pt]
e^{(1)}_{R_2 \to 3} = p_{2,3} e_2, \\[4pt]
e^{(2)}_{R_2 \to 3} = \left(e^{(1)}_{R_1 \to 2} + e^{(2)}_{R_1 \to 2} + e^{(1)}_{R_3 \to 2} + e^{(2)}_{R_3 \to 2} \right) \alpha_{2,3} \pi_{2,K}, \\[4pt]
e_3 = \left(e^{(1)}_{R_2 \to 3} + e^{(2)}_{R_2 \to 3} \right) (1 - \pi_{3,K}), \\[4pt]
e^{(1)}_{R_3 \to 2} = e_3, \\[4pt]
e^{(2)}_{R_3 \to 2} = \left(e^{(1)}_{R_2 \to 3} + e^{(2)}_{R_2 \to 3} \right) \pi_{3,K}.
\end{cases}
\tag{4}
$$

By using $e_1 = 1$, we obtain

$$
\begin{cases}
e_1 = e^{(1)}_{R_1 \to 2} = 1, \\[4pt]
e^{(2)}_{R_1 \to 2} = \dfrac{\pi_{1,K}}{1 - \pi_{1,K}}, \\[6pt]
e_2 = \dfrac{1 - \pi_{2,K}}{(1 - \pi_{1,K})[(\alpha_{2,1} - p_{2,1})\pi_{2,K} + p_{2,1}]}, \\[6pt]
e^{(1)}_{R_2 \to 1} = \dfrac{p_{2,1}(1 - \pi_{2,K})}{(1 - \pi_{1,K})[(\alpha_{2,1} - p_{2,1})\pi_{2,K} + p_{2,1}]}, \\[6pt]
e^{(2)}_{R_2 \to 1} = \dfrac{\alpha_{2,1}\pi_{2,K}}{(1 - \pi_{1,K})[(\alpha_{2,1} - p_{2,1})\pi_{2,K} + p_{2,1}]}, \\[6pt]
e^{(1)}_{R_2 \to 3} = \dfrac{p_{2,3}(1 - \pi_{2,K})}{(1 - \pi_{1,K})[(\alpha_{2,1} - p_{2,1})\pi_{2,K} + p_{2,1}]}, \\[6pt]
e^{(2)}_{R_2 \to 3} = \dfrac{\alpha_{2,3}\pi_{2,K}}{(1 - \pi_{1,K})[(\alpha_{2,1} - p_{2,1})\pi_{2,K} + p_{2,1}]}, \\[6pt]
e_3 = e^{(1)}_{R_3 \to 1} = \dfrac{(1 - \pi_{3,K})[(\alpha_{2,3} - p_{2,3})\pi_{2,K} + p_{2,3}]}{(1 - \pi_{1,K})[(\alpha_{2,1} - p_{2,1})\pi_{2,K} + p_{2,1}]}, \\[6pt]
e^{(2)}_{R_3 \to 1} = \dfrac{\pi_{3,K}[(\alpha_{2,3} - p_{2,3})\pi_{2,K} + p_{2,3}]}{(1 - \pi_{1,K})[(\alpha_{2,1} - p_{2,1})\pi_{2,K} + p_{2,1}]}.
\end{cases}
$$

The routing matrices for more general case

Observing the three examples, it may be easy and convenient to write a routing matrix for a more general bike sharing system. Note that Example Three provides more intuitive understanding on how to write those elements of the routing matrix, thus for a more general bike sharing system we establish the routing matrix $\mathbf{P} = \left(p_{\tilde{i},\tilde{j}} \right)$ as follow:

$$
p_{\tilde{i},\tilde{j}} = \begin{cases}
p_{i,j}, & \text{if } \tilde{i} = \text{Station } i, \tilde{j} = \text{Road } i \to j \\
1 - \pi_{j,K}, & \text{if } \tilde{i} = \text{Road } i \to j, \tilde{j} = \text{Station } j \\
\displaystyle\sum_{l \in \Theta_i \& l \in \Delta_j} \alpha_{l,j} \pi_{l,K}, & \text{if } n_l = K, \tilde{i} = \text{Road } i \to l, \tilde{j} = \text{Road } l \to j \\
0. & \text{otherwise}
\end{cases}
$$

Theorem 1. *The routing matrix* $\mathbf{P}$ *of finite size is irreducible and stochastic, and there exists the unique positive solution to the following system of linear equations*

$$
\begin{cases}
\overrightarrow{e} = \overrightarrow{e}\mathbf{P}, \\
e_1 = 1,
\end{cases}
$$

where $e_1 = 1$ *is the first element of the row vector* $\overrightarrow{e}$, *and* $\overrightarrow{e}$ *is a row vector of the relative arrival rates of this bike sharing system.*

Proof: The outline of this proof is described as follows. It is clear that the size of the routing matrix $\mathbf{P}$ is finite. At the same time, it is well-known that (a) the routing structure of the multiclass closed queueing network indicates that the routing matrix $\mathbf{P}$ is stochastic; and (b) the accessibility of each station node or road node in the bike sharing system shows that the routing matrix $\mathbf{P}$ is irreducible. Thus the routing matrix $\mathbf{P}$ is not only irreducible but also stochastic. For the routing matrix $\mathbf{P}$, applying Theorem 1.1 (a) and (b) of Chap. 1 in Seneta [33], the left eigenvector $\overrightarrow{e}$ of the irreducible stochastic matrix $\mathbf{P}$ of finite sizes corresponding to the maximal eigenvalue 1 is strictly positive, that is, $\overrightarrow{e} > 0$; and $\overrightarrow{e}$ is unique with $e_1 = 1$. This completes this proof. ∎

(c) A joint queue-length process

Let $Q_i^{(v)}(t)$ be the number of bikes parked in Station i with phase v of the MAP at time $t \geq 0$, for $1 \leq i \leq N$, $1 \leq v \leq m$; and $R_{k,l}^{(r)}(t)$ the number of bikes of class r ridden on Road $k \to l$ at time $t \geq 0$, for $r = 1, 2$ and for $l \neq k$ with $1 \leq k, l \leq N$. We write

$$
\mathbf{X}(t) = (\mathbf{L}_1(t), \mathbf{L}_2(t), \ldots, \mathbf{L}_{N-1}(t), \mathbf{L}_N(t)),
$$

where for $1 \leq i \leq N$

$$
\mathbf{L}_i(t) = \left(Q_i^{(1)}(t), Q_i^{(2)}(t), \ldots, Q_i^{(m)}(t); R_{i,j}^{(1)}(t), R_{i,j}^{(2)}(t), \ j \in \Theta_i \right).
$$

Obviously, $\{\mathbf{X}(t) : t \geq 0\}$ is a Markov process due to the exponential and MAP assumptions of this bike sharing system. It is easy to see that the state space of Markov process $\{\mathbf{X}(t) : t \geq 0\}$ is given by

$$\Omega = \left\{ \overrightarrow{n} : 0 \leq n_i^{(v)} \leq K,\ 1 \leq i \leq N, 1 \leq v \leq m, \right.$$

$$0 \leq m_{k,l}^{(r)} \leq NC,\ r = 1,2, l \neq k, 1 \leq k, l \leq N, \tag{5}$$

$$\left. \sum_{i=1}^{N} \sum_{v=1}^{m} n_i^{(v)}(t) + \sum_{k=1}^{N} \sum_{l \in \Theta_k} \sum_{r=1,2} m_{k,l}^{(r)} = NC \right\},$$

where

$$\overrightarrow{n} = (\mathbf{n}_1, \mathbf{n}_2, \ldots, \mathbf{n}_N),$$

for $1 \leq i \leq N$

$$\mathbf{n}_i = \left(n_i^{(1)}, n_i^{(2)}, \cdots, n_i^{(m)}; m_{i,j}^{(1)}, m_{i,j}^{(2)}, j \in \Theta_i \right).$$

It is easy to check that the Markov process $\{\mathbf{X}(t) : t \geq 0\}$ on a finite state space is irreducible, aperiodic and positive recurrent. Therefore, there exists the stationary probability vector

$$\pi = (\pi(\overrightarrow{n}) : \overrightarrow{n} \in \Omega)$$

such that

$$\pi(\overrightarrow{n}) = \lim_{t \to +\infty} P\{\mathbf{X}(t) = \overrightarrow{n}\}.$$

4   A Product-Form Solution and Performance Analysis

In this section, we first provide a product-form solution to the steady-state probabilities of joint queue lengths in the multiclass closed queueing network. Then we provide a nonlinear solution to determine the N undetermined constants: $\pi_{1,K}, \pi_{2,K}, \ldots, \pi_{N,K}$. Also, an example is used to indicate our computational steps. Finally, we analyze performance measures of the bike sharing system by means of the steady-state probabilities of joint queue lengths.

Note that $\{\mathbf{X}(t) : t \geq 0\}$ is an irreducible, aperiodic, positive recurrent and continuous-time Markov process with finite states, thus we have

$$\pi(\overrightarrow{n}) = \lim_{t \to +\infty} P\left\{ Q_i^{(v)}(t) = n_i^{(v)}, 1 \leq i \leq N, 1 \leq v \leq m;\ R_{k,l}^{(1)}(t) = m_{k,l}^{(1)}, R_{k,l}^{(2)}(t) \right.$$

$$\left. = m_{k,l}^{(2)}, 1 \leq k, l \leq N \text{ with } l \neq k, \sum_{i=1}^{N} \sum_{v=1}^{m} n_i^{(v)} + \sum_{k=1}^{N} \sum_{l \in \Theta_k} \sum_{r=1,2} m_{k,l}^{(r)} = NC \right\}.$$

Note that if $\sum_{i=1}^{N} \sum_{v=1}^{m} n_i^{(v)} + \sum_{k=1}^{N} \sum_{l \in \Theta_k} \sum_{r=1,2} m_{k,l}^{(r)} \neq NC$, it is easy to see that $\pi(\overrightarrow{n}) = 0$. In practice, it is a key in the study of bike sharing systems to provide expression for the steady-state probability $\pi(\overrightarrow{n})$, $\overrightarrow{n} \in \Omega$.

4.1   A Product-Form Solution

For the bike sharing system, we establish a multiclass closed queueing network with $N + \sum_{i=1}^{N} |R(i)|$ virtual nodes and with NC virtual customers. As $t \to +\infty$, the multiclass closed queueing network is decomposed into $N + \sum_{i=1}^{N} |R(i)|$ isolated and equivalent queueing systems as follows:

(i) **The ith station node:** An equivalent queue is $M_i/MAP_i/1/K$, where M_i denotes a Poisson process with relative arrival rate e_i, and MAP_i is $MAP(C_i, D_i)$ as a service process.

(ii) **The Road $i \to l$ node:** The two classes of customers correspond to their two queueing processes as follow:

(a) The first queue process on the Road $i \to l$ node is $M_{i \to j}^{(1)}/\sum_{k=1}^{m_{i,j}^{(1)}} M_{i \to j;1}^{(k)}/1$, where $M_{i \to j}^{(1)}$ denotes a Poisson process with relative arrival rate $e_{R_{i \to j}}^{(1)}$, and $\sum_{k=1}^{m_{i,j}^{(1)}} M_{i \to j;1}^{(k)}$ is the random sum of $m_{i,j}^{(1)}$ i.i.d. exponential random variables, each of which is exponential with service rate $\mu_{i,j}$.

(b) The second queue process on the Road $i \to l$ node is $M_{i \to j}^{(2)}/\sum_{k=1}^{m_{i,j}^{(2)}} M_{i \to j;2}^{(k)}/1$, in which $M_{i \to j}^{(2)}$ is a Poisson process with relative arrival rate $e_{R_{i \to j}}^{(2)}$, and $\sum_{k=1}^{m_{i,j}^{(2)}} M_{i \to j;2}^{(k)}$ is the random sum of $m_{i,j}^{(2)}$ i.i.d. exponential random variables, each of which is exponential with service rate $\xi_{i,j}$.

Using the above three classes of isolated queues, the following theorem provides a product-form solution to the steady-state probability $\pi(\overrightarrow{n})$ of joint queue lengths at the virtual nodes for $\overrightarrow{n} \in \Omega$; while its proof is easy by means of Chap. 7 in Bolch et al. [3] and is omitted here.

Theorem 2. *For the two-class closed queueing network corresponding to the bike sharing system, if the undetermined constants $\pi_{1,K}, \pi_{2,K}, \ldots, \pi_{N,K}$ are given, then the steady-state joint probability $\pi(\overrightarrow{n})$ is given by*

$$\pi(\overrightarrow{n}) = \frac{1}{G(NC)} \prod_{i=1}^{N} H(\mathbf{n}_i) H(\mathbf{m}_i), \tag{6}$$

where $\overrightarrow{n} \in \Omega$,

$$H(\mathbf{n}_i) = \frac{\left(n_i^{(1)} + n_i^{(2)} + \cdots + n_i^{(m)}\right)!}{n_i^{(1)}! n_i^{(2)}! \cdots n_i^{(m)}!} \prod_{v=1}^{m} \left(\frac{e_i}{\lambda_i^{(v)}}\right)^{n_i^{(v)}},$$

$$H(\mathbf{m}_i) = \prod_{j \in \Theta_i} \frac{\left(m_{i,j}^{(1)} + m_{i,j}^{(2)}\right)!}{m_{i,j}^{(1)}! m_{i,j}^{(2)}!} \left(\frac{e_{R_{i \to j}}^{(1)}}{m_{i,j}^{(1)} \mu_{i,j}}\right)^{m_{i,j}^{(1)}} \left(\frac{e_{R_{i \to j}}^{(2)}}{m_{i,j}^{(2)} \xi_{i,j}}\right)^{m_{i,j}^{(2)}},$$

and $G(NC)$ is a normalization constant, given by

$$G(NC) = \sum_{\vec{n} \in \Omega} \prod_{i=1}^{N} H(\mathbf{n}_i) H(\mathbf{m}_i).$$

By means of the product-form solution given in Theorem 2, the following theorem further establishes a system of nonlinear equations, whose solution determines the N undetermined constants $\pi_{1,K}, \pi_{2,K}, \ldots, \pi_{N,K}$. Note that $\pi_{i,K}$ is also the steady-state probability of the ith full station for $1 \leq i \leq N$. While its proof is easy by means of the law of total probability and is omitted here.

Theorem 3. *The undetermined constants $\pi_{1,K}, \pi_{2,K}, \ldots, \pi_{N,K}$ can be uniquely determined by the following system of nonlinear equations:*

$$\begin{cases} \pi_{1,K} = \sum_{\substack{\vec{n} \in \Omega \\ \& n_1 = K,}} \pi(\vec{n}), \\ \pi_{2,K} = \sum_{\substack{\vec{n} \in \Omega \\ \& n_2 = K,}} \pi(\vec{n}), \\ \quad\vdots \\ \pi_{N,K} = \sum_{\substack{\vec{n} \in \Omega \\ \& n_N = K,}} \pi(\vec{n}), \end{cases}$$

where $\pi(\vec{n})$ is given by the product-form solution stated in Theorem 2.

To indicate how to compute the undetermined constants $\pi_{1,K}, \pi_{2,K}, \ldots, \pi_{N,K}$, in what follows we give a concrete example.

Example Four: In Example One, we use the product-form solution to determine $\pi_{1,K}$ and $\pi_{2,K}$. By using (2) and (6), we obtain

$$\begin{cases} \pi_{1,K} = \sum_{\substack{\vec{n} \in \Omega \\ \& n_1 = K,}} \pi(\vec{n}), \\ \pi_{2,K} = \sum_{\substack{\vec{n} \in \Omega \\ \& n_2 = K,}} \pi(\vec{n}). \end{cases} \tag{7}$$

We take that $C = 2, K = 3, m = 2$. Thus (7) is simplified as

$$\begin{cases} \pi_{1,K} = \frac{1}{G(NC)} \left(\frac{1}{\lambda_1^{(1)}} + \frac{1}{\lambda_1^{(2)}} \right)^3 \left[\frac{\pi_{1,K}}{\xi_{1,2}\left(1-\pi_{1,K}\right)} + \frac{1}{\mu_{1,2}} + \frac{1-\pi_{2,K}}{1-\pi_{1,K}} \left(\frac{1}{\lambda_2^{(1)}} + \frac{1}{\lambda_2^{(2)}} + \frac{1}{\mu_{1,2}} \right) \right], \\ \pi_{2,K} = \frac{1}{G(NC)} \left(\frac{1}{\lambda_1^{(1)}} + \frac{1}{\lambda_1^{(2)}} \right)^3 \frac{\left(1-\pi_{2,K}\right)^3}{\left(1-\pi_{1,K}\right)^3} \left(\frac{1}{\lambda_1^{(1)}} + \frac{1}{\lambda_1^{(2)}} + \frac{\pi_{2,K}}{\xi_{2,1}\left(1-\pi_{1,K}\right)} \right. \\ \left. \quad + \frac{1}{\mu_{1,2}} + \frac{1-\pi_{2,K}}{\mu_{2,1}\left(1-\pi_{1,K}\right)} \right), \end{cases}$$

$$\tag{8}$$

where the normalization constant $G(NC)$ is given by

$$
\begin{aligned}
&G(NC) \\
&= \frac{1}{256\,(\mu_{1,2})^4} + \left(\frac{1}{\lambda_1^{(1)}} + \frac{1}{\lambda_1^{(2)}}\right)^3 \left[\frac{\pi_{1,K}}{\xi_{1,2}\,(1-\pi_{1,K})} + \frac{1}{\mu_{1,2}}\right. \\
&\quad \left. + \frac{1-\pi_{2,K}}{1-\pi_{1,K}} \left(\frac{1}{\lambda_2^{(1)}} + \frac{1}{\lambda_2^{(2)}} + \frac{1}{\mu_{2,1}}\right)\right] \\
&\quad + \left(\frac{1}{\lambda_1^{(1)}} + \frac{1}{\lambda_1^{(2)}}\right)^2 \left[\frac{1}{4\,(\mu_{1,2})^2} + \frac{(1-\pi_{2,K})^2}{(1-\pi_{1,K})^2} \left(\frac{1}{\lambda_2^{(1)}} + \frac{1}{\lambda_2^{(2)}} + \frac{1}{2\mu_{2,1}}\right)^2\right. \\
&\quad \left. + \frac{1-\pi_{2,K}}{\mu_{1,2}\,(1-\pi_{1,K})} \left(\frac{1}{\lambda_2^{(1)}} + \frac{1}{\lambda_2^{(2)}} + \frac{1}{\mu_{2,1}}\right)\right] \\
&\quad + \left(\frac{1}{\lambda_1^{(1)}} + \frac{1}{\lambda_1^{(2)}}\right) \left\{\frac{1}{27\,(\mu_{1,2})^3} + \frac{1-\pi_{2,K}}{4\,(\mu_{1,2})^2\,(1-\pi_{1,K})} \left(\frac{1}{\lambda_2^{(1)}} + \frac{1}{\lambda_2^{(2)}} + \frac{1}{\mu_{1,2}}\right)\right. \\
&\quad + \frac{(1-\pi_{2,K})^2}{\mu_{1,2}\,(1-\pi_{1,K})^2} \left(\frac{1}{\lambda_2^{(1)}} + \frac{1}{\lambda_2^{(2)}} + \frac{1}{2\mu_{1,2}}\right)^2 \\
&\quad + \frac{(1-\pi_{2,K})^3}{(1-\pi_{1,K})^3} \left[\left(\frac{1}{\lambda_1^{(1)}} + \frac{1}{\lambda_1^{(2)}}\right)^3 + \frac{1}{27\,(\mu_{2,1})^3} + \frac{1}{4\,(\mu_{2,1})^2} \left(\frac{1}{\lambda_2^{(1)}} + \frac{1}{\lambda_2^{(2)}}\right)\right. \\
&\quad \left.\left. + \frac{1}{\mu_{2,1}} \left(\frac{1}{\lambda_2^{(1)}} + \frac{1}{\lambda_2^{(2)}}\right)^2\right]\right\} + \frac{(1-\pi_{2,K})^2}{4\,(\mu_{1,2})^2\,(1-\pi_{1,K})^2} \left(\frac{1}{\lambda_2^{(1)}} + \frac{1}{\lambda_2^{(2)}} + \frac{1}{2\mu_{2,1}}\right)^2 \\
&\quad + \frac{1-\pi_{2,K}}{3\mu_{1,2}\,(1-\pi_{1,K})} \left(\frac{1}{\lambda_2^{(1)}} + \frac{1}{\lambda_2^{(2)}} + \frac{1}{\mu_{2,1}}\right) \\
&\quad + \frac{(1-\pi_{2,K})^3}{\mu_{1,2}\,(1-\pi_{1,K})^3} \left[\left(\frac{1}{\lambda_1^{(1)}} + \frac{1}{\lambda_1^{(2)}}\right)^3 + \frac{1}{27\,(\mu_{2,1})^3} + \frac{1}{4\,(\mu_{2,1})^2} \left(\frac{1}{\lambda_2^{(1)}} + \frac{1}{\lambda_2^{(2)}}\right)\right. \\
&\quad \left. + \frac{1}{\mu_{2,1}} \left(\frac{1}{\lambda_2^{(1)}} + \frac{1}{\lambda_2^{(2)}}\right)^2\right] + \frac{(1-\pi_{2,K})^4}{256\,(\mu_{2,1})^4\,(1-\pi_{1,K})^4} \\
&\quad + \frac{\left(\lambda_2^{(1)} + \lambda_2^{(2)}\right)(1-\pi_{2,K})^4}{27\lambda_2^{(1)}\lambda_2^{(2)}\,(\mu_{2,1})^3\,(1-\pi_{1,K})^4} + \frac{\left(\lambda_2^{(1)} + \lambda_2^{(2)}\right)^2 (1-\pi_{2,K})^4}{4\left(\lambda_2^{(1)}\lambda_2^{(2)}\mu_{2,1}\right)^2 (1-\pi_{1,K})^4} \\
&\quad + \frac{\left(\lambda_2^{(1)} + \lambda_2^{(2)}\right)^3 (1-\pi_{2,K})^3 \left[\xi_{2,1}\,(1-\pi_{2,K}) + \pi_{2,K}\mu_{2,1}\right]}{\xi_{2,1}\mu_{2,1}\left(\lambda_2^{(1)}\lambda_2^{(2)}\right)^3 (1-\pi_{1,K})^4}.
\end{aligned}
\tag{9}
$$

By using (8) and (9), we can compute the two undetermined constants: $\pi_{1,K}$ and $\pi_{2,K}$.

To this end, let $\lambda_1^{(2)} = 7, \lambda_2^{(1)} = 5, \lambda_2^{(2)} = 5, \mu_{1,2} = 2, \mu_{2,1} = 3, \xi_{1,2} = 4, \xi_{2,1} = 5$. When $\lambda_1^{(1)} = 5, 6, 7, 8, 9$, we obtain the values of $\pi_{1,K}$ and $\pi_{2,K}$ which are listed in Table 1.

From Table 1, it is seen that as $\lambda_1^{(1)}$ increases, $\pi_{1,K}$ decreases but $\pi_{2,K}$ increases. This result is the same as the actual intuitive situation. When $\lambda_1^{(1)}$ increases, more bikes are rented from Station 1, so $\pi_{1,K}$ decreases; while when more bikes are rented from Station 1 and are ridden on Road $1 \to 2$, more bikes will be returned to Station 2, so $\pi_{2,K}$ increases.

Table 1. Numerical results of $\pi_{1,K}$ and $\pi_{2,K}$

$\lambda_1^{(1)}$	$\lambda_1^{(2)}$	$\lambda_2^{(1)}$	$\lambda_2^{(2)}$	$\mu_{1,2}$	$\mu_{2,1}$	$\xi_{1,2}$	$\xi_{2,1}$	$\pi_{1,K}$	$\pi_{1,K}$
5	7	5	5	2	3	4	5	0.10434	0.14143
6	7	5	5	2	3	4	5	0.08609	0.14502
7	7	5	5	2	3	4	5	0.07609	0.14815
8	7	5	5	2	3	4	5	0.06424	0.14961
9	7	5	5	2	3	4	5	0.05734	0.15116

Remark 1. *For a large-scale bike sharing system, it is always more difficult and challenging to determine the normalization constant $G(NC)$. Thus it is necessary in the future study to develop some effective algorithms for numerically computing $G(NC)$.*

4.2 Performance Analysis

Now, we consider two key performance measures of the bike sharing system in terms of the steady-state probability $\pi(\overrightarrow{n})$ of joint queue lengths at the virtual nodes for $\overrightarrow{n} \in \Omega$.

(1) *The steady-state probability of problematic stations*

In the study of bike sharing systems, it is a key to compute the steady-state probability of problematic stations. For this bike sharing system, the steady-state probability of problematic stations is given by

$$\Im = \sum_{i=1}^{N} P\{n_i = 0 \text{ or } n_i = K\} = \sum_{i=1}^{N} [P\{n_i = 0\} + P\{n_i = K\}]$$

$$= \sum_{i=1}^{N} \left[\sum_{\substack{\overrightarrow{n} \in \Omega \\ \& n_i = 0}} \pi(\overrightarrow{n}) + \sum_{\substack{\overrightarrow{n} \in \Omega \\ \& n_i = K}} \pi(\overrightarrow{n}) \right].$$

(2) The mean of the steady-state queue length

The steady-state mean of the number of bikes parked at the ith station is given by

$$\mathbf{Q}_i = \sum_{\substack{\overrightarrow{n} \in \Omega \\ \& 1 \le n_i \le K}} n_i \pi(\overrightarrow{n}), \quad 1 \le i \le N,$$

and the steady-state mean of the number of bikes ridden on the Road $k \to l$ for $1 \le k \le N$ and $l \in \Theta_k$ is given by

$$\mathbf{Q}_{R_{k \to l}} = \sum_{r=1,2} \sum_{\substack{\overrightarrow{n} \in \Omega \\ \& 1 \le m_{k,l}^{(r)} \le NC}} m_{k,l}^{(r)} \pi(\overrightarrow{n}).$$

Remark 2. *In the practical bike sharing systems, arrivals of bike users often have some special important behavior and characteristics, such as, time-inhomogeneity, space-heterogeneity, and arrival burstiness. To express such behavior and characteristics, this paper uses the MAPs to express non-Poisson (and non-renewal) arrivals of bike users. It is seen that such a MAP-based study is a key to generalize and extend the arrivals of bike users to a more general arrival process in practice, for example, a renewal process, a periodic MAP, a periodic time-inhomogeneous arrival process and so on. In fact, the methodology of this paper may be applied to deal with more general arrivals of bike users. Thus it is very interesting for our future study to analyze the space-heterogeneous or time-inhomogeneous arrivals of bike users in the bike sharing systems.*

5   Concluding Remarks

In this paper, we first propose a more general bike sharing system with Markovian arrival processes and under an irreducible path graph. Then we establish a multiclass closed queueing network by means of some virtual ideas, including, virtual customers, virtual nodes and virtual service times. Furthermore, we set up the routing matrix, which gives a nonlinear solution to computing the relative arrival rates. Based on this, we give the product-form solution to the steady-state probabilities of joint queue lengths at the virtual nodes. Finally, we compute the steady-state probability of problematic stations, and also deal with other interesting performance measures of the bike sharing system. Along these lines, there are a number of interesting directions for potential future research, for example:

– Analyzing bike sharing systems with phase type (PH) riding-bike times on the roads;
– discussing repositioning bikes by trucks in bike sharing systems with information technologies;
– developing effective algorithms for establishing the routing matrix, and for computing the relative arrival rates;

- developing effective algorithms for computing the product-form steady-state probabilities of joint queue lengths at the virtual nodes, and further for calculating the steady-state probability of problematic stations; and
- applying periodic MAPs, periodic PH distributions, or periodic Markov processes to study time-inhomogeneous bike sharing systems. This is a very interesting but challenging topic in the future study of bike sharing system.

Acknowledgements. This work was supported in part by the National Natural Science Foundation of China under grant Nos. 71671158 and 71471160; and by the Natural Science Foundation of Hebei province under grant No. G2017203277.

References

1. Adelman, D.: Price-directed control of a closed logistics queueing network. Oper. Res. **55**, 1022–1038 (2007)
2. Baskett, F., Chandy, K.M., Muntz, R.R., Palacios, F.G.: Open, Closed, and Mixed Networks of Queues with Different Classes of Customers. J. ACM **22**, 248–260 (1975)
3. Bolch, G., Greiner, S., de Meer, H., Trivedi, K.S.: Queueing Networks and Markov Chains: Modeling and Performance Evaluation with Computer Science Applications. Wiley, New York (2006)
4. Bruell, S.C., Balbo, G.: Computational Algorithms for Closed Queueing Networks. Elsevier Science Ltd., Oxford (1980)
5. Chakravarthy, S.R.: The batch markovian arrival process: a review and future work. Adv. Probab. Theory Stochast. Process. **1**, 21–49 (2001)
6. DeMaio, P.: Bike-sharing: history, impacts, models of provision, and future. J. Public Transp. **12**, 41–56 (2009)
7. Fishman, E., Washington, S., Haworth, N.: Bike share: a synthesis of the literature. Transp. Rev. **33**, 148–165 (2013)
8. Fricker, C., Gast, N., Mohamed, A.: Mean field analysis for inhomogeneous bike-sharing systems. In: DMTC Proceedings of AQ, AofA 2012, pp. 365–376 (2012)
9. Fricker, C., Gast, N.: Incentives and redistribution in homogeneous bike-sharing systems with stations of finite capacity. Euro. J. Transp. Logistics **5**, 261–291 (2013)
10. Fricker, C., Tibi, D.: Equivalence of ensembles for large vehicle-sharing models. Annal. Appl. Probab. **27**, 883–916 (2017)
11. George, D.K., Xia, C.H.: Fleet-sizing and service availability for a vehicle rental system via closed queueing networks. Eur. J. Oper. Res. **211**, 198–207 (2011)
12. Henderson, W., Pearce, C.E.M., Taylor, P.G., van Dijk, N.M.: Closed queueing networks with batch services. Queueing Syst. **6**, 59–70 (1990)
13. Kelly, F.P.: Reversibility and Stochastic Networks. Wiley, New York (1979)
14. Kelly, F.P.: Reversibility and Stochastic Networks. Cambridge University Press, Cambridge (2001)
15. Kelly, F.P., Pollett, P.K.: Sojourn times in closed queueing networks. Adv. Appl. Probab. **15**, 638–656 (1983)
16. Labadi, K., Benarbia, T., Barbot, J.P., Hamaci, S., Omari, A.: Stochastic petri net modeling, simulation and analysis of public bicycle sharing systems. IEEE Trans. Autom. Sci. Eng. **12**, 1380–1395 (2015)
17. Leurent, F.: Modelling a Vehicle-Sharing Station as a Dual Waiting System: Stochastic Framework and Stationary Analysis. HAL Id: hal-00757228, pp. 1–19 (2012)

18. Li, Q.L.: Constructive Computation in Stochastic Models with Applications: The RG-factorizations. Springer, Heidelberg (2010). https://doi.org/10.1007/978-3-642-11492-2
19. Li, Q.L., Fan, R.N.: Bike-Sharing Systems under Markovian Environment. arXiv preprint arXiv:1610.01302, pp. 1–44 (2016)
20. Li, Q.L., Fan, R.N., Ma, J.Y.: A unified framework for analyzing closed queueing networks in bike sharing systems. In: Dudin, A., Gortsev, A., Nazarov, A., Yakupov, R. (eds.) ITMM 2016. CCIS, vol. 638, pp. 177–191. Springer, Cham (2016). https://doi.org/10.1007/978-3-319-44615-8_16
21. Li, Q.L., Chen, C., Fan, R.N., Xu, L., Ma, J.Y.: queueing analysis of a large-scale bike sharing system through mean-field theory. arXiv preprint arXiv:1603.09560, pp. 1–51 (2016)
22. Lucantoni, D.M.: New results on the single server queue with a batch markovian arrival process. Stoch. Models 7, 1–46 (1991)
23. Meddin, R., DeMaio, P.: The bike-sharing world map (2012). http://www.metrobike.net
24. Neuts, M.F.: Structured Stochastic Matrices of M/G/1 Type and Their Applications. Marcel Decker Inc., New York (1989)
25. Neuts, M.F.: Matrix-analytic methods in the theory of queues. In: Dhashalow, J. (ed.) Advances in Queueing, pp. 265–292. CRC Press, Boca Raton (1995)
26. Onvural, R.O.: Survey of closed queueing networks with blocking. ACM Comput. Surv. 22, 83–121 (1990)
27. Ramaswami, V.: The N/G/1 queue and its detailed analysis. Adv. Appl. Probab. 12, 222–261 (1980)
28. Raviv, T., Kolka, O.: Optimal inventory management of a bikesharing station. IIE Trans. 45, 1077–1093 (2013)
29. Raviv, T., Tzur, M., Forma, I.A.: Static repositioning in a bike-sharing system: models and solution approaches. EURO J. Transp. Logistics 2, 187–229 (2013)
30. Reiser, M., Kobayashi, H.: Queuing networks with multiple closed chains: theory and computational algorithms. IBM J. Res. Dev. 19, 283–294 (1975)
31. Reiser, M.: Mean-value analysis and convolution method for queue-dependent servers in closed queueing networks. Perform. Eval. 1, 7–18 (1981)
32. Schuijbroek, J., Hampshire, R., van Hoeve, W.J.: Inventory rebalancing and vehicle routing in bike-sharing systems. Technical report 2013–2, Tepper School of Business, Carnegie Mellon University, pp. 1–27 (2013)
33. Seneta, E.: Non-Negative Matrices and Markov Chains. Springer, New York (2006). https://doi.org/10.1007/0-387-32792-4
34. Serfozo, R.: Introduction to Stochastic Networks. Springer, New York (1999). https://doi.org/10.1007/978-1-4612-1482-3
35. Shu, J., Chou, M.C., Liu, Q., Teo, C.P., Wang, I.L.: Models for effective deployment and redistribution of bicycles within public bicycle-sharing systems. Oper. Res. 61, 1346–1359 (2013)
36. Waserhole, A., Jost, V.: Vehicle sharing system pricing regulation: transit optimization of intractable queuing network. HAL Id: hal-00751744, pp. 1–20 (2012)
37. Waserhole, A., Jost, V.: Vehicle sharing system pricing regulation: a fluid approximation. HAL Id: hal-00727041, pp. 1–35 (2013)
38. Waserhole, A., Jost, V., Brauner, N.: Pricing techniques for self regulation in vehicle sharing systems. Electron. Notes Discrete Math. 41, 149–156 (2013)
39. Waserhole, A., Jost, V.: Pricing in Vehicle Sharing Systems: Optimization in Queuing Networks with Product Forms. EURO J. Transp. Logistics 5, 293–320 (2016)

Network Models

Energy-Efficient Strategy with a Speed Switch and a Multiple-Sleep Mode in Cloud Data Centers

Shunfu Jin[1]([⊠]), Shanshan Hao[1], and Wuyi Yue[2]

[1] School of Information Science and Engineering, Yanshan University,
Qinhuangdao 066004, China
`jsf@ysu.edu.cn`, `ashanshan931106@126.com`
[2] Department of Intelligence and Informatics, Konan University,
Kobe 658-8501, Japan
`yue@konan-u.ac.jp`

Abstract. Due to the rapid growth of energy costs and increasingly strict environmental standards, energy consumption has become a significant expenditure for the operating and maintaining of a cloud data center. To improve the energy efficiency of cloud data centers, in this paper, we propose an energy-efficient strategy with a speed switch and a multiple-sleep mode. According to current traffic loads, a proportion of Virtual Machines (VMs) operate at a low speed or a high speed, while the remaining VMs either sleep or operate at a high speed. In our strategy, we develop a continuous-time queueing model with an adaptive service rate and a partial synchronous vacation. We construct a two-dimensional Markov chain based on the total number of requests in the system and the state of all the VMs. By using the method of a matrix geometric solution, we mathematically estimate the energy saving level of the system. Numerical experiments with analysis and simulation show that our proposed energy-efficient strategy can effectively reduce the energy consumption on the premise of guaranteeing the Quality of Service of CDCs.

Keywords: Cloud data center · Energy efficiency · Speed switch
Multiple-sleep · Matrix geometric solution

1 Introduction

The rapid development of information technology and the explosive growth in global data has generated enormous demand for cloud computing. Consequently, Cloud Data Centers (CDCs) are growing exponentially, both in number and in size, to provide universal service. International Data Corporation (IDC) predicts that the total number of CDCs deployed worldwide will peak at 8.6 million in 2017 [1]. Currently, high energy consumption and serious environmental pollution are significant factors restricting the development of CDCs.

The energy consumption of a VM is approximately in line with the CPU utilization, so the most direct method of conserving energy is to operate all the

© Springer International Publishing AG 2017
W. Yue et al. (Eds.): QTNA 2017, LNCS 10591, pp. 143–154, 2017.
https://doi.org/10.1007/978-3-319-68520-5_9

VMs at lower voltage and frequency. One of the common techniques for optimizing energy consumption in CDCs is dynamic power management (DPM). DPM refers to dynamic CPU energy consumption and CPU processing speed adjustment according to current traffic load. In [2], Li proved that if the application environment and average energy consumption are given, there is an optimal speed scheme that minimizes the average response time of requests. In [3], Wang *et al.* presented a workload predictor based on online Bayes classifier and an novel DPM technique based on adaptive reinforcement learning algorithm to reduce the energy consumption in stochastic dynamic systems. In [4], Chen *et al.* proposed a Dynamic Voltage and Frequency Scaling (DVFS) scheme based on DPM technique, by which the best fitting voltage and frequency for a multi-core embedded system is dynamically predicted. All the methods based on DPM technique mentioned above can improve energy efficiency from the perspective of reducing the energy consumption of each VM in CDCs. However, all the VMs in the CDCs remain open all the time, even though there are no requests in CDCs. Even when operating at low-speed and in low-voltage mode, the accumulated energy consumption by thousands of VMs in CDCs can not be ignored.

In respect to the low utilization of VMs in CDCs, pushing part of VMs to enter a sleep state or a power-off state during lower workload hours can also save energy. In [5], Chou *et al.* proposed a DynSleep scheme. DynSleep dynamically postpones the processing of some requests, creating longer idle periods, which allows the use of deep sleep mode that save more energy. In [6], Dabbagh *et al.* proposed an integrated energy-aware resource provisioning framework for CDCs. This framework first predicts the number of cloud users that will arrive at CDCs in the near future, then estimates the number of VMs that are needed to serve cloud users. In [7], Liao *et al.* proposed an energy-efficient strategy, which dynamically switches two backup groups of servers on and off according to different thresholds. Using the methods above, energy can be conserved by decreasing the number of VMs running in the system. However, few methods can accurately estimate the behavior of requests. Their arrival and departure are stochastic. Pushing VMs to enter a sleep state or a power-off state based on only the predicted behavior of requests is very risky, and might lead to significant sacrifice of the Quality of Service (QoS).

In this paper, by applying DPM technique and introducing a sleep mode, we propose an energy-efficient strategy with a speed switch and a multiple-sleep mode. Accordingly, we establish a continuous-time queueing model with an adaptive service rate and a partial synchronous vacation to investigate the behavior of cloud users and all the VMs in CDCs with the proposed energy-efficient strategy. From the perspective of the total number of cloud users in the CDC and the state of all the VMs, we construct a two-dimensional Markov chain to analyze the queueing model. Finally, we mathematically and numerically evaluate the energy saving level of the system.

The rest of this paper is organized as follows. In Sect. 2, we propose a strategy for improving the energy efficiency in cloud data centers (CDCs), and develop a continuous-time queueing model accordingly. Section 3 derives the system model in steady state with the proposed strategy. In Sect. 4, we investigate the performance measures in terms of energy saving level. Moreover, numerical results are provided to show the influence of parameters on the proposed strategy in Sect. 5. Finally, conclusions are summarized in Sect. 6.

2   Energy-Efficient Strategy and System Model

2.1   Energy-Efficient Strategy

In conventional CDCs, all the VMs remain open regardless of traffic load. This results in a large amount of energy to be wasted, which is referred to as idle energy consumption. Furthermore, inappropriate VM scheduling also generates additional energy consumption, referred to as luxury energy consumption. In order to improve the energy efficiency of CDCs, we propose an energy-efficient strategy with a speed switch and a multiple-sleep mode to reduce both the idle energy consumption and the luxury energy consumption.

In the proposed energy-efficient strategy, all the VMs in the CDC are divided into two modules, namely, the base-line module and the reserve module. The VMs in the base-line module are always active, and its processing speed can be switched between a low speed and a high speed in accordance with the traffic load. The VMs in the reserve module can be awakened from multiple sleeps.

Based on the stochastic behavior of cloud users, as well as the operational characteristics of sleep timers, the CDC will be converted among the following three cases:

Case I: The VMs in the base-line module operate at a low speed while the VMs in the reserve module are asleep. The level of energy-conservation in the CDC is the most significant in this case.

Case II: The VMs in the base-line module operate at a high speed while the VMs in the reserve module are asleep. The level of energy-conservation in the CDC is relatively obvious in this case.

Case III: The VMs in the base-line module operate at a high speed while the VMs in the reserve module are awake and operate at a high speed. The QoS of the CDC is the most ideal in this case.

To avoid frequently switching the processing speed of VMs in the base-line module, we use a dual-threshold, marked as θ_1 ($\theta_1 = 0, 1, 2, \ldots$) and θ_2 ($\theta_2 = 0, 1, 2, \ldots$), to jointly control the VMs processing speed in the base-line module, in which we set $0 < \theta_2 < \theta_1$. When the number of cloud users in the CDC exceeds the threshold θ_1, all the VMs in the base-line module will operate at a high speed. When the number of cloud users in the CDC is less than the threshold θ_2, all the VMs in the base-line module will operate at a low speed. To guarantee the QoS in the CDC even when the traffic load is heavy, we use another threshold, called the activation threshold θ_3, to wake up the VMs in the reserve module. If the number of cloud users waiting in the CDC buffer exceeds the threshold θ_3, all the VMs in the reserve module will be awakened and operate at a high speed after the sleep timer expires. Otherwise, the sleep timer will be restarted with a random duration, and all the VMs in the reserve module will go to sleep again.

For convenience of presentation, we denote the number of VMs in the base-line module as n, and the number of VMs in the reserve module as m. To avoid the appearance that all the VMs in the reserve module are awake while the VMs in the base-line module operate at a low speed, we set $(n - \theta_2) \geq m$. To ensure all the cloud users in the CDC buffer can be served once the VMs in the reserve module are awakened, we set $0 < \theta_3 < m$.

In **Case I**, each cloud user is served immediately on arrival at the CDC by a VM available in the base-line module at a low speed. However, with the arrival of the cloud users, more VMs in the base-line module will be occupied. Here, we call the VMs being occupied by cloud users as busy VMs. When the number of busy VMs in the base-line module exceeds the threshold θ_1, all the VMs in the base-line module will be switched to a high speed, i.e., the CDC will be converted to the **Case II** state. The cloud users that have not received service will be served continuously by the same VM, but at a high speed. In this CDC case, there are no cloud users waiting in the CDC buffer. Therefore, when the sleep timer expires, this sleep timer will be restarted with a random duration, and all the VMs in the reserve module will go to sleep again.

In **Case II**, if there are idle VMs in the base-line module, the following cloud users will be served immediately in the base-line module at a high speed. Otherwise, the cloud users have to wait in the CDC buffer. On the one hand, as cloud users arrive, more cloud users will wait in the CDC buffer. When the sleep timer expires, if the number of cloud users waiting in the CDC buffer exceeds the activation threshold θ_3, all the VMs in the reserve module will be awakened and directly operate at a high speed, i.e., the CDC will be converted to the **Case III** state. Then all the cloud users in the CDC buffer will be served immediately in the reserve module at a high speed. Otherwise, the CDC will remain in the **Case II** state. On the other hand, as cloud users that have received service depart, fewer VMs in the base-line module will be busy. When the number of busy VMs in the base-line module decreases below the threshold θ_2, all the VMs in the base-line module will be switched to a low speed, i.e., the CDC will be converted to the **Case I** state. The cloud users that have not received service will be served continuously by the same VM, but at a low speed.

In **Case III**, if there are idle VMs in either the base-line module or the reserve module, the following cloud users will be served immediately at a high speed. Otherwise, the cloud users will have to wait in the CDC buffer. However, as cloud users that have received service depart, fewer VMs in both the base-line module and the reserve module will be busy. When the number of idle VMs in the base-line module is equal to the number of busy VMs in the reserve module, the cloud users that have not received service in the reserve module will be migrated to the idle VMs in the base-line module, and continuously served at a high speed. Then the sleep timer will be restarted with a random duration, and all the VMs in the reserve module will go to sleep again, i.e., the CDC will be converted to the **Case II** state.

2.2 System Model

A continuous-time queueing model with an adaptive service rate and a partial synchronous vacation is established to capture the related performance measures of the CDC using the proposed energy-efficient strategy. In this queueing model, the requests from cloud users are regarded as customers, each VM is regarded as an independent server that can only serve one request at a time. A sleep is abstracted as a vacation [8]. The system buffer is supposed to be infinite.

We assume that the arrival intervals of requests follow an exponential distribution with parameter λ ($0 < \lambda < 1$). We assume that the service time of a request when the system is in the **Case I** state follows an exponential distribution with parameter μ_l ($0 < \mu_l < 1$). The service time of a request when the system is in either the **Case II** state or the **Case III** state follows an exponential distribution with parameter μ_h ($\mu_l < \mu_h < 1$). Furthermore, we assume that the energy consumption level of each VM during the sleep state is J_v ($J_v > 0$), the energy consumption level of each idle VM is J_o ($J_o > J_v$), the energy consumption level of each busy VM operating at the low speed and the high speed are J_l and J_h ($J_h > J_l$), respectively. In addition, we assume that the time length of a sleep timer follows an exponential distribution with parameter ψ ($\psi > 0$). Here, we refer to the parameter ψ as the sleep parameter.

Let random variable $N(t) = i$, $i \in \{0, 1, 2, \ldots\}$ be the total number of requests in the system at instant t, which is called the system level. Let random variable $C(t) = j$, $j \in \{1, 2, 3\}$ be the system case at instant t. $j = 1, 2, 3$ represents the system is in the states of **Case I**, **Case II** and **Case III**, respectively. $\{N(t), C(t), t \geq 0\}$ constitutes a two-dimensional continuous-time Markov chain (CTMC). The state-space Ω of the CTMC is given as follows:

$$\Omega = \{(i, j) \mid i \in \{0, 1, 2, \ldots\}, \ j \in \{1, 2, 3\}\}. \tag{1}$$

For the two-dimensional CTMC, we define $\pi_{i,j}$ as the steady-state probability when the system level is i and the system case is j. $\pi_{i,j}$ is given as follows:

$$\pi_{i,j} = \lim_{t \to \infty} P\{N(t) = i, C(t) = j\}, \ i \in \{0, 1, 2, \ldots\}, \ j \in \{1, 2, 3\}. \tag{2}$$

We define $\boldsymbol{\pi}_i$ as the steady-state probability vector when the system level is i. $\boldsymbol{\pi}_i$ can be given by

$$\boldsymbol{\pi}_i = \begin{cases} (\pi_{i1}, \pi_{i2}), \ i \in \{0, 1, 2, \ldots, n\} \\ (\pi_{i2}, \pi_{i3}), \ i \in \{n+1, n+2, n+3, \ldots\}. \end{cases} \tag{3}$$

The steady-state probability distribution $\boldsymbol{\Pi}$ of the CTMC is composed of $\boldsymbol{\pi}_i$ ($i \geq 0$). $\boldsymbol{\Pi}$ is given by

$$\boldsymbol{\Pi} = (\boldsymbol{\pi_0}, \boldsymbol{\pi_1}, \boldsymbol{\pi_2}, \ldots). \tag{4}$$

3 Model Analysis

3.1 Transition Rate Matrix

According to the proposed energy-efficient strategy, the system case is related to the system level. The state transition with the transition rate of the system model is illustrated in Fig. 1.

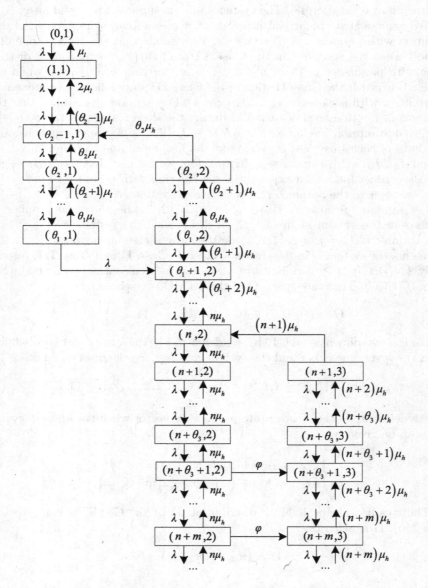

Fig. 1. The state transition of the system model.

Let $Q_{u,v}$ be the one step state transition rate sub-matrix for the system level changing from u ($u = 0, 1, 2, \ldots$) to v ($v = 0, 1, 2, \ldots$). For clarity, $Q_{u,u-1}$, $Q_{u,u}$ and $Q_{u,u+1}$ are abbreviated as B_u, A_u and C_u, respectively. Then, Q is given as follows:

$$
Q = \begin{pmatrix}
A_0 & C_0 & & & & & & \\
B_1 & A_1 & C_1 & & & & & \\
& \ddots & \ddots & \ddots & & & & \\
& & B_{n-1} & A_{n-1} & C_{n-1} & & & \\
& & & B_n & A_n & C_n & & \\
& & & & B_{n+1} & A_{n+1} & C_{n+1} & \\
& & & & & B_{n+2} & A_{n+2} & C_{n+2} \\
& & & & & & \ddots & \ddots & \ddots \\
& & & & & & & B_{n+m} & A_{n+m} & C_{n+m} \\
& & & & & & & & \ddots & \ddots & \ddots
\end{pmatrix} . \tag{5}
$$

The block-tridiagonal structure of Q shows that the state transitions occur only between adjacent system levels. Hence, the two-dimensional CTMC $\{N(t), C(t), t \geq 0\}$ can be seen as a type of Quasi Birth-and-Death (QBD) process.

3.2 Steady-State Probability Distribution

For the CTMC $\{N(t), C(t), t \geq 0\}$ with the one step state transition rate matrix Q, the necessary and sufficient conditions for positive recurrence are that the matrix quadratic equation:

$$
R^2 B_{n+m} + R A_{n+m} + C_{n+m} = 0 \tag{6}
$$

has a minimal non-negative solution R and that the spectral radius $SP(R) < 1$.

We assume the rate matrix $R = \begin{pmatrix} r_{11} & r_{12} \\ 0 & r_{22} \end{pmatrix}$, then substitute R, B_{n+m}, A_{n+m}, and C_{n+m} into Eq. (6), we have

$$
\begin{pmatrix} n\mu_h r_{11}^2 & (n+m)\mu_h(r_{11}+r_{22})r_{12} \\ 0 & (n+m)\mu_h r_{22}^2 \end{pmatrix}
$$
$$
+ \begin{pmatrix} -(\lambda + n\mu_h + \psi)r_{11} & r_{11}\psi - (\lambda + (n+m)\mu_h)r_{12} \\ 0 & -(\lambda + (n+m)\mu_h)r_{22} \end{pmatrix} + \begin{pmatrix} \lambda & 0 \\ 0 & \lambda \end{pmatrix} = \begin{pmatrix} 0 & 0 \\ 0 & 0 \end{pmatrix} . \tag{7}
$$

By solving Eq. (7), we can get r_{11} and r_{22} as follows:

$$
\begin{cases}
r_{11} = \dfrac{(\lambda + n\mu_h + \psi) - \sqrt{(\lambda + n\mu_h + \psi)^2 - 4n\lambda\mu_h}}{2n\mu_h} \\
r_{22} = \dfrac{\lambda}{(n+m)\mu_h},
\end{cases} \tag{8}
$$

and r_{12} can be given by

$$r_{12} = \frac{r_{11}\phi}{\lambda + (n + m)(1 - r_{11} - r_{22})\mu_h}. \tag{9}$$

The rate matrix R has been given in closed-form. Note that $SP(R) = max\{r_{11}, r_{22}\}$, and r_{11} can be proved mathematically less than 1. Therefore, the necessary and sufficient condition for positive recurrence of the CTMC $\{N(t), C(t), t \geq 0\}$ is equivalent to $r_{22} < 1$, that is, $\lambda < (n + m)\mu_h$.

With the rate matrix R obtained, we construct a square matrix $B[R]$ as follows:

$$B[R] = \begin{pmatrix} A_0 & C_0 & & & & & & \\ B_1 & A_1 & C_1 & & & & & \\ & \ddots & \ddots & \ddots & & & & \\ & & B_{n-1} & A_{n-1} & C_{n-1} & & & \\ & & & B_n & A_n & C_n & & \\ & & & & B_{n+1} & A_{n+1} & C_{n+1} & \\ & & & & & B_{n+2} & A_{n+2} & C_{n+2} \\ & & & & & & \ddots & \ddots & \ddots \\ & & & & & & & B_{n+m-1} & A_{n+m-1} & C_{n+m-1} \\ & & & & & & & & B_{n+m} & R \times B_{n+m} + A_{n+m} \end{pmatrix}. \tag{10}$$

By using the geometric-matrix method [9], we can give an equation set as follows:

$$\begin{cases} (\pi_0, \pi_1, \dots, \pi_{n+m})B[R] = \underbrace{(0, 0, \dots, 0)}_{2(n+m+1)} \\ (\pi_0, \pi_1, \dots, \pi_{n+m-1})e + \pi_{n+m}(I - R)^{-1}e_1 = 1 \end{cases} \tag{11}$$

where e is a $2(n + m) \times 1$ vector with ones, and e_1 is a 2×1 vector with ones.

We further construct an augmented matrix as follows:

$$(\pi_0, \pi_1, \dots, \pi_{n+m}) \left(B[R] \vdots \begin{matrix} e \\ (I - R)^{-1}e_1 \end{matrix} \right) = \underbrace{(0, 0, \dots, 0, 1)}_{2(n+m+1)}. \tag{12}$$

Applying the Gauss-Seidel method [10] to solve Eq. (12), we can obtain π_0, π_1, $\dots$, π_{n+m}. From the structure of the transition rate matrix Q, we know π_i $(i = n + m + 1, n + m + 2, n + m + 3, \dots)$ satisfies the matrix geometric solution form as follows:

$$\pi_i = \pi_{n+m}R^{i-(n+m)}, \; i \geq (n + m). \tag{13}$$

Substituting π_{n+m} obtained in Eq. (12) into Eq. (13), we can obtain π_i $(i = n+m+1, n+m+2, n+m+3, \dots)$. Then the steady-state probability distribution $\Pi = (\pi_0, \pi_1, \pi_2, \dots)$ of the system can be given mathematically.

4   Performance Measures

We define the energy saving level as the difference between the energy consumption level of the CDC with the proposed energy-efficient strategy and that of the conventional CDC.

The energy consumption level C of the CDC with the proposed energy-efficient strategy is given as follows:

$$C = C_1 + C_2 + C_3 \tag{14}$$

where C_1, C_2 and C_3 are the average energy consumption level when the system is in the states of **Case I**, **Case II** and **Case III**, respectively.

$$C_1 = \sum_{i=0}^{\theta_1} \pi_{i,1}(iJ_l + (n-i)J_o + mJ_v),$$

$$C_2 = \sum_{i=\theta_2}^{n} \pi_{i,2}(iJ_h + (n-i)J_o + mJ_v) + \sum_{i=n+1}^{\infty} \pi_{i,2}(nJ_h + mJ_v),$$

$$C_3 = \sum_{i=n+1}^{n+m} \pi_{i,3}(iJ_h + (n+m-i)J_o) + \sum_{i=n++m+1}^{\infty} \pi_{i,3}((n+m)J_h). \tag{15}$$

The energy consumption level C' in the conventional CDC is given as follows:

$$C' = (n+m)J_h \left(\frac{\lambda}{(n+m)\mu_h} \right) + (n+m)J_o \left(1 - \frac{\lambda}{(n+m)\mu_h} \right)$$

$$= \frac{\lambda J_h}{\mu_h} + J_o \left(n+m - \frac{\lambda}{\mu_h} \right). \tag{16}$$

Combining Eqs. (14) and (16), the energy saving level S of the CDC with the proposed energy-efficient strategy is given as follows:

$$S = C' - C. \tag{17}$$

5   Numerical Experiments

In order to evaluate the energy saving level of the CDC with the proposed energy-efficient strategy, we provide numerical experiments with analysis and simulation. The analysis results are obtained based on Eq. (11) using Matlab 2011a. The simulation results are obtained by averaging over 10 independent runs using MyEclipse 2014. In the numerical experiments, the parameters are set in Table 1.

To elucidate the better energy saving effect of the proposed energy-efficient strategy, a comparison between the proposed energy-efficient strategy and the conventional DPM strategy is given. In conventional DPM strategy, all the VMs

Table 1. Numerical parameters.

Parameters	Value
Total number $(n+m)$ of VMs in the system	50
Service rate μ_l when VM operates at the low speed	0.01 ms^{-1}
Service rate μ_h when VM operates at the high speed	0.02 ms^{-1}
Dual-threshold θ_1, θ_2	20, 10
Activation threshold θ_3	15
Energy consumption level J_v of a sleeping VM	0.2 mJ
Energy consumption level J_o of an idle VM	0.4 mJ
Energy consumption level J_l of a busy VM operating at the low speed	0.45 mJ
Energy consumption level J_h of a busy VM operating at the high speed	0.5 mJ

are open all the time, but their processing speed can be switched between a low speed and a high speed according to current traffic load of system.

By setting the number of VMs in the reserve module $m = 20$ as an example, we examine the influence of the arrival rate λ of requests on the energy saving level S of the system for different sleep parameters ϕ in Fig. 2(a). By setting the sleep parameter $\phi = 0.05$ as an example, we examine the influence of the arrival rate λ of requests on the energy saving level S of the system for different numbers m of VMs in the reserve module in Fig. 2(b). In Fig. 2(a) and (b), the solid line represents the analysis results with the proposed energy-efficient strategy, the dotted line represents the analysis results with the conventional DPM strategy.

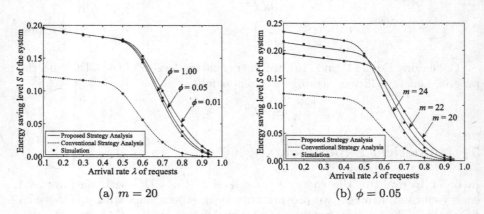

(a) $m = 20$ (b) $\phi = 0.05$

Fig. 2. Energy saving level S of the system.

In Fig. 2, we observe that for the same sleep parameter ϕ and the same number m of VMs in the reserve module, the energy saving level S of the system will initially decrease gradually then decrease sharply as the arrival rate λ of requests increases. When λ is smaller (such as $\lambda < 0.5$ for $\phi = 0.05$ and $m = 20$), as λ increases, it becomes more possible that the number of requests in the system will exceed the threshold θ_1. That is, all the VMs in the base-line module will be switched to the high speed from the low speed. Note that the energy consumption of a VM operating at the high speed is greater than that operating at the low speed. Thus, in this situation, the energy consumption will increase, and the energy saving level will decrease gradually as the arrival rate of requests increases. When λ is larger (such as $\lambda > 0.5$ for $\phi = 0.05$ and $m = 20$), all the VMs in the base-line module will be busy, and the upcoming requests will have to wait in the system buffer. As λ increases, the number of requests waiting in the system buffer is more likely to exceed the activation threshold θ_3, so the VMs in the reserve module will be awakened after the sleep timer expires. Note that the energy consumption of a VM operating at the high speed is greater than that when asleep. Obviously, the energy saving due to sleep mode is greater than that due to switching from the high speed to the low speed. Thus, in this situation, the energy saving level will decrease sharply as the arrival rate of requests increases.

From Fig. 2(a), we notice that for the same arrival rate λ of requests, the energy saving level S of the system will decrease as the sleep parameter ϕ increases. The larger the value of ϕ is, the more likely the VMs in the reserve module will be awake. Thus, the energy consumption of the system will increase, and the energy saving level will decrease.

From Fig. 2(b), we notice that for a smaller arrival rate λ of requests (such as $\lambda < 0.5$ for $\phi = 0.05$), the energy saving level S of the system will increase as the number m of VMs in the reserve module increases. When λ is smaller, no matter how small the value of n is, all the VMs in the reserve module will be more likely to go to sleep again after the sleep timer expires. Thus, as the value of m increases, the energy saving level of the system will increase.

On the other hand, for a larger arrival rate λ of requests (such as $\lambda > 0.55$ for $\phi = 0.05$), the energy saving level S of the system will decrease as the number m of VMs in the reserve module increases. When λ is larger, as the value of n decreases, the number of requests waiting in the system buffer will increase, and the probability that all the VMs in the reserve module awakened will be greater. Thus, as the value of m increases, the energy saving level of the system will decrease.

From the numerical experiments shown in Fig. 2, we see that the analysis results match well with the simulation results. What's more, the contrast experiments show that the energy saving effect of the proposed energy-efficient strategy performs better than that of the conventional DPM strategy.

6   Conclusions

In this paper, we proposed a novel energy-efficient strategy with a speed switch and a multiple-sleep mode. By applying DPM technology and introducing a sleep mode, our energy-efficient strategy can improve energy efficiency significantly by reducing both the luxury energy consumption and the idle energy consumption. We established a continuous-time queueing model with an adaptive service rate and a partial synchronous vacation, and constructed a two-dimensional continuous-time Markov chain to analyze the queueing model. By using the method of a matrix geometric solution, we derived the expression of the system performance measure in terms of the energy saving level of the system. Numerical experiments show that on the premise of guaranteeing the QoS of CDCs, the energy saving effect of CDCs when using our proposed energy-efficient strategy is remarkable when compared with conventional CDCs.

As a future research, we will investigate the delay performance and optimize the proposed strategy with an intelligent optimization algorithm.

Acknowledgments. This work was supported by National Natural Science Foundation (No. 61472342) and Hebei Province Natural Science Foundation (No. F2017203141), China.

References

1. Villars, R.: Worldwide Datacenter Installation Census and Construction Forecase, 2015–2019. Internet Data Center (2015)
2. Li, K.: Improving multicore server performance and reducing energy consumption by workload dependent dynamic power management. IEEE Trans. Cloud Comput. **4**, 122–137 (2016)
3. Wang, Y., Xie, Q., Ammari, A., Pedram, M.: Deriving a near-optimal power management policy using model-free reinforcement learning and bayesian classification. In: 48th IEEE Design Automation Conference, pp. 41–46. IEEE Press, New York (2011)
4. Chen, Y., Chang, M., Liang, W., Lee, C.: Performance and energy efficient dynamic voltage and frequency scaling scheme for multicore embedded system. In: 6th IEEE International Conference on Communications and Electronics, pp. 58–59. IEEE Press, New York (2016)
5. Chou, C., Wong, D., Bhuyan, L.: DynSleep: fine-grained power management for a latency-critical data center application. In: 16th International Symposium on Low Power Electronics and Design, pp. 212–217. IEEE Press, New York (2016)
6. Dabbagh, M., Hamdaoui, B., Guizani, M., Rayes, A.: Energy-efficient resource allocation and provisioning framework for cloud data centers. IEEE Trans. Netw. Serv. Manage. **12**, 377–391 (2015)
7. Liao, D., Li, K., Sun, G., Anand, V., Gong, Y., Tan, Z.: Energy and performance management in large data centers: a queuing theory perspective. In: 4th International Conference on Computing. Networking and Communications, pp. 287–291. IEEE Press, New York (2015)
8. Tian, N., Zhang, Z.: Vacation Queueing Models Theory and Applications. Springer, America (2006).
9. Latouche, G., Ramaswami, V.: Introduction to Matrix Analytic Methods in Stochastic Modeling. Society for Industrial and Applied Mathematics, America (1999)
10. Greenbaum, A.: Iterative Methods for Solving Linear Systems. Society for Industrial and Applied Mathematics, America (1997)

Performance Analysis of Broadcast Packets in Vehicular Ad Hoc Networks

Jaedeok Kim[1] and Ganguk Hwang[2(✉)]

[1] Artificial Intelligence Team, Samsung Electronics, Suwon, Republic of Korea
05jaedeok@gmail.com
[2] Department of Mathematical Sciences,
Korea Advanced Institute of Science and Technology,
Daejeon, Republic of Korea
guhwang@kaist.edu

Abstract. In this paper we analyze the performance of a broadcast packet in a VANET with the slotted ALOHA protocol where locations of vehicles are modeled by a one-dimensional stochastic geometry. We consider the packet delivery probability under a broadcast delay constraint. Since the successful transmission of a broadcast packet is significantly affected by interferences at receivers which are spatially correlated, it is important to capture the spatial correlations properly in order to obtain an accurate expression of the packet delivery probability. However, the exact analysis of the spatial correlations in interference is not mathematically tractable. In this paper we provide an accurate approximation of the spatial correlations in interference and derive the packet delivery probability with the help of the approximation. Numerical and simulation results are provided to validate our analysis and to investigate the performance of a VANET.

Keywords: Performance evaluation · Broadcast · VANET Packet delivery probability

1 Introduction

With a growing interest in the intelligent transportation system (ITS), vehicular ad hoc networks (VANETs) have been developed and many relevant industries have been rapidly growing. To support wireless communications in a vehicular environment, dedicated short range communication (DSRC) is projected and several standards such as the IEEE 802.11p and the IEEE 1609 family are developed [1,2]. In a VANET, a broadcast packet is widely used for various purposes. For instance, an application in a VANET aiming at the safety by reducing accidents on roads, uses broadcast packets in order to inform the current status of a vehicle to neighbor vehicles [3]. Since there are no acknowledgement messages in the broadcast mode, it is hard to know whether or not a packet is successfully transmitted to a node. So a broadcast packet requires more stringent constraints on its performance than a unicast packet with acknowledgement, and it is important to accurately analyze the performance of a broadcast packet.

© Springer International Publishing AG 2017
W. Yue et al. (Eds.): QTNA 2017, LNCS 10591, pp. 155–167, 2017.
https://doi.org/10.1007/978-3-319-68520-5_10

In this paper, we analyze the performance of a broadcast packet in a one-dimensional VANET with the slotted ALOHA protocol. The motivation of a one-dimensional network comes from its application to vehicular networks where vehicles are mostly located in a linear form. Moreover, the slotted ALOHA protocol is a good approximation of the CSMA protocol, e.g., a recent simulation study in [4] shows that the behavior of CSMA appears similar to that of the slotted ALOHA as the network density becomes high. So the analysis of dense VANETs with the slotted Aloha is performed to approximate the CSMA-based MAC protocol in [5], ad hoc networks with the slotted ALOHA protocol are widely considered and analyzed, e.g., [6–8].

We consider the packet delivery probability (PDP) as a performance metric of a broadcast packet. The PDP is important for safety-related applications such as the stopped vehicle hazard warning, in which a broadcast packet is requested to be delivered to all neighbor vehicles. Since the locations of vehicles significantly affect the PDP, we use stochastic geometry theory to model the locations of vehicles in the network. In addition, bearing in mind a safety or control related broadcast packet, we consider a broadcast delay constraint in the computation of the performance metrics [9].

The PDP of a broadcast packet is significantly affected by channel fading and interference to nodes (i.e. vehicles). Considering the path loss in the signal power, interference from a transmitter to a receiver is determined by the distance between them and the locations of the other transmitters. So there exist spatial correlations in interference at receivers and accordingly successful receptions of a broadcast packet are spatially correlated. While the stochastic properties of interference have been studied with the help of stochastic geometry theory [10,11], the exact analysis of the spatial correlations in interference is not mathematically tractable. As a result, it is hard to obtain the joint probability of successful receptions of a broadcast packet at all receivers of interest. To overcome the difficulty, we propose an approximation method to get the PDP as follows. Since the signal power is exponentially decreasing in distance, the main interferer is the nearest transmitter. So it is plausible to assume that successful receptions at receivers are independent when the location of the nearest transmitter is given. By using our approximation assumption, we derive an analytical expression of the PDP which is easy to compute numerically. We show its accuracy through numerical and simulation results. The details are explained in Sects. 3 and 5.

The performance of a VANET such as the PDP has been extensively studied. In [3,12], Campolo et al. analyze the PDP of a broadcast packet that is broadcasted on the control channel interval defined in the IEEE 1609 standard. In [13], Hassan et al. study the impact of retransmissions on the PDP in the IEEE 802.11p standard. Hassanabadi and Valaee in [14] propose a rebroadcasting protocol using network coding in order to improve the PDP of a broadcast packet on the control channel interval. In these works, the transmission range and the impact of locations of nodes are not considered. They assume that all receivers cannot receive a broadcast packet from a transmitter if there exists any other transmitter, although they are able to receive due to finite transmission power. With a consideration of interference, Ma et al. in [15] investigate the PDP by using a one-dimensional Poisson point process. They also consider a carrier sensing range so as to investigate the impact of hidden terminals. However, they assume a static transmission range of a broadcast packet while the transmission ranges of a broadcast packet in practice are different from node to node due to independent channel fading and path loss. Such random characteristic of transmission ranges is considered in [16] where Liu and Andrews analyze the PDP in a sensor network by using a clustered PPP. They also consider the impact of interference from other transmitters in their analysis. However, the spatial correlations in interference are not considered and they assume that the successes of transmissions are independent from node to node. Note that the spatial correlations significantly affect the PDP, but the consideration of the spatial correlations makes the analysis of the PDP almost intractable. So we propose an approximation method in this paper to obtain the PDP while capturing the spatial correlations.

Our main contributions are summarized as follows.

- We propose an approximation method to obtain the PDP of a broadcast packet with a consideration of the spatial correlations in interference and a broadcast delay constraint.
- We derive an analytical expression of the PDP which is easy to compute numerically. It is greatly helpful in solving an optimization problem on the optimal access probability.
- Through numerical and simulation results, we show that the PDP is significantly affected by the spatial correlations in interference.

The remainder of this paper is organized as follows. In Sect. 2, we describe our system model. In Sects. 3 and 4, we derive an analytical expression of the PDP by considering the spatial correlations in interference and a broadcast delay constraint. In Sect. 5, we investigate the performance of a VANET through numerical and simulation results. Finally, we summarize our conclusions in Sect. 6.

2   System Modeling

We consider a one-dimensional vehicular ad hoc network (VANET). The time axis is slotted and each slot duration is equal to a packet transmission time.

For modeling and analysis, we tag an arbitrary node in the network and call it the tagged node. The tagged node is assumed to be located at the origin by translating the axis. The locations of untagged nodes are modeled by the one dimensional Poisson point process (PPP) $\Phi = \{X_i\}$ with intensity λ. So the distance between two nodes follows the exponential distribution with mean $1/\lambda$. Experimental measurements show that the inter-arrival time at an arbitrary observation point is exponentially distributed in a free flow network [17,18], which provides a good reason for the use of the PPP in our modeling.

Each node adopts the p-persistent slotted ALOHA. At each slot each node independently tries to broadcast its packet with an access probability p. No acknowledgement message is used in the broadcast mode. A node broadcasts a packet only once and removes the packet regardless of the broadcasting result. In many vehicular applications a packet has an expiry time depending on applications [2,9]. So we assume that the expiry time of a packet is fixed to T_{exp} slots. After T_{exp} slots from the generation of a packet, the packet is expired and dropped if it is still waiting for transmission.

We call a node who broadcasts a packet an active node and denote the set of all active nodes by $\Phi_a := \{X_i \in \Phi : a_i = 1\}$ where a_i indicates the channel access of node X_i. a_i has the value 1 with probability p and has the value 0 with probability $1 - p$. In addition, we let $\Phi_{ia}^+ := \{Y_i \in \Phi : Y_i \geq 0, a_i = 0\}$ be the set of all inactive nodes in the positive direction. Inactive nodes try to receive broadcast packets from active nodes. The reason why we consider only the positive direction is explained as follows. In most vehicular applications, information generated by a node is useful for the following nodes. So it is more meaningful to investigate how a packet is broadcasted in the positive direction. In fact, the broadcasting in the negative direction can be analyzed in the same way as given for the positive direction. We therefore do not provide the analysis for the negative direction in this paper.

An inactive node can successfully decode the packet from the tagged node, called the tagged packet, only when the SINR exceeds a given threshold θ. The channel is modeled by a block Rayleigh fading channel, that is, the channel state remains invariant during a slot and changes slot-by-slot. Since the channel has the Rayleigh fading, the fading from node i and node j is exponentially distributed with mean $1/\mu$. Note that the fading is independent of the node indices i and j. Then the interference to an inactive node located at u is given by

$$I_{\Phi_a}(u) = \sum_{X_i \in \Phi_a} \frac{F_i}{|X_i - u|^\alpha} \tag{1}$$

and the received SINR of the inactive node at u is given by

$$\text{SINR}(u, \Phi_a) = \frac{F/u^\alpha}{W + I_{\Phi_a}(u)}, \tag{2}$$

where α is the pass loss exponent, W denotes the noise power at the inactive node, and F, F_i are the fading at the inactive node from the tagged node and node $X_i \in \Phi_a$, respectively. When the channel is assume to be the additive white Gaussian noise (AWGN) channel, W becomes an exponential random variable with mean $1/\sigma$ [19].

It is worth mentioning that $\text{SINR}(u, \Phi_a) > \theta$ does not always guarantee that $\text{SINR}(u', \Phi_a) > \theta$ for $0 < u' < u$ due to the independence in fading, while the previous works, e.g., [13,15,20] assume that all nodes at $u'(< u)$ successfully receive a packet if $\text{SINR}(u, \Phi_a) > \theta$. So the transmission range of a packet is not only a random variable, but also differs from node to node even for the same packet, which should be considered in the analysis. Hence, we define the transmission range of a receiver by a random variable R such that $[R > u] = [\text{SINR}(u, \Phi_a) > \theta]$.

3   Analysis of the Packet Delivery Probability

In this section, we analyze the PDP of a broadcast packet. The concept of performance may differ from application to application since each application has its own objectives. So a performance metric should be adapted to the objectives of each application. In addition, since an expired packet is dropped, the expiry time of a packet has to be considered in defining a performance metric. Bearing these observations in mind, we carefully define the packet delivery probability for a broadcast packet as follows.

The packet delivery probability (PDP) is defined by the probability that the tagged packet is broadcasted and no node within the range L_{rd} fails to receive the tagged packet. Obviously, the PDP is a function of the access probability of the ALOHA protocol and we denote the PDP for a given access probability p by $P_D(p)$.

The PDP is important for safety-related applications such as the stopped vehicle hazard warning, in which a broadcast packet of the tagged node is requested to be delivered to all neighbor nodes of the tagged node with high probability. To explain it more precisely, we consider the following scenario. Suppose that there exists a node who fails to receive a safety packet, e.g., a packet with the stopped vehicle hazard warning, from the tagged node. The node has no information about the status of the tagged node in this case and accordingly the node is likely to crash the tagged node or its neighbor. Hence, such a packet is strongly requested to be delivered to all neighbor nodes.

In order to define neighbor nodes of the tagged node, we need a neighbor range based on the reaction distance. For instance, the *pre-crash sensing* application recommends to have at least 50 m as the reaction distance [9]. In our analysis, we use L_{rd} as the reaction distance and consider all nodes in $[0, L_{rd}]$ in the analysis of $P_D(p)$.

We now derive $P_D(p)$. Let T denote the broadcast delay defined by the time until the tagged packet is broadcasted from its generation. Since the tagged node independently decides whether to be active with probability p at each slot, T is a geometric random variable with mean $1/p$.[1] Since the expiry time of the packet is given by T_{exp} slots, it is required that $T \leq T_{exp}$. Otherwise the tagged packet is dropped due to its expiration. Then the probability $P_{tx}(p)$ that the tagged packet is broadcasted is given by

$$P_{tx}(p) = 1 - (1 - p)^{T_{exp}}.$$

---

[1] In this paper, we consider a scenario as follows. The arrival rate of a broadcast packet is low and each node has a queue of capacity 1. So a generated packet in the queue can be served just after its generation and hence the broadcast delay T follows a geometric distribution as explained. However, it can be extended to the case of having a general queue and a general arrival process of broadcast packets. For instance, when the arrival process follows the Bernoulli process, the broadcast delay T can be analyzed by using the Geom/G/1 queueing model, and the following analysis is not changed to obtain the performance metrics. For simplicity in our analysis, we thus consider a simple version as explained before.

Let $\mathcal{D}$ denote the event that all untagged nodes in $[0, L_{rd}]$ receives the tagged packet. Note that an inactive node located at u successfully receives the tagged packet when $R > u$, whereas any active nodes cannot receive the tagged packet because we assume that a node cannot transmit and receive a packet simultaneously. Let X_1^+ denote the location of the first active node in the positive direction in Φ_a. Then the event $\mathcal{D}$ is given by

$$\mathcal{D} = [X_1^+ > L_{rd}, R_i > Z_i, Z_i \in \mathcal{N}_0],$$

where $\mathcal{N}_0$ is the set of all inacitve nodes in $[0, L_{rd}]$ and R_i is the transmission range with respect to the node Z_i.

From the definition of the PDP where the delay constraint T_{exp} is also considered, we have

$$\begin{aligned} P_D(p) &= \mathbb{P}\{T \leq T_{exp}, \mathcal{D}\} \\ &= \mathbb{P}\{T \leq T_{exp}\}\mathbb{P}\{\mathcal{D} \mid T \leq T_{exp}\} \\ &= P_{tx}(p)P_D^0(p), \end{aligned} \tag{3}$$

where $P_D^0(p) = \mathbb{P}\{\mathcal{D} \mid T \leq T_{exp}\}$. Since T is independent of the transmission range R_i and the locations of nodes Φ, we have $P_D^0(p) = \mathbb{P}\{\mathcal{D}\}$. It then follows that

$$\begin{aligned} P_D^0(p) &= \mathbb{E}[\mathbb{P}\{\mathcal{D} \mid X_1^+\}] \\ &= \mathbb{E}[\mathbf{1}_{[X_1^+ > L_{rd}]}\mathbb{P}\{R_i > Z_i, Z_i \in \mathcal{N}_0 \mid X_1^+\}] \\ &= \int_{L_{rd}}^{\infty} \mathbb{P}\{R_i > Z_i, Z_i \in \mathcal{N}_0 \mid X_1^+ = x_1\}\lambda p e^{-\lambda p x_1} \, dx_1 \end{aligned} \tag{4}$$

where we use the fact that X_1^+ is an exponential random variable with parameter λp.

To derive the joint probability $\mathbb{P}\{R_i > Z_i, Z_i \in \mathcal{N}_0 \mid X_1^+ = x_1\}$, first recall that $[R_i > Z_i] = [\text{SINR}(Z_i, \Phi_a) > \theta]$. Due to the interference term $I_{\Phi_a}(Z_i)$ in $\text{SINR}(Z_i, \Phi_a)$, the transmission ranges R_i are not independent and the joint distribution is very complicated to evaluate directly. We therefore need to approximate the joint distribution. Note that R_i are independent when Φ_a is given and the main interferer is the nearest active node located at X_1^+. So, it is plausible to assume that R_i are independent when X_1^+ is given. When $N_0 := |\mathcal{N}_0| = n$ (≥ 0), the locations Z_i are independent and uniformly distributed over $[0, L_{rd}]$ by the property of a PPP. Then,

$$\begin{aligned} &\mathbb{P}\{R_i > Z_i, Z_i \in \mathcal{N}_0 \mid N_0 = n, X_1^+ = x_1\} \\ &\approx \left(\mathbb{P}\{R > U \mid X_1^+ = x_1\}\right)^n \\ &= \left(\frac{1}{L_{rd}} \int_0^{L_{rd}} \mathbb{P}\{R > u \mid X_1^+ = x_1\} \, du\right)^n \\ &=: (p_{suc,x_1})^n, \end{aligned} \tag{5}$$

where U is the uniform random variable on $[0, L_{rd}]$ and R is a random variable having the same distribution as R_i. Here, we can evaluate $\mathbb{P}\{R > u \mid X_1^+ = x_1\}$ by using the following proposition.

Proposition 1. *Let X_1^+ be the location of the nearest active node in the positive direction in Φ_a. The distribution of the transmission range R conditioned on $X_1^+ = x_1$ is given by*

$$\mathbb{P}\{R > u \mid X_1^+ = x_1\} = \mathbb{E}[e^{-\mu\theta u^\alpha W}] \frac{|x_1 - u|^\alpha}{|x_1 - u|^\alpha + \theta u^\alpha} e^{-2\lambda pu\theta^{1/\alpha}\pi/\alpha \sin(\pi/\alpha)}$$

$$\times \exp\left(\lambda p \int_0^{x_1} \frac{\theta u^\alpha}{\theta u^\alpha + |t - u|^\alpha}\, dt\right). \tag{6}$$

In particular, when the channel has the AWGN with mean $1/\sigma$,

$$\mathbb{P}\{R > u \mid X_1^+ = x_1\} = \frac{\sigma}{\sigma + \mu\theta u^\alpha} \frac{|x_1 - u|^\alpha}{|x_1 - u|^\alpha + \theta u^\alpha} e^{-2\lambda pu\theta^{1/\alpha}\pi/\alpha \sin(\pi/\alpha)}$$

$$\times \exp\left(\lambda p \int_0^{x_1} \frac{\theta u^\alpha}{\theta u^\alpha + |t - u|^\alpha}\, dt\right).$$

Proof. We omit the proof due to space limitation.

By using (5), we have, for $x_1 > L_{rd}$

$$\mathbb{P}\{R_i > Z_i, Z_i \in \mathcal{N}_0 \mid X_1^+ = x_1\}$$
$$= \mathbb{E}[\mathbb{P}\{R_i > Z_i, Z_i \in \mathcal{N}_0 \mid N_0, X_1^+ = x_1\} \mid X_1^+ = x_1]$$
$$= \mathbb{E}[(p_{suc,x_1})^{N_0} \mid X_1^+ = x_1].$$

Since $\mathcal{N}_0 = \Phi_{ia}^+ \cap [0, L_{rd}]$, N_0 $(= |\mathcal{N}_0|)$ follows the Poisson distribution with parameter $\lambda(1 - p)L_{rd}$ regardless of the value of x_1. So

$$\mathbb{E}[z^{N_0} \mid X_1^+ = x_1] = \exp\left(-\lambda(1 - p)L_{rd}(1 - z)\right)$$

and it follows that

$$\mathbb{P}\{R_i > Z_i, Z_i \in \mathcal{N}_0 \mid X_1^+ = x_1\}$$
$$= \exp\left(-\lambda(1 - p)L_{rd}(1 - p_{suc,x_1})\right)$$
$$= \exp\left(-\lambda(1 - p) \int_0^{L_{rd}} \mathbb{P}\{R \leq u \mid X_1^+ = x_1\}\, du\right).$$

Substituting this into (4), we obtain

$$P_D^0(p) = \lambda p \int_{L_{rd}}^{\infty} e^{-\lambda px_1 - \lambda(1-p)\int_0^{L_{rd}} \mathbb{P}\{R \leq u \mid X_1^+ = x_1\}du}\, dx_1.$$

Therefore, the PDP is given by

$$P_D(p) = P_{tx}(p) \times P_D^0(p)$$

$$= (1 - (1-p)^{T_{exp}}) \times \lambda p \int_{L_{rd}}^{\infty} e^{-\lambda p x_1 - \lambda(1-p) \int_0^{L_{rd}} \mathbb{P}\{R \le u \mid X_1^+ = x_1\} du} \, dx_1.$$

$$(7)$$

The numerical evaluation of $P_D(p)$ is not easy if we directly use (7) because the expression in (7) contains an integral in the exponent and the integrand $\mathbb{P}\{R \le u \mid X_1^+ = x_1\}$ of the integral also has another integral. So for an easy numerical evaluation we provide in Sect. 4 an approximated expression of the integrand $\mathbb{P}\{R \le u \mid X_1^+ = x_1\}$ (equivalently, $\mathbb{P}\{R > u \mid X_1^+ = x_1\}$) based on a polynomial. Using our approximation, we can easily compute the value of $P_D(p)$ that is necessary for our optimization problem given below.

In the design of the p-persistent ALOHA, it is important to determine the optimal access probability p_D^* for a reliable broadcasting. From our analysis, the optimal access probability p_D^* is given by

$$p_D^* = \underset{0 \le p \le 1}{\operatorname{argmax}} P_D(p). \tag{8}$$

We call $P_D^* := P_D(p_D^*)$ the optimal PDP.

4   Approximation of the Distribution of R

We know from the previous section that the performance metrics are closely related to the distribution of R. However, evaluating the value of $P_D(p)$ is not an easy task as explained before. To explain it more precisely, recall that from (7) $P_D(p)$ contains the integral

$$\int_{L_{rd}}^{\infty} e^{-\lambda p x_1 - \lambda(1-p) \int_0^{L_{rd}} \mathbb{P}\{R \le u \mid X_1^+ = x_1\} du} \, dx_1.$$

The integrand contains an integral at the exponent whose integrand is a function of x_1. In addition, we know from Proposition 1 that the integrand $\mathbb{P}\{R > u \mid X_1^+ = x_1\}$ also contains the exponent having the integral of the function of x_1

$$\int_0^{x_1} \frac{\theta u^\alpha}{\theta u^\alpha + |t - u|^\alpha} \, dt.$$

Due to the integral exponent, the numerical evaluation is very complicated and slow. It then follows that solving the optimization problem (8) becomes extremely hard from the numerical evaluation viewpoint.

We therefore need to approximate the exponent of the conditional distribution $\mathbb{P}\{R > u \mid X_1^+ = x_1\}$. By the same argument used in the proof of Proposition 1, the integral is changed into

$$\int_0^{x_1} \frac{\theta u^\alpha}{\theta u^\alpha + |t - u|^\alpha} \, dt = \begin{cases} u\theta^{\frac{1}{\alpha}} \left(\int_0^{\theta^{-\frac{1}{\alpha}}} \frac{1}{1+s^\alpha} \, ds + \int_0^\zeta \frac{1}{1+s^\alpha} \, ds \right), & \text{if } 0 \le u < x_1, \\ u\theta^{\frac{1}{\alpha}} \left(\int_0^{\theta^{-\frac{1}{\alpha}}} \frac{1}{1+s^\alpha} \, ds - \int_0^\zeta \frac{1}{1+s^\alpha} \, ds \right), & \text{if } u \ge x_1, \end{cases}$$

where $\zeta = |x_1 - u|/u\theta^{1/\alpha}$. Note that two terms have the same form of integral. So it is enough to approximate each integral, and the following proposition provides a polynomial approximation with its error bound.

Proposition 2. *Suppose that* $\alpha > 1$. *For any* $m \in \mathbb{N}$,

$$\left| \int_0^\beta \frac{1}{1+s^\alpha} \, ds - \tilde{I}_m(\beta) \right| \leq \begin{cases} \frac{\beta^{\bar{m}\alpha+1}}{\bar{m}\alpha+1}, & if \beta < 1, \\ \frac{1}{\bar{m}^2\alpha^2-1} - \frac{\beta^{-\bar{m}\alpha+1}}{\bar{m}\alpha-1}, & if \beta \geq 1. \end{cases}$$

Here $\bar{m} = 2\lceil m/2 \rceil$ *and*

$$\tilde{I}_m(\beta) := \begin{cases} \sum_{k=0}^m \frac{(-1)^k}{k\alpha+1} \beta^{k\alpha+1}, & if \beta < 1, \\ 1 - 2\sum_{k=1}^m \frac{(-1)^k}{k^2\alpha^2-1} + \sum_{k=1}^m \frac{(-1)^k}{k\alpha-1} \beta^{-k\alpha+1}, & if \beta \geq 1. \end{cases}$$

Proof. We omit the proof due to space limitation.

By Proposition 2, the conditional distribution of R is approximated as follows.

$$\mathbb{P}\{R > u \mid X_1^+ = x_1\}$$
$$\approx \mathbb{E}[e^{-\mu\theta u^\alpha W}] \frac{|x_1 - u|^\alpha}{|x_1 - u|^\alpha + \theta u^\alpha} e^{-\lambda p u\theta^{1/\alpha} \{C_0 - \tilde{I}_m(\theta^{-\frac{1}{\alpha}}|\frac{x_1}{u}-1|)\}}, \tag{9}$$

where $C_0 = 2\pi/\alpha \sin(\pi/\alpha) - \tilde{I}_m(\theta^{-\frac{1}{\alpha}})$.

The experimental measurements show that the range of the value of α in a VANET is not less than 1.5 [21–23]. So the assumption $\alpha > 1$ always holds in practice. In addition, the error function provided in Proposition 2 goes to 0 as $m \to \infty$. However, we find that choosing $m = 1$ is enough to approximate in practical scenarios, which will be shown through numerical results in Sect. 5.

5   Numerical Results

This section validates our mathematical model and investigates the performance of a broadcast packet. To this end, we provide both numerical and simulation results using Matlab. Throughout this section, the system parameter values are fixed as follows unless otherwise mentioned. We use the path loss exponent $\alpha = 1.77$ adopted from the measurement data [22]. The success threshold θ is set to $5\,\text{dB}$. The AWGN channel is considered in our examples. Since the constant multiplication of the noise W and the fading F does not affect the value of the SINR, the average of W is assumed to be normalized by $1/\sigma = 1$ and the average of F is assumed to be $1/\mu = 10^7$.

We first validate the approximation of the tail probability of the transmission range R based on Proposition 2. The intensity λ is chosen to be $0.05\,(nodes/m)$ and the location X_1^+ of the nearest active node is fixed by $X_1^+ = 200\,\text{m}$. Figure 1 plots the approximated values by (9) with the access probabilities $p = 0.1, 0.4, 0.7$. When we evaluate the approximated values, we use $m = 1$ in Proposition 2 which is the simplest approximation. For a comparison purpose,

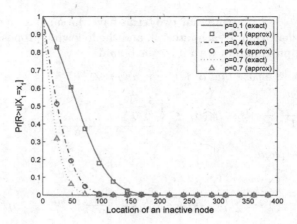

Fig. 1. Validation of our approximation

the exact values by (6) are also plotted. As we expect the approximated values are well matched with the exact values. From now on, we use (9) instead of (6) for numerical evaluation.

In the following examples, we investigate the PDP via numerical and simulation results. To this end, we use the traffic intensities $\lambda = 0.05 \, (nodes/m)$ and $\lambda = 0.1 \, (nodes/m)$, the expiration time $T_{exp} = 10 \, slots$, and the minimum requirement of transmission range $L_{rd} = 50 \, m$. For a comparison purpose, we also provide analytical results when R_i are assumed to be independent as in the existing works, e.g., [16,24]. In the following figures, numerical results of our analysis are denoted by *ana*, while numerical results with the independence assumption are denoted by *indep*.

The PDP $P_D(p)$ is plotted in Fig. 2 as the access probability p varies from 0.01 to 0.95. As seen in the figure, our analytic results are relatively well matched with simulation results, which validates our approximation on the joint success probability used in Sect. 3. However, when R_i are assumed to be independent, the resulting PDP are significantly underestimated in both cases. The reason is explained as follows. The main contributor of interference at a receiver is the nearest active node. Since the neighbor nodes of the tagged node have the common nearest active node located at X_1^+, interferences at the neighbor nodes are strongly correlated. In addition, since the signal power is exponentially decreasing in distance, the other active nodes contribute less significantly to interference in the SINR at a receiver than the nearest active node. So the PDP are not significantly affected by the correlations induced by the other active nodes except the nearest active node. From the above reasons, the PDP is well estimated by our approximation that considers only the nearest active node.

We now investigate the performances of a broadcast packet with the optimal access probabilities which are obtained by solving (8). For a comparison purpose, two different values of the fading power $1/\mu$ are considered. $1/\mu \gg 1/\sigma$ implies

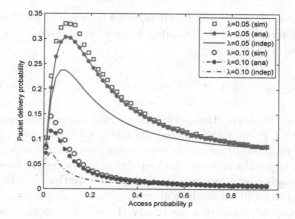

Fig. 2. The packet delivery probability $P_D(p)$ when p varies

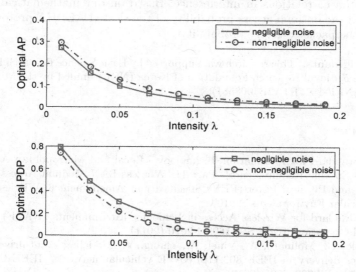

Fig. 3. The optimal access probability p_D^* and the optimal PDP P_D^* when λ varies

that the transmission power is so large that the noise can be negligible, while otherwise the noise should not be negligible. Since we consider the normalized noise, we change the values of μ to represent each case. In our examples, we use $1/\mu = 10^7$ for the negligible noise case and $1/\mu = 10^3$ for the non-negligible noise case.

In Fig. 3, the optimal PDP P_D^* is plotted with the corresponding optimal access probability p_D^* as the intensity varies from 0.01 to 0.2. We see that the PDP is significantly degraded as λ increases. A large value of λ implies that there are many nodes within the range L_{rd}. In order to make all neighbor nodes inactive, the optimal access probability p_D^* should be small enough as we see in the figure. However, the use of a small value of p increases the broadcast delay T, which makes more packets discarded before broadcasting. As a result, the PDP goes to 0 as λ increases.

6   Conclusions

In this paper, we analyzed the performance of a broadcast packet in a one-dimensional VANET with the slotted ALOHA protocol. We considered the packet delivery probability as our performance metric with a delay constraint. We proposed an approximation method to obtain the packet delivery probability of a broadcast packet with a consideration of the spatial correlations in interference. We then derived an analytical expression of the packet delivery probability which is easy to compute numerically. Through numerical and simulation results, we showed that the packet delivery probability is significantly affected by the spatial correlations in interference. Based on our mathematical model, we obtained the optimal access probability of the slotted ALOHA protocol that optimizes the packet delivery probability.

Acknowledgments. This research was supported by Basic Science Research Program through the National Research Foundation of Korea (NRF) funded by the Ministry of Education (NRF-2017R1A2B4008581).

References

1. IEEE Standard for Information Technology - Local and Metropolitan Area Networks - Specific Requirements - Part 11: Wireless LAN Medium Access Control (MAC) and Physical Layer (PHY) Specifications Amendment 6: Wireless Access in Vehicular Environments (2010)
2. IEEE Standard for Wireless Access in Vehicular Environments (WAVE) - Multi-channel Operation, IEEE Std 1609.4-2010 (2011)
3. Campolo, C., Molinaro, A., Vinel, A., Zhang, Y.: Modeling event-driven safety messages delivery in IEEE 802.11p/WAVE vehicular networks. IEEE Commun. Lett. **17**(12), 2392–2395 (2013)
4. Subramanian, S., Werner, M., Liu, S., Jose, J., Lupoaie, R., Wu, X.: Congestion control for vehicular safety: synchronous and asynchronous MAC algorithms. In: The Nineth ACM International Workshop on Vehicular Inter-networking (VANET) (2012)
5. Nguyen, T.V., Baccelli, F., Zhu, K., Subramanian, S., Wu, Z.Z.: A performance analysis of CSMA based broadcast protocol in VANETs. In: Proceedings of the IEEE INFOCOM, pp. 2805–2813 (2013)
6. Blaszczyszyn, B., Mühlethaler, P., Toor, Y.: Maximizing throughput of linear vehicular ad-hoc networks (AVNETs) - a stochastic approach. In: European Wireless Conference, pp. 32–36 (2009)
7. Blaszczyszyn, B., Mühlethaler, P., Toor, Y.: Stochastic analysis of ALOHA in vehicular ad-hoc networks. Ann. Telecommun. **68**(1–2), 95–106 (2013)
8. Zhang, L., Valaee, S.: Congestion control for vehicular networks with safety-awareness. IEEE/ACM Trans. Netw. **24**(6), 3290–3299 (2016)

9. CAMP Vehicle Safety Communications Consortium: Vehicle Safety Communications Project: Task 3 Final Report: Identify Intelligent Vehicle Safety Applications Enabled by DSRC. National Highway Traffic Safety Administration, US Department of Transportation (2005)
10. Ganti, R.K., Haenggi, M.: Spatial and temporal correlation of the interference in ALOHA ad hoc networks. IEEE Commun. Lett. **13**(9), 631–633 (2009)
11. Gong, Z., Haenggi, M.: Interference and outage in mobile random networks: expectation, distribution, and correlation. IEEE Trans. Mob. Comput. **13**(2), 337–349 (2014)
12. Campolo, C., Vinel, A., Molinaro, A., Koucheryavy, Y.: Modeling broadcasting in IEEE 802.11p/WAVE vehicular networks. IEEE Commun. Lett. **15**(2), 199–201 (2011)
13. Hassan, M.I., Vu, H.L., Sakurai, T., Andrew, L.L.H.: Effect of retransmissions on the performance of the IEEE 802.11 MAC protocol for DSRC. IEEE Trans. Veh. Technol. **61**(1), 22–34 (2012)
14. Hassanabadi, B., Valaee, S.: Reliable periodic safety message broadcasting VANETs using network coding. IEEE Trans. Wireless Commun. **13**(3), 1284–1297 (2014)
15. Ma, X., Zhang, J., Wu, T.: Reliability analysis of one-hop safety-critical broadcast services in VANETs. IEEE Trans. Veh. Technol. **60**(8), 3933–3946 (2011)
16. Liu, C.-H., Andrews, J.G.: Multicast outage probability and transmission capacity of multihop wireless networks. IEEE Trans. Inf. Theory **57**(7), 4344–4358 (2011)
17. Neelakantan, P.C., Babu, A.V.: Connectivity analysis of vehicular ad hoc networks from a physical layer perspective. Wireless Pers. Commun. **71**(1), 45–70 (2013)
18. Roess, R.P., Prassas, E.S., Mcshane, W.R.: Traffic Engineering. Pearson Prentice Hall, Englewood Cliffs (2004)
19. Tse, D., Viswanath, P.: Fundamentals of Wireless Communication. Cambridge University Press, Cambridge (2005)
20. Miorandi, D., Altman, E.: Connectivity in one-dimensional ad hoc networks: a queueing theoretical approach. Wireless Netw. **12**(6), 573–587 (2006)
21. Abbas, T., Sjöberg, K., Karedal, J., Tufvesson, F.: A measurement based shadow fading model for vehicle-to-vehicle network simulations. Int. J. Antennas Propag. **2015**, Article ID 190607 (2015)
22. Karedal, J., Czink, N., Paier, A., Tufvesson, F., Molisch, A.F.: Path loss modeling for vehicle-to-vehicle communications. IEEE Trans. Veh. Technol. **60**(1), 323–328 (2011)
23. Zang, Y., Stibor, L., Orfanos, G., Guo, S., Reumerman, H-J.: An error model for inter-vehicle communications in highway scenarios at 5.9 GHz. In: Proceeding on PE-WASUN 2005, pp. 49–56 (2005)
24. Stamatiou, K., Haenggi, M.: Delay characterization of multihop transmission in a poisson field of interference. IEEE/ACM Trans. Netw. **22**(6), 1794–1807 (2014)

Performance Evaluation of Cognitive Radio Networks with a Finite Buffer Setting for Primary Users

Yuan Zhao[1(✉)] and Wuyi Yue[2]

[1] School of Computer and Communication Engineering,
Northeastern University at Qinhuangdao, Qinhuangdao 066004, China
yuanzh85@163.com
[2] Department of Intelligence and Informatics, Konan University,
Kobe 658-8501, Japan
yue@konan-u.ac.jp

Abstract. In this paper, in order to reduce possible packet loss of the primary users (PUs) in cognitive radio networks, we assume there is a buffer with a finite capacity for the PU packets. If a PU packet arrives during the transmission period of another PU packet, this newly arriving PU packet can initially access the PU buffer. In order to evaluate the influence of the buffer setting of the PU packets on the secondary users (SUs), we construct and analyze a discrete-time Markov chain model. Accordingly, we determine the expressions of some important performance measures of the PU packets and the SU packets, including the average queue length, the blocking rate, the throughput of PU packets and the average queue length, the interrupted rate, and the average delay of SU packets, respectively. Finally, we show numerical results to evaluate how the buffer setting of the PU packets influences the system performance of the PU packets and the SU packets.

Keywords: Cognitive radio networks · Finite buffer setting Markov chain · Performance evaluation

1 Introduction

The demand for wireless spectrum resources is increasing rapidly. Cognitive radio networks, which are proposed as one of the mobile communication systems of the future, have been attracting more and more attention in both academic and industrial fields of research [1,2].

Spectrum resources in cognitive radio networks are shared by two types of users, namely, primary users (PUs) and secondary users (SUs). The SUs can make opportunistic use of the spectrum when the spectrum is not used by PUs [3,4].

Considering the absolute right of the PUs, the transmission of SUs can not be guaranteed. During the transmission of SUs, PUs can arrive at any instant

© Springer International Publishing AG 2017
W. Yue et al. (Eds.): QTNA 2017, LNCS 10591, pp. 168–179, 2017.
https://doi.org/10.1007/978-3-319-68520-5_11

and interrupt the transmission of the SUs. Therefore, from the perspective of SUs, it is necessary to adjust their behaviors based on the history of spectrum utilization of PUs [5–7].

In recent years, as stated in [8], some researchers have dealt with the influence of PUs actions on the system performance of cognitive radio networks.

In [9], Jin et al. introduced an energy saving strategy to cognitive radio networks. Via the Matrix-Geometric Solution method, they derived the average latency of SU packets and the system energy saving ratio. With numerical results, they illustrated how the arrival rate of PU packets influenced the system performance of SU packets.

In [10], Li and Han considered socially optimal queuing control for the SU packets in cognitive radio networks. They assumed the emergence of PUs was a server interruption. Based on an assumption of Markovian service interruptions, the authors analyzed the individually and socially optimal threshold strategies for the SUs.

As mentioned above, we find that most of the available literature relating to cognitive radio networks does not consider the buffer setting for the PUs. However, in order to reduce the possible packet loss of PUs, it is necessary to set a buffer for the PU packets, especially for the networks that have a large number of newly arriving PU packets. In [11], Asheralieva and Miyanaga proposed a spectrum allocation algorithm to assign the spectrum resources based on the buffer sizes of the PUs and SUs. With numerical results, they showed that the buffer setting for the PUs had an important impact on both the PUs and SUs.

On the other hand, in most of the available literature relating to cognitive radio networks, in order to avoid possible computational complexity, some continues-time Markov models were built to analyze the system performance. However, considering the digital nature of model communication networks, the discrete-time models are more suitable for the system analysis of cognitive radio networks. Therefore, in this paper, overcoming the computational complexity, we build and analyze a discreet-time Markov chain model.

In this paper, we analyze a spectrum resource allocation strategy in cognitive radio networks with a slotted time structure. In order to reduce the possible packet loss of the PUs, we assume that there is a buffer with a finite capacity set for PUs. Different from the model analysis in [11], considering the change in the number of PU packets and SU packets in the system, we construct and analyze a two-dimensional discrete-time Markov chain model. Moreover, we derive the expressions of some important performance measures of the PU packets and the SU packets. Using numerical results, we compare the system performance between a finite buffer setting and a zero buffer setting for the PUs.

The remainder of this paper is organized as follows. A Markov chain model considering finite buffer setting for PUs is demonstrated in Sect. 2. The model analysis is carried out in Sect. 3. In Sect. 4, the expressions for some important performance measures are derived. In Sect. 5, numerical results are provided to compare the system performance between a finite buffer setting and a zero buffer setting for the PUs. Finally, conclusions are drawn in Sect. 6.

2   System Model

In this paper, we focus on the spectrum allocation strategy in a cognitive radio network with a slotted time structure. We focus on a single channel in the spectrum. This single channel is authorized to one PU. The packets generated from this PU can access the system and can be transmitted on the channel always. SU packets generated from SUs can access the system and be transmitted opportunistically.

We assume the packet arrivals occur during the beginning instant of a slot, and the packet departures occur during the ending instant of a slot.

In order to reduce the possible packet loss, we set buffers for the PU packets and the SU packets, respectively. In practice, the buffers can accommodate the packets that can not access the channel directly. In this paper, we assume that the PU buffer is allocated for the newly arriving PU packets, and the SU buffer is allocated for both of the newly arriving SU packets and the interrupted SU packets. The capacity of the PU buffer is finite and denoted as K ($K > 0$). Considering the lower priority of SU packets in cognitive radio networks, there may be a large number of SU packets waiting in the system. Therefore, we assume the capacity of the SU buffer is infinite.

In order to demonstrate the system actions of PU packets and SU packets more intuitively, we depict Fig. 1 as follows.

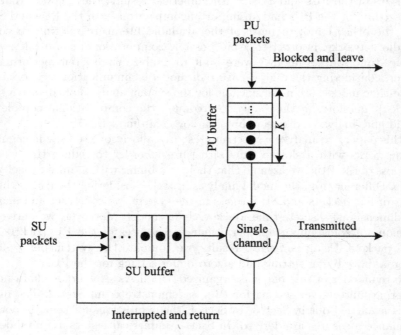

Fig. 1. System actions of PU packets and SU packets.

As shown in Fig. 1, in order to avoid complexity, we assume the PU packets and the SU packets will access the system with one queue respectively.

When a PU packet is being transmitted on the channel, we assume no other packet (a PU packet or an SU packet) can access the channel until the transmission of this PU packet is complete.

A PU packet that cannot access the channel enters the PU buffer. A newly arriving PU packet can wait in the PU buffer to be transmitted. However, a newly arriving PU packet has to leave the system when the PU buffer is full.

Considering the higher priority of PU packets, a PU packet can interrupt an SU packet's transmission. The SU buffer is prepared for both of the newly arriving SU packets and the interrupted SU packets. We assume that an interrupted SU packets can return to the SU buffer and wait for the next transmission.

The intervals for the PU packets and the SU packets' arrivals are supposed to follow geometric distributions with arrival rates λ_1 and λ_2, where $\bar{\lambda}_1 = 1 - \lambda_1$ ($\bar{\lambda}_2 = 1 - \lambda_2$).

The transmission times for the PU packets and the SU packets are supposed to follow geometric distributions with transmission rates μ_1 and μ_2, where $\bar{\mu}_1 = 1 - \mu_1$ ($\bar{\mu}_2 = 1 - \mu_2$).

We introduce two notations to denote different packet numbers at the instant $t = n^+$ as follows:

$S_n \triangleq$ the number of SU packets in the system,
$P_n \triangleq$ the number of PU packets in the system.

Based on the behaviors of the two kinds of packets in the system, we can build a discrete-time Markov chain $\{S_n, P_n\}$. The state space Ω for $\{S_n, P_n\}$ can be given as follows:

$$\Omega = \{(i, j) : 0 \leq i < \infty, 0 \leq j \leq K + 1\}. \tag{1}$$

3   Model Analysis

In this section, based on the Markov chain model built in Sect. 2, we derive the system steady-state distribution.

In order to analyze the system model, we firstly define the system level as the number of SU packets in the system, and further define the system stage as the number of PU packets in the system.

From the transitions of the system level and stage, we show the state transition probability matrix $\boldsymbol{Q}$ of $\{S_n, P_n\}$ as follows:

$$Q = \begin{pmatrix} C_0 & B_0 & & & & \\ D & C & B & & & \\ & D & C & B & & \\ & & \ddots & \ddots & \ddots & \\ & & & D & C & B \\ & & & & \ddots & \ddots & \ddots \end{pmatrix}. \tag{2}$$

As shown in Eq. (2), the state transition probability matrix $\boldsymbol{Q}$ presents as a block structure. Each non-zero block in $\boldsymbol{Q}$ is defined as follows:

(1) $\boldsymbol{C}_0$ is a $(K+2) \times (K+2)$ block matrix when the number of SU packets in the system is fixed at 0.

$$\boldsymbol{C}_0 = \begin{pmatrix} \bar{\lambda}_1\bar{\lambda}_2 & \lambda_1\bar{\lambda}_2 & & & \\ \bar{\lambda}_1\bar{\lambda}_2\mu_1 & \bar{\lambda}_2(\bar{\lambda}_1\bar{\mu}_1 + \lambda_1\mu_1) & \lambda_1\bar{\lambda}_2\bar{\mu}_1 & & \\ \ddots & & \ddots & & \ddots \\ & & \bar{\lambda}_1\bar{\lambda}_2\mu_1 & \bar{\lambda}_2(\bar{\lambda}_1\bar{\mu}_1 + \lambda_1\mu_1) & \lambda_1\bar{\lambda}_2\bar{\mu}_1 \\ & & & \bar{\lambda}_1\bar{\lambda}_2\mu_1 & \bar{\lambda}_2(\bar{\mu}_1 + \lambda_1\mu_1) \end{pmatrix}. \tag{3}$$

(2) $\boldsymbol{B}_0$ is a $(K+2) \times (K+2)$ block matrix when the number of SU packets in the system changes from 0 to 1.

$$\boldsymbol{B}_0 = \begin{pmatrix} \bar{\lambda}_1\lambda_2 & \lambda_1\lambda_2 & & & \\ \bar{\lambda}_1\lambda_2\mu_1 & \lambda_2(\bar{\lambda}_1\bar{\mu}_1 + \lambda_1\mu_1) & \lambda_1\lambda_2\bar{\mu}_1 & & \\ \ddots & & \ddots & & \ddots \\ & & \bar{\lambda}_1\lambda_2\mu_1 & \lambda_2(\bar{\lambda}_1\bar{\mu}_1 + \lambda_1\mu_1) & \lambda_1\lambda_2\bar{\mu}_1 \\ & & & \bar{\lambda}_1\lambda_2\mu_1 & \lambda_2(\bar{\mu}_1 + \lambda_1\mu_1) \end{pmatrix}. \tag{4}$$

(3) $\boldsymbol{D}$ is a $(K+2) \times (K+2)$ block matrix when the number of SU packets in the system is reduced by one.

$$\boldsymbol{D} = \begin{pmatrix} \bar{\lambda}_1\bar{\lambda}_2\mu_2 & \lambda_1\bar{\lambda}_2\mu_2 & 0 & \cdots & 0 \\ 0 & 0 & 0 & \cdots & 0 \\ \vdots & \vdots & \vdots & & \vdots \\ 0 & 0 & 0 & \cdots & 0 \end{pmatrix}. \tag{5}$$

(4) $\boldsymbol{C}$ is a $(K+2) \times (K+2)$ block matrix when the number of SU packets in the system is fixed.

$$\boldsymbol{C} = \begin{pmatrix} \bar{\lambda}_1(\bar{\lambda}_2\bar{\mu}_2 + \lambda_2\mu_2) & \lambda_1(\bar{\lambda}_2\bar{\mu}_2 + \lambda_2\mu_2) & & & \\ \bar{\lambda}_1\bar{\lambda}_2\mu_1 & \bar{\lambda}_2(\bar{\lambda}_1\bar{\mu}_1 + \lambda_1\mu_1) & \lambda_1\bar{\lambda}_2\bar{\mu}_1 & & \\ \ddots & & \ddots & & \ddots \\ & & \bar{\lambda}_1\bar{\lambda}_2\mu_1 & \bar{\lambda}_2(\bar{\lambda}_1\bar{\mu}_1 + \lambda_1\mu_1) & \lambda_1\bar{\lambda}_2\bar{\mu}_1 \\ & & & \bar{\lambda}_1\bar{\lambda}_2\mu_1 & \bar{\lambda}_2(\bar{\mu}_1 + \lambda_1\mu_1) \end{pmatrix}. \tag{6}$$

(5) $\boldsymbol{B}$ is a $(K+2) \times (K+2)$ block matrix when the number of SU packets in the system is increased by one.

$$\boldsymbol{B} = \begin{pmatrix} \bar{\lambda}_1\lambda_2\bar{\mu}_2 & \lambda_1\lambda_2\bar{\mu}_2 & & & \\ \bar{\lambda}_1\lambda_2\mu_1 & \lambda_2(\bar{\lambda}_1\bar{\mu}_1 + \lambda_1\mu_1) & \lambda_1\lambda_2\bar{\mu}_1 & & \\ \ddots & & \ddots & & \ddots \\ & & \bar{\lambda}_1\lambda_2\mu_1 & \lambda_2(\bar{\lambda}_1\bar{\mu}_1 + \lambda_1\mu_1) & \lambda_1\lambda_2\bar{\mu}_1 \\ & & & \bar{\lambda}_1\lambda_2\mu_1 & \lambda_2(\bar{\mu}_1 + \lambda_1\mu_1) \end{pmatrix}. \tag{7}$$

From each non-zero block in the transition probability matrix Q, we conclude that the two-dimensional Markov chain $\{S_n, P_n\}$ follows a Quasi Birth and Death (QBD) process [12].

By referencing [13], we note that the stability condition for the system model is the average arrival rate being less than the average service rate. Moreover, by referencing the analysis results presented in [14], the stability condition of the system model can be calculated as follows:

$$\lambda_2 - \mu_2 \left[1 + \frac{\lambda_1}{\bar{\lambda}_1 \mu_1} \frac{1 - \alpha^{K+1}}{1 - \alpha} \right]^{-1} < 0 \tag{8}$$

where $\alpha = \lambda_1 \bar{\mu}_1 (\bar{\lambda}_1 \mu_1)^{-1}$.

Let $\pi_{i,j}$ be the steady-state distribution of the two-dimensional Markov chain, where $\pi_{i,j} = \lim_{n \to \infty} P\{S_n = i, P_n = j\}$. By using the Matrix-Geometric Solution method and the recursive algorithm shown in [9], we can obtain the system steady-state distribution $\pi_{i,j}$.

4   Performance Measures

In this section, considering the finite buffer setting for the PU packets, with the steady-state distribution $\pi_{i,j}$ obtained in Sect. 3, we give some performance measures for the PU packets and the SU packets, respectively.

4.1   Performance Measures of PU Packets

The average queue length E_{PU} of the PU packets is the average number of PU packets in the system when in the steady state. The number of PU packets in the system is closely related to the buffer capacity K of the PU packets. So the average queue length E_{PU} of the PU packets can be given as follows:

$$E_{PU} = \sum_{i=0}^{\infty} \sum_{j=0}^{K+1} j\pi_{i,j}. \tag{9}$$

The blocking rate β_{PU} of the PU packets is the number of PU packets that are blocked per slot due to the overflow of the PU buffer. Considering the capacity K of the PU buffer, the blocking rate β_{PU} of the PU packets is given as follows:

$$\beta_{PU} = \bar{\mu}_1 \lambda_1 \sum_{i=0}^{\infty} \pi_{i,K+1}. \tag{10}$$

The throughput θ_{PU} of the PU packets is defined as the probability that one PU packet can be transmitted completely. Considering the priority of the PU packets in cognitive radio networks, a PU packet can be transmitted completely so long as it can access the system without being blocked. Therefore, the throughput θ_{PU} of the PU packets can be given as follows:

$$\theta_{PU} = \frac{\lambda_1 - \beta_{PU}}{\lambda_1}. \tag{11}$$

4.2   Performance Measures of SU Packets

The average queue length E_{SU} of the SU packets is the number of SU packets per slot in the steady state. The average queue length E_{SU} of the SU packets can be given as follows:

$$E_{SU} = \sum_{i=0}^{\infty} \sum_{j=0}^{K+1} i \pi_{i,j}. \tag{12}$$

The interrupted rate γ_{SU} of the SU packets is the number of interrupted SU packets in the system per slot. The interrupted rate γ_{SU} of the SU packets can be given as follows:

$$\gamma_{SU} = \sum_{i=1}^{\infty} \pi_{i,0} \bar{\mu}_2 \lambda_1. \tag{13}$$

The average delay δ_{SU} of the SU packets is the average time period for an SU packet sojourning in the system. By using Little's law [15], δ_{SU} can be given as follows:

$$\delta_{SU} = \frac{E_{SU}}{\lambda_2} \tag{14}$$

where E_{SU} is the average queue length of the SU packets defined in Eq. (12).

5   Numerical Results

We present the numerical results in this section to show how the capacity of the PU buffer influences the system performance. In the following numerical results, $K = 0$ indicates the case where the buffer setting for the PU packets is set at zero, and as an example, $K = 2$ indicates the case where there is a finite buffer setting for the PU packets.

Figure 2 demonstrates the average queue length E_{PU} of the PU packets versus the arrive rate λ_1 of the PU packets.

From Fig. 2, we observe that as the PU packet arrival rate λ_1 increases or the PU packet transmission rate μ_1 deceases, the average queue length E_{PU} of the PU packets will increase. We know that as the PU packet arrival rate increases or the PU packet transmission rate deceases, more PU packets will wait in the system to be transmitted. As a result, the average queue length of the PU packets will increase.

Besides this, from Fig. 2, we see that compared with the zero buffer setting case ($K = 0$), the average queue length E_{PU} is higher for $K = 2$. It is obvious that a greater PU buffer can accommodate more PU packets, and the average queue length of the PU packets will be correspondingly higher.

We examine the change trend for the throughput θ_{PU} for the PU packets in Fig. 3.

In Fig. 3, we conclude that as the arrival rate λ_1 of the PU packets increases, the throughput θ_{PU} of the PU packets decreases. The reason is that the higher

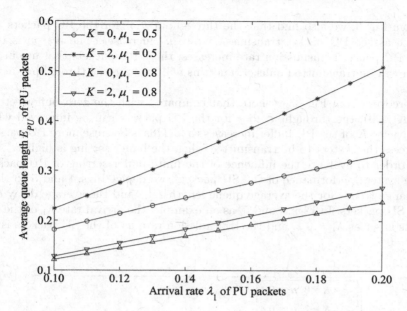

Fig. 2. Average queue length E_{PU} of PU packets.

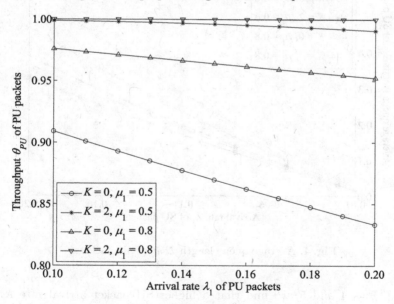

Fig. 3. Throughput θ_{PU} of PU packets.

the PU packet arrival rate is, the higher the possibility for the PU packets being blocked, then the possibility for the PU packets being transmitted completely will decrease. As a result, a lower throughput of the PU packets can be realized.

From Fig. 3, we also find that the throughput θ_{PU} for the PU packets will increase as the PU packet transmission rate μ_1 increases. The reason is that as the PU packet transmission rate increases, those PU packets that are in the system will be transmitted quicker, and this will increase the throughput of PU packets.

Moreover, from Fig. 3, we note that compared with the zero buffer setting case ($K = 0$), the throughput θ_{PU} for the PU packets can be improved when the capacity K of the PU buffer increases to 2. This is because more PU packets can access the system to be transmitted when the buffer setting is higher.

In order to evaluate the influence of the finite buffer setting of PU packets on the system performance of the SU packets, we depict Figs. 4 and 5 to show the change trends for the average queue length E_{SU} and the average delay δ_{SU} of the SU packets. In Figs. 4 and 5, as an example, the arrival rate λ_1 of the PU packets is set as $\lambda_1 = 0.1$, and the transmission rate μ_2 of the SU packets is set as $\mu_2 = 0.8$.

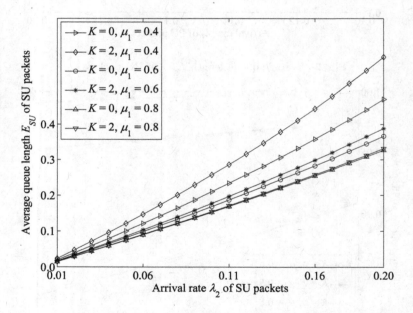

Fig. 4. Average queue length E_{SU} of SU packets.

From Figs. 4 and 5, we find that a higher SU packet arrival rate λ_2 can increase both the average queue length E_{SU} and the average delay δ_{SU} of the SU packets. We know that as the SU packet arrival rate increases, considering the infinite buffer setting for SU packets, more SU packets can access and wait in the SU buffer. Obviously, this will increase the average queue length and the average delay of the SU packets.

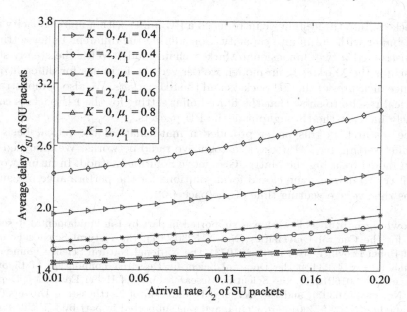

Fig. 5. Average delay δ_{SU} of SU packets.

On the other hand, from Figs. 4 and 5, we find that the average queue length E_{SU} and the average delay δ_{SU} of the SU packets will decrease as the PU packet transmission rate μ_1 increases. The reason is that as the PU packet transmission rate increases, the transmissions of the PU packets will be quicker and the possibility for the SU packets to be transmitted will be higher, and this will decrease the average queue length and the average delay of the SU packets.

From Figs. 4 and 5, we can also note that compared with the zero buffer setting case ($K = 0$), the average queue length E_{SU} and the average delay δ_{SU} of the SU packets will increase when the capacity K of the PU buffer increases to 2. This is because the greater the PU buffer capacity is, and the more the PU packets there are in the system, then the possibility for the SU packets being transmitted will be lower, and so larger numbers of SU packets have to sojourn in the SU buffer. As a result, both the average queue length and the average delay of the SU packets will be increased.

From Figs. 2, 3, 4 to 5, we find that compared with the zero buffer setting, the throughput of the PU packets can be improved by setting a finite PU buffer capacity. However, the average queue length of the PU packets and the SU packets, and the average delay of the SU packets will increase. Therefore, in practice, cognitive radio networks should set different PU buffer capacities according to different network environments.

6 Conclusions

In order to reduce possible packet loss of the primary users (PUs) in cognitive radio networks, this paper assumed there was a buffer with a finite capacity for

PU packets. Into the system we introduced a PU buffer with a finite capacity and an SU buffer with an infinite capacity. Considering the different buffer settings, we constructed a two-dimensional Markov chain model. With the steady-state analysis for the Markov chain model, we derived the formulas for different performance measures of the PU packets and the SU packets. We also demonstrated numerical results to show that the finite buffer setting for the PU packets could intensely increase the throughput of the PU packets.

The research in this paper provided a mathematical theoretical basis for the buffer setting for PU packets in cognitive radio networks. We analyzed the system model by using the Matrix-Geometric Solution method. In future works, we will try to derive some closed-form solutions for the performance measures, such as the average waiting time of the packets.

Acknowledgments. This work was supported in part by the Fundamental Research Funds for the Central Universities (No. N152303007), the Natural Science Foundation of Hebei Province (No. F2016501073), the National Natural Science Foundation of China (No. 61472342), the Doctoral Scientific Research Foundation of Liaoning Province (No. 201601016), the Scientific Research Fund of Hebei Education Department (No. QN2016307), and the Doctoral Foundation of Northeastern University at Qinhuangdao (No. XNB201606), China, and was supported in part by MEXT, Japan.

References

1. Zhao, Q., Sadler, B.M.: A survey of dynamic spectrum access. IEEE Signal Process. Mag. **24**, 79–89 (2007)
2. Marinho, J., Monteiro, E.: Cognitive radio: survey on communication protocols, spectrum decision issues, and future research directions. Wirel. Netw. **18**, 147–164 (2012)
3. Naeem, M., Anpalagan, A., Jaseemuddin, M., Lee, D.C.: Resource allocation techniques in cooperative cognitive radio networks. IEEE Commun. Surv. Tutor. **16**, 729–744 (2014)
4. Benmammar, B., Amraoui, A., Krief, F.: A survey on dynamic spectrum access techniques in cognitive radio networks. Int. J. Commun. Netw. Inf. Secur. **5**, 68–79 (2013)
5. Wang, B., Liu, K.J.R.: Advances in cognitive radio networks: a survey. IEEE J. Sel. Top. Signal Process. **5**, 5–23 (2011)
6. Tragos, E.Z., Zeadally, S., Fragkiadakis, A.G., Siris, V.A.: Spectrum assignment in cognitive radio networks: a comprehensive survey. IEEE Commun. Surv. Tutor. **15**, 1108–1135 (2013)
7. Zhang, Z., Long, K., Wang, J.: Self-organization paradigms and optimization approaches for cognitive radio technologies: a survey. IEEE Wirel. Commun. **20**, 36–42 (2013)
8. Saleem, Y., Rehmani, M.H.: Primary radio user activity models for cognitive radio networks: a survey. J. Netw. Comput. Appl. **43**, 1–16 (2014)
9. Jin, S., Chen, S., Zhang, J.: Social optimization and pricing policy in cognitive radio networks with an energy saving strategy. Mob. Inf. Syst. **2016**, 1–10 (2016). Article ID 2426580

10. Li, H., Han, Z.: Socially optimal queuing control in cognitive radio networks subject to service interruptions: to queue or not to queue? IEEE Trans. Wirel. Commun. **10**, 1656–1666 (2011)

11. Asheralieva, A., Miyanaga, Y.: Joint bandwidth and power allocation for LTE-based cognitive radio network based on buffer occupancy. Mob. Inf. Syst. **2016**, 1–23 (2016). Article ID 6306580

12. Alfa, A.S.: Queueing Theory for Telecommunications: Discrete Time Modelling of a Single Node System. Springer, New York (2010). https://doi.org/10.1007/978-1-4419-7314-6

13. Do, C.T., Tran, N.H., Nguyen, M.V., Hong, C.S.: Social optimization strategy in unobserved queueing systems in cognitive radio networks. IEEE Commun. Lett. **16**, 1944–1947 (2012)

14. Tian, N., Xu, X., Ma, Z.: Discrete-Time Queueing Theory. Science Press, Beijing (2008)

15. Tian, N., Zhang, G.: Vacation Queueing Models: Theory and Applications. Springer, New York (2006). https://doi.org/10.1007/978-0-387-33723-4

Dynamic Lot-Sizing in Sequential Online Auction Systems with Reserve Price

Shuren Liu[1]([✉]), Licai Lei[2], and Pei Tang[1]

[1] School of Mathematics and Computational Science, Xiangtan University,
Xiangtan 411105, China
`liushuren@xtu.edu.cn`, `735389365@qq.com`
[2] Business School, Xiangtan University, Xiangtan 411105, China
`leilicai0735@126.com`

Abstract. We study an optimal inventory scrapping and allocation problem for a retailer who uses sequential online auctions with reserve price. It is assumed that the buyers arrive according to a Poisson process or constant case with uniform bid distributions and multi-unit Vickrey auction mechanism is employed in each auction. We build a Markov decision process model for the retailer's lot-size decision in these auctions with reserve price. It is shown that the single-auction expected revenue function satisfies a second order condition, leading to the retailer's optimal lot-size policy. That is, the optimal inventory-scraping policy is a threshold policy and the optimal lot-size policy is a monotone staircase with unit jumps policy when we introduce the reserve price in sequential online auction systems. We also show these results for the case where the reserve price is a decision variable.

Keywords: Auctions/bidding · Reserve price · Lot size
Markov decision processes

1 Introduction

With the rise of electronic commerce, sequential online auctions of identical items are increasingly used as a viable mechanism for pricing goods in retail businesses. These auctions are conducted either on the seller's own website such as Dell and Sam's Club, or through auction giant eBay.com and other similar firms such as Ubid. On the other hand, large retailers also use sequential online auctions as an inventory clearing tool which combines scrapping excess inventories and lot-size decisions, such as Overstock. In this paper, we study the optimal inventory-scraping and lot-size decisions for an online retailer who scraps, allocates, and sells homogeneous goods through sequential online auctions with reserve price.

Lot-sizes are one of the key decision variables in revenue management and sequential online auctions, which mean the number of units to be allocated in

The project was supported in part by National Natural Science Foundation of China (Grant No. 71401150).

W. Yue et al. (Eds.): QTNA 2017, LNCS 10591, pp. 180–195, 2017.
https://doi.org/10.1007/978-3-319-68520-5_12

each period. Lot-sizes in revenue management represent seat inventory control. The research on seat inventory control, pioneered by [1], has received a good deal of attention. We refer interested readers to [2] for a comprehensive review. Recently, Kim [3] examines the impact of customer buying behavior on the optimal airline seat inventory allocation decisions and develops an efficient heuristic algorithm to reduce computation time. Wen et al. [4] examine dynamic capacity management with uncertain demand and dynamic price. By using Markov decision processes, they show the optimality of a so-called save-up-to level policy. That is, it is optimal to allocate as much as the save-up-to level to future periods if possible in each period.

In sequential online auctions, a small lot size in each auction may result in bidder competition thus increasing the clearing-price. But the number of units may remain too high and the future holding costs will increase, which may also decrease the total revenue. A large lot size in each auction may cause some negative effects and fail due to the insufficient demand and the lower price of units auctioned off. So there is a problem for the retailer to allocate the optimal amount of units to each auction so as to maximize the total revenue. There is a wide variety of work on sequential online auctions with lot-size decisions. Segev et al. [5] first analyze sequential online auctions. They focus on how many units to optimally offer in each auction, formulating the optimization problem as a dynamic programming and showing a solution using existing data taken from Onsale. Vulcano et al. [6] consider sequential online auctions for selling a fixed quantity of a product. They prove that dynamic variants of the first-price and second-price auction mechanisms maximize the seller's expected revenue, conditioned on that the number of units auctioned off is determined at the end of the auction after all buyers' bids are submitted. Du et al. [7] consider a similar problem. They assume that the seller determines the number of units allocated to each auction at the beginning of the auction, but the number of units auctioned off in each auction is determined by arriving buyers' bids and the reserve price. Chou and Parlar [8] study an optimal quota allocation for a revenue-maximizing auction holder facing a random number of bidders. Tripathi, et al. [9] study the optimal lot size policies for sequential online auctions under the restrictions: a fixed number of participating bidders in each auction, uniform bid distribution, and the use of multi-unit Dutch auction mechanism. Assuming that the lot size does not change over time, they derive a simple close-form lot-size expression that resembles the well-known Economic Order Quantity (EOQ) formula in inventory management.

There are two papers that have studied inventory-scrapping and lot-size decisions in sequential online auctions. Pinker et al. [10] assume a fixed number of participating bidders with uniform bid distributions in each auction and employ multi-unit Vickrey auction mechanism. By using a deterministic dynamic programming, they show that it is optimal to scrap inventory only one time before beginning the sequential online auctions, and derive a closed-form lot-sizing policy. Furthermore, Chen et al. [11] build a stochastic dynamic programming model for the seller's inventory-scrapping and lot-size decisions in sequential online auctions. The model incorporates a random number of participating bidders in each auction, allows for any bid distribution, and is not restricted to any specific price-determination mechanism. When the single-auction expected revenue function satisfies a second order condition, they show that the optimal inventory-scraping policy is a scrap-down-to level policy and the optimal lot-sizing policy is a monotone staircase with unit jumps policy.

In general, the retailer sets a reserve price on the units to ensure his profit and prohibit collusion from buyers. The reserve price set by the retailer may be either private, or public (i.e., announced on the website). Rosenkranz and Schmitz [12] provide theoretical explanations for when reserve prices should be public or secret. Choi et al. [13] present evidence from 260,000 online auctions of second-hand cars to identify the impact of public reserve prices on auction outcomes. Meanwhile, there is a lot of literature that has studied the sequential online auctions with the reserve prices. Caillauda and Mezzetti [14] study the equilibrium reserve prices in sequential ascending auctions. Vulcano et al. [6] present a sequential online auction model for selling a fixed quantity of a product. Each unit has a hidden reserve price that depends on the inventory on hand. The model in [6] does not consider inventory costs or scrapping but is further generalized to include inventory holding and ordering with no fixed setup costs in [15] and with fixed setup costs in [16]. Recently, Ghate [17] studies the optimal minimum bids (i.e., the optimal public reserve price) and inventory scrapping in sequential, single-unit, Vickey auctions with demand learning. In contrast to our paper, there is no lot-size decisions in [17] due to a single-unit auction in each period.

The basic setting in our paper is similar to [11] in that we consider a seller who conducts a sequence of online auctions of retail goods. However, we incorporate the seller's reserve price in each auction in contrast to their work. Chen et al. [11] point out "Unfortunately, simultaneous dynamic optimization of multiple design variables such as scrapped inventory, lot-size, and minimum bid, and especially a structural analysis of the corresponding optimal policies, will be difficult. It may be viable under more restrictive assumptions and should provide an interesting avenue for future research" (page 264). In this paper, we try on it. For the Poisson arrival process or constant arrival case with uniform bid distributions, we show that a threshold inventory-scrapping policy, and a monotone staircase with unit jumps lot-sizing policy are optimal. We also show these results for the case where the reserve price is a decision variable.

The remainder of this paper is organized as follows. In Sect. 2, we introduce the basic model with its notations and build a Markov decision process model for the retailer's lot-size decisions in sequential online auctions with reserve price. In Sect. 3, we discuss the multi-unit Vickery auction mechanism and give the single-auction expected revenue function. In Sect. 4, we show that the single-auction expected revenue function satisfies a second order condition and obtain the retailer's optimal inventory-scraping and lot-size policy. Finally, Sect. 5 is concluding section.

2   The Basic Model

Suppose that a retailer has some initial identical items on hand. He conducts non-overlapping sequential online auctions, i.e., auctions one after another on the website. We call each auction a period auction. The duration of each period auction, t_0, is pre-determined. Each period auction is a multi-unit auction with reserve price v. As in [11], at the beginning of each period, the retailer determines how many units to scrap from his stocks on hand and how many units to offer for a period auction. Therefore, at the beginning of auction t with inventory level i on hand, the retailer must make two decisions: (1) scrapping y units for a value $s \geq 0$ per unit; (2) determining lot size x from the remaining units $i - y$ for auction t.

In online auctions, bidders generally arrive stochastically, one after another. So their arrivals form a stochastic point process. For simplicity, it is often assumed to be a Poisson process, which is checked statistically and is true in many cases (see [18, 19] for instance). Bidders arrive according to a Poisson process with rate λ, and each bidder is risk-neutral. Moreover, each bidder wishes to purchase at most one unit and she has a valuation on each unit. This valuation is private and symmetric, that is, each bidder knows her own valuation deterministically, yet only knows other bidders' valuations as random variables which are drawn independently from the same distribution function $F(.)$. We call this type of valuation independent and private valuation, IPV for short. Analytical literature in auctions uses uniform bid distributions (see [11, 20] for instance), we follow this trend that it is assumed that $F(.)$ is the uniform distribution function on an interval $[\underline{v}, \overline{v}]$.

For multi-unit auctions, the main two mechanisms are discriminatory auctions (i.e., first-price auctions) and Vickery auctions (i.e., second-price auctions). It is assumed that each period auction has an exogenous reserve price v, which is announced at the beginning of the auction. So, each arriving buyer whose valuation is greater than v will submit a bid. For details, in a multi-unit auction with lot size x, each of the x highest bidders will win a unit if her bid is greater than the reserve price; other buyers will lose. Hence, all buyers will wait until the end of the auction. The winners are determined and get units after knowing all bids. The price a winner pays is her own bid for the unit in discriminatory auctions, while it is the maximum of the reserve price v and the $(x + 1)$-th bid in Vickery auctions. Here, we let the $(x + 1)$-th bid be zero if there is no $(x + 1)$-th bid. The retailer's revenue from such an auction is the sum of all winners' payments. Weber [21] shows that the retailer will get the same expected revenue under these two auction mechanisms. This is the well-known Revenue Equivalence Theorem in the multi-unit auctions, which is also supported by some empirical research (see [22] for instance). It is apparent that this result is still true in our setting where bidders arrive sequentially and stochastically.

Furthermore, we assume that buyers across auctions are independent, which excludes repeat bidders. This assumption is reasonable if unsuccessful bidders are impatient and simply leave to buy elsewhere and if bidders do not "wait around" to bid in future auctions (see [11, 15, 16] for instance). On the other hand, Krishna [23] points out that relaxing this assumption would make the analysis more complicated for sequential auctions. Pinker et al. [10] and Goes et al. [24] include repeat bidders for sequential auctions, but behavior of repeat bidders in later auctions is not yet well-understood.

The total expected profit of the retailer from the sequential auctions is the (discounted) sum of the expected profit gained from each period auction. The retailer's objective is to maximize his total expected profit.

The decision epoch is defined as the beginning of each period auction. The notations are given as follows.

t: index for horizons;

i: state variable denoting inventory level at the beginning of a period auction;

y: decision variable denoting the number of units to be scrapped;

x: decision variable denoting the lot size offered to the period auction, $0 \leq x \leq i - y$;

h: unit holding cost per period auction; we assume that units are assigned to the winners at the end of the auction, and so the total holding cost is $h(i - y)$ for a period auction with state i and scrapping units y;

s: the value per unit for scrapped inventory;

v: the reserve price set by the retailer for all period auctions;

$\alpha \in (0, 1]$: one period discounted factor;

q_m: probability of exactly m bidders arriving in one period auction whose bids are larger than or equal to the reserve price v;

$\pi(x)$: expected revenue gained by the retailer in one period auction with the reserve price v as a function of the lot size x. The problem how to compute $\pi(x)$ will be studied in Sect. 3 under the multi-unit Vickery auctions.

Note that both the probability q_m and revenue function $\pi(x)$ depend on the number of arriving buyers, the distribution function $F(.)$, and the reserve price v set by the retailer. We will see this point in the next section.

As in [11], for the auction lot size x, let

$$\partial \Pi(x) \overset{\Delta}{=} \frac{\Delta \pi(x)}{\sum\limits_{m=x+1}^{\infty} q_m},$$

where the numerator $\Delta \pi(x) = \pi(x + 1) - \pi(x)$ is the first difference of $\pi(x)$, which represents the marginal single-auction expected revenue; the denominator $\sum\limits_{m=x+1}^{\infty} q_m$ represents the probability that the auction is successful due to a sufficient number of bidders participating. Furthermore, let

$$\partial^2 \Pi(x) = \partial \Pi(x + 1) - \partial \Pi(x).$$

We introduce the following condition in [11] on the single-auction revenue function.

Condition A. Let $\widehat{x}$ be the smallest maximizer of $\phi(.)$. Suppose

$$\partial^2 \Pi(x) \leq 0, \, for \, x \in \{0, 1, \cdots, \widehat{x} - 2\}.$$

Condition A implies $\pi(.)$ is concave and unimodal by Lemmas 2.2 and 2.4 in [11].

Let $\xi_x = min\{x, m\}$, where x is lot size and m is the number of bidders arriving in one period auction whose bids are larger than or equal to the reserve price v. Thus, when $m > x$, $\xi_x = x$ represents that the period auction is successful; when $m \leq x$, $\xi_x = m$ represents that the period auction fails. Let $V_t(i)$ denote the profit-to-go function at the beginning of auction t with the inventory level i on hand. Therefore, by Markov decision processes (see [25] for instance), $V_t(i)$ satisfies the following optimality equation:

$$V_t(i) = \max_{\substack{0 \leq y \leq i \\ 0 \leq x \leq i-y}} [sy - h(i - y) + \pi(x)$$
$$+ \alpha E[V_{t+1}(i - y - \xi_x)]], i \geq 1, 1 \leq t \leq T, \tag{1}$$

with the boundary condition $V_{T+1}(i) = 0$ for $i \geq 0$ and $V_t(0) = 0$ for all $1 \leq t \leq T$. Though the above optimality equation is the same as the optimality Eq. (3) in [11], it should be noted that both the probability q_m and single-auction revenue function $\pi(x)$ depend on the reserve price v in this paper. The concrete expressions of q_m and $\pi(x)$ will be studied in Sect. 3.

Furthermore, following [11], we rewrite the optimality Eq. (1) as:

$$V_t(i) = \max_{0 \leq y \leq i} [sy + W_t(i - y)], i \geq 1, 1 \leq t \leq T, \tag{2}$$

where,

$$W_t(j) = \max_{0 \leq x \leq j} [-hj + \pi(x) + \alpha E[V_{t+1}(j - \xi_x)]], j \geq 1, 1 \leq t \leq T. \tag{3}$$

Note that $W_t(0) = 0$ for all $1 \leq t \leq T$.

3   Multi-unit Vickery Auctions

In this section, we study bidding behavior and the retailer's revenue function under a multi-unit Vickery auction with lot-size x and reserve price v.

For online auctions on eBay, Ockenfels and Roth [26] study late and multiple bidding in second-price auctions and point out that bidders often submit bids very close to or just at the end of auctions. Further, Peters and Severinov [27] reach the conclusion of a truth revealing bidding behavior for a stylized eBay auction setting with multiple simultaneous auctions. Dang et al. [20] obtain the dominant strategy for buyers is to report their true valuations under IPV for second-price online auctions. Thus, we adopt multi-unit Vickery auctions to determine a winning price for each auction (see also [28] for instance).

Suppose that bidders arrive according to a Poisson process with arrival rate λ during the auction with length of t_0; and each bidder is risk-neutral. Moreover, each bidder wishes to purchase at most one unit and she has a valuation on each unit. Bidders' valuations are IPV with the uniform distribution function $F(\cdot)$ on an interval $[\underline{v}, \overline{v}]$. It is noted the bidders will bid her true valuation of the product when the auction mechanism is the multi-unit Vickery auction. At the end of the auction, each of the x highest bidders will win a unit if her bid is greater than the reserve price. The price a winner pays is the maximum of the reserve price and the $(x+1)$-th bid under multi-unit Vickery auctions. Here, we let the $(x+1)$-th bid be zero if there is no $(x+1)$-th bid.

Under the Vickery auction mechanism, it is assumed the the retailer sets a reserve price v. When the retailer announces his reserve price v, the bidders whose intended bids are less than the reserve price v will obviously not bid and leave. The bidders who believe that the value of the unit is more than v will be willing to place a bid. We call bidders who arrive and bid "active bidders". Then, by Proposition 4.1 in [7], the active bidders arrive according to a Poisson distribution function with rate $\lambda t_0 \bar{F}(v)$, where $\bar{F}(v) = 1 - F(v)$. Furthermore, let v_i be the valuation of i-th active bidder. It is easy to see that $v_i \in [v, \overline{v}]$, and its distribution function is

$$F_v(z) \triangleq (F(z) - F(v))/\bar{F}(v) = \frac{z - v}{\overline{v} - v}$$

for $v \leq z \leq \overline{v}$. Let $f_v(\cdot)$ be its probability density function.

For any period auction, if we let N denote the number of the arriving active bidders, then the probability of the event "*there are exactly m active bidders whose valuations are over the reserve price v*" is obviously given by

$$q_m = P_{rob}\{N = m\} = \frac{1}{m!}(\lambda t_0 \bar{F}(v))^m e^{-\lambda t_0 \bar{F}(v)}, \ m \geq 0. \tag{4}$$

Suppose that m active bidders participate in the period auction and lot-size x is offered with $x < m$. The bids of m bidders are their own valuations under the multi-unit Vickery auction and so drawn from the probability density $f_v(\cdot)$, and denoted by $v_1, v_2, \cdots, v_m$, respectively. The reverse order statistics of $v_1, v_2, \cdots, v_m$ are denoted by respectively $v_{(1)} \geq v_{(2)} \geq \cdots \geq v_{(m)}$.

Next, we introduce Euler integration to obtain the expected pay-off (price) under the multi-unit Vickery auction. Let $\Gamma(\kappa) = \int_0^\infty e^{-z} z^{\kappa-1} dz$ be the gamma function for any $\kappa \geq 0$. Let $B(\kappa, \gamma) = \int_0^1 z^{\kappa-1}(1-z)^{\gamma-1} dz$ be the beta function evaluated at (κ, γ). It is easy to see that $B(\kappa, \gamma) = \frac{\Gamma(\kappa)\Gamma(\gamma)}{\Gamma(\kappa+\gamma)}$. It is well known (see [29]) that the density function of l-th reverse order statistics $v_{(l)}$ is

$$d_{l|m}(z) = \frac{1}{B(l, m-l+1)} F_v(z)^{m-l}[1 - F_v(z)]^{l-1} f_v(z), \ z \geq 0, \ l = 1, 2, \cdots, m,$$

where $\frac{1}{B(l,m-l+1)} = \frac{m!}{(l-1)!(m-l)!}$.

According to multi-unit Vickery auction principle [21], the active bidders with the private valuations $v_{(1)} \geq v_{(2)} \geq \cdots \geq v_{(x)}$ win x units, and all of them pay a uniform price that equals to the $(x+1)$-th highest bid $v_{(x+1)}$. Therefore, the expected pay-off (price) is

$$E(v_{(x+1)}) = \int_v^{\overline{v}} z d_{x+1|m}(z) dz$$

$$= \int_v^{\overline{v}} z \frac{1}{B(x+1, m-x)} F_v(z)^{m-x-1} [1 - F_v(z)]^x f_v(z) dz$$

$$= \frac{1}{B(x+1, m-x)} \int_v^{\overline{v}} z F_v(z)^{m-x-1} [1 - F_v(z)]^x f_v(z) dz.$$

By taking a variable transformation $u = F_v(z) = \frac{z-v}{\overline{v}-v}$, we obtain

$$E(v_{(x+1)}) = \frac{1}{B(x+1, m-x)} \int_0^1 [u(\overline{v}-v) + v] u^{m-x-1} (1-u)^x du$$

$$= \frac{\overline{v}-v}{B(x+1, m-x)} B(m-x+1, x+1)$$

$$+ v \frac{1}{B(x+1, m-x)} B(m-x, x+1)$$

$$= (\overline{v}-v) \frac{\Gamma(m-x+1)\Gamma(x+1)}{\Gamma(m+2)} \frac{\Gamma(m+1)}{\Gamma(x+1)\Gamma(m-x)} + v$$

$$= v + (\overline{v}-v) \frac{m-x}{m+1}$$

$$= \overline{v} - (\overline{v}-v) \frac{x+1}{m+1}.$$

Hence, the revenue gained by the retailer from the multi-unit Vickery auction with the lot-size x and reserve price v (not including the holding cost) is

$$\pi(x) = \sum_{m=0}^{x} mv q_m + x \sum_{m=x+1}^{\infty} q_m \left[\overline{v} - (\overline{v}-v) \frac{x+1}{m+1} \right]. \tag{5}$$

4   Structural Analysis of Optimal Policies

Chen et al. [11] study an optimal lot-size problem for a retailer who sells a fixed number of units through sequential online auctions. For each period auction without reserve price, they show that Condition A is true under multi-unit Vickery auctions, multi-unit Dutch auctions and Yankee auctions for the Poisson or constant arrival process with uniform bid distributions (Proposition 3.12 in their paper). It should be noted that after checking the proof of results in [11], all their results are true under Condition A, irrespective of the concrete expression of $\pi(x)$. Thus, we will show that Condition A is true under certain conditions in the following.

Theorem 1. *For the Poisson arrival process with uniform bid distributions, Condition A holds under the multi-unit Vickery auction with reserve price v.*

Proof. Under multi-unit Vickery auctions for the Poisson arrival process with uniform bid distributions, the expected revenue gained by the retailer in one period auction with lot-size x and reserve price v is as follows by (5):

$$\pi(x) = \sum_{m=0}^{x} mvq_m + x \sum_{m=x+1}^{\infty} q_m \left(\overline{v} - (\overline{v} - v)\frac{x+1}{m+1} \right).$$

Thus,

$$\Delta\pi(x) = \pi(x+1) - \pi(x)$$

$$= v\left(\sum_{m=0}^{x+1} mq_m - \sum_{m=0}^{x} mq_m \right) + (x+1) \sum_{m=x+2}^{\infty} q_m \left(\overline{v} - (\overline{v} - v)\frac{x+2}{m+1} \right)$$

$$- x \sum_{m=x+1}^{\infty} q_m \left(\overline{v} - (\overline{v} - v)\frac{x+1}{m+1} \right)$$

$$= v(x+1)q_{x+1} + (x+1)\overline{v} \sum_{m=x+2}^{\infty} q_m - x\overline{v} \sum_{m=x+1}^{\infty} q_m$$

$$+ x(x+1)(\overline{v} - v) \sum_{m=x+1}^{\infty} \frac{q_m}{m+1} - (x+1)(x+2)(\overline{v} - v) \sum_{m=x+2}^{\infty} \frac{q_m}{m+1}.$$

Let $Q(x) = 1 - \sum_{m=x+1}^{\infty} q_m$, then $\sum_{m=x+2}^{\infty} q_m = 1 - Q(x+1)$. Note that $q_m = \frac{m+1}{\lambda t_0 \bar{F}(v)} q_{m+1}$ by (4), then

$$\Delta\pi(x) = v(x+1)q_{x+1} + (x+1)\overline{v}\left(1 - Q(x+1)\right) - x\overline{v}\left(1 - Q(x)\right)$$

$$+ \frac{x(x+1)(\overline{v} - v)\left(1 - Q(x+1)\right)}{\lambda t_0 \bar{F}(v}$$

$$- \frac{(x+1)(x+2)(\overline{v} - v)\left(1 - Q(x+2)\right)}{\lambda t_0 \bar{F}(v)}$$

$$= v(x+1)q_{x+1} + (x+1)\overline{v}\left(1 - Q(x) - q_{x+1}\right) - x\overline{v}\left(1 - Q(x)\right)$$

$$+ \frac{(x+1)(\overline{v} - v)}{\lambda t_0 \bar{F}(v)}\left(x(1 - Q(x) - q_{x+1}) \right.$$

$$\left. - (x+2)(1 - Q(x) - q_{x+1} - q_{x+2}) \right)$$

$$= v(x+1)q_{x+1} + \overline{v}\left(1 - Q(x)\right) - \frac{2(x+1)(\overline{v} - v)}{\lambda t_0 \bar{F}(v)}\left(1 - Q(x)\right)$$

$$+ \frac{(x+1)(\overline{v} - v)}{\lambda t_0 \bar{F}(v)}\left((x+2)q_{x+1} + (x+2)q_{x+2} - xq_{x+1} \right) - (x+1)\overline{v}q_{x+1}$$

$$= v(x+1)q_{x+1} + \overline{v}\left(1 - Q(x)\right) - \frac{2(x+1)(\overline{v} - v)}{\lambda t_0 \bar{F}(v)}\left(1 - Q(x)\right)$$

$$+ \frac{(x+1)q_{x+1}}{\lambda t_0 \bar{F}(v)}[2\overline{v} - 2v - v\lambda t_0 \bar{F}(v)].$$

Hence,

$$\partial\Pi(x) = \frac{\Delta\pi(x)}{\sum\limits_{m=x+1}^{\infty} q_m} = \frac{\Delta\pi(x)}{1-Q(x)}$$

$$= \bar{v} - \frac{2(x+1)(\bar{v}-v)}{\lambda t_0 \bar{F}(v)} + \frac{v(x+1)q_{x+1}}{1-Q(x)}$$

$$+ \frac{(x+1)\Big(2\bar{v} - 2v - v\lambda t_0 \bar{F}(v)\Big)}{\Big(1-Q(x)\Big)\lambda t_0 \bar{F}(v)} q_{x+1}$$

$$= \bar{v} - \frac{2(x+1)(\bar{v}-v)}{\lambda t_0 \bar{F}(v)} + \frac{\lambda t_0 \bar{F}(v)v q_x}{1-Q(x)} + \frac{\Big(2\bar{v} - 2v - v\lambda t_0 \bar{F}(v)\Big)}{\Big(1-Q(x)\Big)} q_x.$$

Therefore,

$$\partial^2\Pi(x) = \partial\Pi(x+1) - \partial\Pi(x)$$

$$= \frac{2(x+1)(\bar{v}-v)}{\lambda t_0 \bar{F}(v)} - \frac{2(x+2)(\bar{v}-v)}{\lambda t_0 \bar{F}(v)} + \frac{v\lambda t_0 \bar{F}(v)q_{x+1}}{1-Q(x+1)} - \frac{v\lambda t_0 \bar{F}(v)q_x}{1-Q(x)}$$

$$+ \big(2\bar{v} - 2v - v\lambda t_0 \bar{F}(v)\big)\Big(\frac{q_{x+1}}{1-Q(x+1)} - \frac{q_x}{1-Q(x)}\Big)$$

$$= \frac{2(v-\bar{v})}{\lambda t_0 \bar{F}(v)} + 2(\bar{v}-v)\Big(\frac{q_{x+1}}{1-Q(x+1)} - \frac{q_x}{1-Q(x)}\Big). \tag{6}$$

To prove that $\partial^2\Pi(x) \leq 0$, we define

$$h(x) = \frac{q_x}{1-Q(x)}.$$

Its first difference is $\Delta h(x) = \frac{q_{x+1}}{1-Q(x+1)} - \frac{q_x}{1-Q(x)}$. Thus, it is equivalent to show by (6)

$$\Delta h(x) \leq \frac{1}{\lambda t_0 \bar{F}(v)},$$

that is,

$$LHS = \lambda t_0 \bar{F}(v)\Big(\frac{q_{x+1}}{1-Q(x+1)} - \frac{q_x}{1-Q(x)}\Big) \leq 1.$$

Then, replacing λ of (A.12) in Appendix A in [11] with $\lambda t_0 \bar{F}(v)$ and implementing the same analysis, we have $LHS \leq 1$. Therefore,

$$\partial^2\Pi(x) = \partial\Pi(x+1) - \partial\Pi(x) \leq 0.$$

□

Theorem 1 show that Condition A holds. Thus, we have the following important results by Theorem 2.1 in [11].

Theorem 2. *(1) A threshold inventory-scrapping policy is optimal. That is, for any inventory $i \geq 1$, let $0 \leq i_t^* = \min\{i \mid s \geq W_t(i+1) - W_t(i)\}$, then we scrap all units above the time-dependent threshold i_t^*, and not to scrap any inventory below the threshold i_t^*.*

(2) A monotone staircase with unit jumps lot-sizing policy is optimal. That is, if x is the smallest optimal lot size in post-scraping inventory j, either x or $x+1$ is the smallest optimal lot size in post-scraping inventory $j+1$. Furthermore, the smallest optimal lot size jumps from x to $x+1$ at $j+1 \leq k^* = \min\{k \mid \Delta\pi(x) > \alpha(V_{t+1}(k-x) - V_{t+1}(k-x-1))\}$.

Under the multi-unit Vickery auction with reserve price v, for the Poisson arrival process with uniform bid distributions, we show that both Theorems 1 and 2 are adapted from [11], that is, the reserve price does not play a role in optimal lot-sizing in sequential online auction systems.

In the following, we assume that reserve price v is also an endogenous decision variable for each period auction. For example, in the auctions on eBay, retailers specify reserve prices for their auctions. Then, besides the inventory-scrpping quantity and the amount allocated for the period auction, the retailer has to determine the reserve price. In this case, we write the probability q_m as $q_m(v)$ and the revenue function $\pi(x)$ as $\pi(x,v)$ for allocating lot-size x in one period, respectively. Hence, the optimality equation for the sequential online auctions is as follows:

$$V_t(i) = \max_v V_t(i,v), \tag{7}$$

$$V_t(i,v) = \max_{\substack{0 \leq y \leq i \\ 0 \leq x \leq i-y}} [sy - h(i-y) + \pi(x,v) + \alpha E[V_{t+1}(i-y-\xi_x)]],$$

$$i \geq 1, 1 \leq t \leq T, \tag{8}$$

with the boundary condition $V_{T+1}(i) = 0$ for $i \geq 0$ and $V_t(0) = 0$ for all $1 \leq t \leq T$. Clearly, $V_1(i)$ is the maximum total discounted expected profit and $V_1(i,v)$ is the maximum total discounted expected profit under the current reserve price v, when there are T horizons remaining and the current state is i.

Note that the proof of Theorem 1 does not concern the reserve price, Theorem 1 holds when the reserve price v is an endogenous decision variable for each period auction. Therefore, Theorem 2 also holds with the optimal threshold and lot size depend on the reserve price v.

When buyers arrive according to a general stochastic process with uniform bid distributions, we give the following Remark 1.

Remark 1. We assume that the number N of arriving bidders in one period auction is a random variable with a probability mass function $g(n)$. We first compute q_m, the probability of the event "there are exactly m bidders whose valuations are over reserve price v in one period auction". It is easy to see that

$$q_m = \sum_{n=m}^{\infty} g(n) C_n^m \bar{F}(v)^m F(v)^{n-m}, \quad m = 0, 1, \ldots. \tag{9}$$

Specially, for the constant arrival case, i.e., the number N of arriving bidders in one period auction is a fixed integer, we have

$$q_m = C_N^m \bar{F}(v)^m F(v)^{N-m}, \quad m = 0, 1, \ldots, N. \tag{10}$$

According to Sect. 3, the single-auction revenue gained by the retailer with lot-size x and reserve price v (not including the holding cost) is the same as (5) in form, that is,

$$\pi(x) = \sum_{m=0}^{x} mvq_m + x \sum_{m=x+1}^{\infty} q_m \left[\bar{v} - (\bar{v} - v) \frac{x+1}{m+1} \right],$$

where q_m is defined by (9). The first difference of $\pi(x)$ is

$$\Delta\pi(x) = \pi(x+1) - \pi(x)$$

$$= v \left(\sum_{m=0}^{x+1} mq_m - \sum_{m=0}^{x} mq_m \right) + (x+1) \sum_{m=x+2}^{\infty} q_m \left(\bar{v} - (\bar{v} - v) \frac{x+2}{m+1} \right)$$

$$- x \sum_{m=x+1}^{\infty} q_m \left(\bar{v} - (\bar{v} - v) \frac{x+1}{m+1} \right)$$

$$= (x+1)vq_{x+1} + (x+1) \sum_{m=x+2}^{\infty} q_m \left(\bar{v} - (\bar{v} - v) \frac{x+2}{m+1} \right)$$

$$- x \sum_{m=x+2}^{\infty} q_m \left(\bar{v} - (\bar{v} - v) \frac{x+1}{m+1} \right) - xq_{x+1}(\bar{v} - (\bar{v} - v) \frac{x+1}{x+2})$$

$$= (x+1)vq_{x+1} + \sum_{m=x+2}^{\infty} q_m(\bar{v} - (\bar{v} - v) \frac{2(x+1)}{m+1})$$

$$- xq_{x+1}(\bar{v} - (\bar{v} - v) \frac{x+1}{x+2}).$$

Then,

$$\partial\Pi(x) = \frac{\Delta\pi(x)}{\sum_{m=x+1}^{\infty} q_m}$$

$$= \frac{1}{\sum_{m=x+1}^{\infty} q_m} [(x+1)vq_{x+1} - (\bar{v} - v) \sum_{m=x+1}^{\infty} q_m \frac{2(x+1)}{m+1} - \bar{v}q_{x+1}$$

$$+ \bar{v} \sum_{m=x+1}^{\infty} q_m + (\bar{v} - v)q_{x+1} \frac{2(x+1)}{x+2} - xq_{x+1}\bar{v} + q_{x+1}(\bar{v} - v) \frac{x^2+x}{x+2}]$$

$$= \overline{v} + \frac{1}{\sum\limits_{m=x+1}^{\infty} q_m} [(x+1)vq_{x+1} - (\sum_{m=x+1}^{\infty} q_m \frac{2(\overline{v}-v)(x+1)}{m+1} + q_{x+1}v(x+1))]$$

$$= \overline{v} - \frac{\sum\limits_{m=x+1}^{\infty} q_m \frac{2(\overline{v}-v)(x+1)}{m+1}}{\sum\limits_{m=x+1}^{\infty} q_m}.$$

Furthermore,

$$\partial^2 \Pi(x) = \partial \Pi(x+1) - \partial \Pi(x)$$

$$= \frac{\Delta \pi(x+1)}{\sum\limits_{m=x+2}^{\infty} q_m} - \frac{\Delta \pi(x)}{\sum\limits_{m=x+1}^{\infty} q_m}$$

$$= \frac{\sum\limits_{m=x+1}^{\infty} q_m \frac{2(x+1)(\overline{v}-v)}{m+1}}{\sum\limits_{m=x+1}^{\infty} q_m} - \frac{\sum\limits_{m=x+2}^{\infty} q_m \frac{2(x+2)(\overline{v}-v)}{m+1}}{\sum\limits_{m=x+2}^{\infty} q_m}$$

$$= \frac{2(x+1)(\overline{v}-v) \sum\limits_{m=x+1}^{\infty} \frac{q_m}{m+1} \sum\limits_{m=x+2}^{\infty} q_m}{\sum\limits_{m=x+1}^{\infty} q_m \sum\limits_{m=x+2}^{\infty} q_m}$$

$$- \frac{2(x+2)(\overline{v}-v) \sum\limits_{m=x+2}^{\infty} \frac{q_m}{m+1} \sum\limits_{m=x+1}^{\infty} q_m}{\sum\limits_{m=x+1}^{\infty} q_m \sum\limits_{m=x+2}^{\infty} q_m}$$

$$= \frac{[q_{x+1}(\sum\limits_{m=x+1}^{\infty} q_m - (x+1) \sum\limits_{m=x+1}^{\infty} \frac{q_m}{m+1}) - \sum\limits_{m=x+1}^{\infty} \frac{q_m}{m+1} \sum\limits_{m=x+1}^{\infty} q_m]}{\sum\limits_{m=x+1}^{\infty} q_m \sum\limits_{m=x+2}^{\infty} q_m}$$

$$*2(\overline{v}-v). \tag{11}$$

For a general stochastic arrival process with uniform bid distributions, we can not show that $\partial^2 \Pi(x) \leq 0$ due to the complexity of (11) above and q_m in (9). Specially for the constant arrival case with uniform bid distributions, numerical computations show that $\partial^2 \Pi(x) \leq 0$. Let

$$Z(x,v) = q_{x+1}(\sum_{m=x+1}^{N} q_m - (x+1) \sum_{m=x+1}^{N} \frac{q_m}{m+1})$$

$$- \sum_{m=x+1}^{N} \frac{q_m}{m+1} \sum_{m=x+1}^{N} q_m, \tag{12}$$

where $q_m = C_N^m \bar{F}(v)^m F(v)^{N-m} (m=0,1,\ldots,N)$, $F(v) = \frac{v-\underline{v}}{\overline{v}-\underline{v}}$, $\bar{F}(v) = \frac{\overline{v}-v}{\overline{v}-\underline{v}}$. Note that $Z(x,v)$ is the main part of the numerator in (11) for the constant case

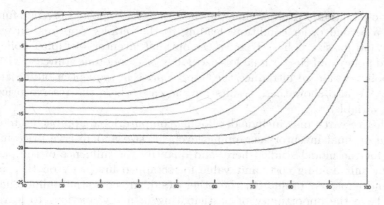

Fig. 1. The image of $Z(x, v)$

with uniform bid distributions. It is assumed that there are $N = 20$ bidders in each period auction. The bids are distributed uniformly on the interval $[10, 100]$, and so $\underline{v} = 10$, $\overline{v} = 100$. For given $x = 1, 2, \cdots, N - 1$ (note that $\sum\limits_{m=x+1}^{N} q_m = \sum\limits_{m=N+1}^{N} q_m$ in (12) is meaningless as $x = N$), the reserve price v discretely fetches on the interval $[10, 100]$, we compute $Z(x, v)$ and obtain Fig. 1. In Fig. 1, the horizontal axis represents the reserve price v; the vertical axis represents the value of $Z(x, v)$. There are 19 curves in Fig. 1 due to $x = 1, 2, \cdots, 19$. It is easy to see that $Z(x, v) \leq 0$, thus $\partial^2 \Pi(x) = \dfrac{2(\overline{v}-v)Z(x,v)}{\sum\limits_{m=x+1}^{N} q_m \sum\limits_{m=x+2}^{N} q_m} \leq 0$. So, Theorem 1 holds for the constant arrival case with uniform bid distributions.

Finally, we give the following Remark 2 on the infinite horizon case.

Remark 2. It is easy to extend all the results in this paper to the infinite horizon case with stationary models, where all parameters are stationary independently of the period index. Then, the optimal inventory-scrapping, lot-size and reserve price decisions for the retailer are all stationary, i.e., irrespective of the period index t. In this case, the retailer will become more convenient to manage his inventory. The details are omitted here.

5   Conclusions

In this paper, we study an optimal inventory-scrapping and lot-size decision problem for a retailer who uses sequential online auctions with reserve price. We build a Markov decision process model for the retailer's problem. It is assumed that the buyers arrive according to a Poisson process or constant case with uniform bid distributions and multi-unit Vickrey auction mechanism is employed in each period auction. We show that the single-auction expected revenue function satisfies a second order condition, leading to the retailer's optimal inventory-scraping and lot-size policy. That is, the optimal inventory-scraping policy is a threshold policy and the optimal lot-size policy is a monotone staircase with unit jumps policy when we introduce the reserve price in sequential online auction systems. We also show these results for the case where the reserve price is a decision variable.

Further research may include developing computational approaches to determine the optimal inventory-scrapping level, lot size and reserve price in each auction for the model studied here, and quantify the influence of the parameters, e.g., unit holding cost, unit value for scrapped inventory on the optimal policy and expected profit. On the other hand, in sequential online auctions, bidders have the opportunity of participating in many auctions to learn and choose the bidding strategy that best fits their preferences. Thus, an extension of our model may include incorporating information acquired in early auctions into lot-size decisions in later auctions through Bayesian updates (see [10, 24, 30] for instance).

References

1. Littlewood, K.: Forecasting and control of passenger booking. AGIFORS Symp. Proc. **12**, 95–117 (1972)
2. Mcgill, J.I., van Ryzin, G.: Revenue management: research overview and prospects. Trans. Sci. **33**, 233–256 (1999)
3. Kim, S.W.: The impact of customer buying behavior on the optimal allocation decisions. Int. J. Prod. Econ. **163**, 71–88 (2015)
4. Wen, X.Q., Xu, C., Hu, Q.Y.: Dynamic capacity management with uncertain demand and dynamic price. Int. J. Prod. Econ. **175**, 121–131 (2016)
5. Segev, A., Beam, C., Shanthikumar, J.: Optimal design of internet-based auctions. Inf. Technol. Manag. **2**, 121–163 (2001)
6. Vulcano, G., van Ryzin, G., Maglaras, C.: Optimal dynamic auctions for revenue management. Manag. Sci. **48**, 1388–1407 (2002)
7. Du, L., Hu, Q.Y., Yue, W.Y.: Analysis and evaluation for optimal allocation in sequential internet auction systems with reserve price. Dyn. Contin. Discret. Implus. Syst. Ser. B: Appl. Algorithms **12**, 617–631 (2005)
8. Chou, F.S., Parlar, M.: Optimal quota allocation for a revenue-maximizing auction holder facing a random number of bidders. Int. Trans. Oper. Res. **12**, 559–580 (2005)
9. Tripathi, A.K., Nair, S.K., Karuga, G.: Optimal lot sizing policies for sequential online auctions. IEEE Trans. Knowl. Data Eng. **21**, 554–567 (2009)
10. Pinker, E., Seidmann, A., Varkrat, Y.: Using bid data for the management of sequential, multi-unit, online auctions with uniformly distributed bidder valuations. Eur. J. Oper. Res. **202**, 574–583 (2010)
11. Chen, X., Ghate, A., Tripathi, A.: Dynamic lot-sizing in sequential online retail auctions. Eur. J. Oper. Res. **215**, 257–267 (2011)
12. Rosenkranz, S., Schmitz, P.W.: Reserve prices in auctions as reference points. Econ. J. **117**, 637–653 (2007)

13. Choi, S., Nesheim, L., Rasul, I.: Reserve price effects in auctions: estimates from multiple regression-discontinuity designs. Econ. Inq. **54**, 294–314 (2016)
14. Caillauda, B., Mezzetti, C.: Equilibrium reserve prices in sequential ascending auctions. J. Econ. Theor. **117**, 78–95 (2004)
15. van Ryzin, G., Vulcano, G.: Optimal auctioning and ordering in an infinite horizon inventory-pricing system. Oper. Res. **52**, 346–367 (2004)
16. Huh, W.T., Janakiraman, G.: Inventory management with auctions and other sales channels: optimality of (s, S) policies. Manag. Sci. **54**, 139–150 (2008)
17. Ghate, A.: Optimal minimum bids and inventory scrapping in sequential, single-unit, vickey auctions with demand learning. Eur. J. Oper. Res. **245**, 555–570 (2015)
18. Pinker, E., Seidmann, A., Varkrat, Y.: Managing online auctions: current business and research issues. Manag. Sci. **49**, 1457–1484 (2003)
19. Etizon, H., Pinker, E.J., Seidmann, A.: Analyzing the simultaneous use of auctions and posted prices for online selling. Manu. Serv. Oper. Manag. **8**, 68–91 (2006)
20. Dang, C.Y., Hu, Q.Y., Liu, J.: Bidding strategies in online auctions with different ending rules and value assumptions. Electron. Commer. Res. Appl. **14**, 101–111 (2015)
21. Weber, R.J.: Multiple-object auctions. In: Engelbrechtp-Wiggans, R., Martin, S., Robert, M.S. (eds.) Auctions, Bidding and Contracting: Uses and Theory, pp. 165–191. New York University Press, New York (1983)
22. Lucking-Reiley, D.: Using field experiments to test equivalence between auction formats: magic on the internet. Am. Econ. Rev. **89**, 1063–1080 (1999)
23. Krishna, V.: Auction Theory. Academic Press, New York (2002)
24. Goes, P., Karuga, G., Tripathi, A.K.: Bidding behavior evolution in sequential auctions: characterization and analysis. Manag. Inf. Syst. Q. **36**, 1021–1042 (2012)
25. Hu, Q.Y., Yue, W.Y.: Markov Decision Processes with Their Applications. Springer, New York (2008). https://doi.org/10.1007/978-0-387-36951-8
26. Ockenfels, A., Roth, A.E.: Late and multiple bidding in second price internet auctions: theory and evidence concerning different rules for ending an auction. Games Econ. Behav. **55**, 297–320 (2006)
27. Peters, M., Severinov, S.: Internet auctions with many traders. J. Econ. Theor. **130**, 220–245 (2006)
28. Jiang, Z.Z., Fang, S.C., Fan, Z.P., Wang, D.W.: Selecting optimal selling format of a product in B2C online auctions with boundedly rational customers. Eur. J. Oper. Res. **226**, 139–153 (2013)
29. David, H.A.: Order Statistics, 2nd edn. Wiley, New York (1981)
30. Lorentziadis, P.L.: Optimal bidding in auctions from a game theory perspective. Eur. J. Oper. Res. **248**, 347–371 (2016)

Queueing Models II

The Priority of Inbound Calls over Outbound Calls Modeled as a Discrete-Time Retrial/Delay System

Rein Nobel[✉] and Maik Dekker

Department of Econometrics and Operations Research, Vrije Universiteit
Amsterdam, Amsterdam, Netherlands
r.d.nobel@vu.nl, m.h.r.dekker@student.vu.nl

Abstract. A one-server discrete-time queueing model is studied with two arrival streams. Both arrival streams are in batches and we distinguish between a stream of low-priority customers, who are put in a queue which is served on a first-come-first-served basis, and a stream of (primary) high-priority customers, who are served uninterruptedly when the batch of high-priority customers finds the server idle upon arrival. High-priority customers are treated as retrial customers, but once in the orbit they lose their high-priority status. The Late Arrival Setup is chosen with Delayed Access. The high-priority retrial customers can be interpreted as inbound calls, and the low-priority customers as outbound calls in a call-center. The joint steady-state distribution of the queue length of the low-priority customers and the orbit size of secondary retrial customers is studied using probability generating functions. Several performance measures will be calculated, such as the mean queue length of the low-priority customers and the orbit size of the secondary retrial customers.

Keywords: Inbound and outbound calls · Discrete-time retrial queue · Priority customers · Generating functions

1 Introduction

In call-centers inbound calls have priority over outbound calls. Outbound calls will be handled only when after the end of a call no inbound calls are coming in, i.e. when a server would stay idle if he would not start answering outbound calls. Inbound calls do not wait in a queue and when upon arrival they find a busy tone they will try to call again some random time later. Outbound calls, for instance in the form of e-mails sent to the call-center with a request to be called back, will be handled by the center in the order of their arrival, when time is available due to the absence of incoming calls.

To model this priority-scheme for inbound calls over requests for being called back by the center we study a mixed retrial/delay model in discrete time with

---

This paper is based on the second author's Bachelor thesis [4].

W. Yue et al. (Eds.): QTNA 2017, LNCS 10591, pp. 199–216, 2017.
https://doi.org/10.1007/978-3-319-68520-5_13

one server. More specifically, we consider a one-server queueing model in discrete time with two types of customers. Time is divided in slots, and all events [arrivals, start of a service and departures] are considered to occur at the slot boundaries only. The high-priority customers [primary inbound calls] arrive in batches following a general probability distribution. When upon arrival of a batch of high-priority customers the server is idle, the complete batch is accepted for an uninterrupted (batch-)service. When upon arrival of a batch of high-priority customers the server is busy, the complete batch will be sent into orbit, and the individual customers lose their high-priority-status. They will approach the server individually [so-called secondary arrivals] some random time later, independently from the other customers in the orbit.

The low-priority customers [outbound calls] also arrive in batches, possibly following a different probability distribution, and they are put in a queue which is served in the order of arrival [within a batch in random order]. The low-priority customers are served individually and a low-priority customer is selected for service only when the server is idle and no batch of primary high-priority customers arrives in the idle slot. In case neither primary high-priority customers arrive nor low-priority customers are present in the queue, then a possible secondary arrival is selected for [an individual] service. The non-selected secondary customers are resent into the orbit. When neither low-priority customers are present in the queue at the end of the idle slot, nor any primary or secondary retrial customers will have arrived in the idle slot, the server stays idle also the following slot.

Notice that the modeling assumption is made that in the time slot following a (batch-)service completion the server always stays idle, even when the queue of low-priority customers is not empty, to enable the start of the service of an incoming batch of high-priority primary customers.

The service times of the high-priority [inbound calls] and the low-priority [outbound calls] customers are all independent and follow [possibly] a different general distribution. To resolve the conflict of simultaneous arrivals and departures we have chosen for the *late arrival setup with delayed access*, i.e. arrivals have precedence over departures and a service of newly arrived customers can only start at the time slot following the slot of the arrival at the earliest. For an overview of discrete-time retrial queues with the late arrival setup we refer to Nobel [7] and for the most complete monograph on retrial queues we refer to Artalejo and Gómez-Corral [1].

So, in this paper we will extend the classical discrete-time one-server retrial model of Nobel and Moreno [9] by adding a second type of customers [the outbound calls] who upon arrival are put in a queue. These low-priority customers will be served one by one on a first-come-first-served basis. The retrial primary customers [inbound calls] are given *non-preemptive* priority over the queued customers [the outbound calls]. Rejected inbound calls lose their priority, but they continue to act as retrial customers, and their service time remains unaltered. In Sharkawy [10] the high-priority retrial customers maintained their high-priority status in the orbit, but it turned out to be impossible to derive a closed-form expression for the probability generating function of the joint steady-state distribution of the queue size and the orbit size. For this technical reason in this paper we made the modeling assumption that high-priority customers lose their priority status once sent into the orbit.

In previous papers (Nobel and Moreno [8] and Nobel [6]) the priority has been mainly modeled the other way around: non-preemptive priority of the queued customers over the retrial customers. This is a natural hierarchy in mobile telephony for modeling handover calls [high priority] versus new calls [low priority] competing for the same target channel, see Nobel [6]. As pointed out before, giving priority to the retrial customers over the queued customers leads to an intractable model (see again Sharkawy [10]), and only to guarantee tractability we made the admittedly somewhat awkward assumption that high-priority retrial customers lose their priority once they have been sent into the orbit. In Artalejo et al. [2] a [somewhat simplified] continuous-time counterpart of our model with single arrivals is discussed, in which the retrial customers do not lose their high-priority status once they are sent into the orbit, but the authors only consider exponential service times, introduce a finite buffer size for the low-priority customers and, most importantly, they give *preemptive* priority to the retrial customers. These three characteristics of their model enable an algorithmic analysis. We think that in a call center outbound calls should not be interrupted by incoming inbound calls, and for that reason we have chosen for *non-preemptive* priority for the inbound calls, but to get an analytic solution we have to pay a price! Of course, it is also possible to give a practical application in which our modeling assumption that the high-priority retrial customers lose their priority status is more natural than in the call-center environment. Take for instance a small military field hospital with one operation unit where regularly scheduled patients [outbound calls!] and incoming emergency patients [inbound calls!] have to be operated. When an ambulance with a group of emergency patients arriving from the battlefield finds the operation unit busy they will be sent away (maybe after some necessary minimal treatment), and subsequently they will compete individually with the regular patients, i.e. they lose their high-priority [emergency] status. Although we had in mind a call-center application when we started this paper, the above hospital example illustrates that our technical assumption is quite realistic in another environment!

A discrete-time model with the easier priority setup, i.e. the queued customers have priority over the retrial customers, has been studied in Choi and Kim [3], but also they discuss only single arrivals and all customers follow the same service-time distribution. Further, they have chosen the early arrival setup. A continuous-time retrial model with priority for the queued customers has been studied by Falin et al. [5], but also in that paper only single arrivals have been considered. The model discussed in this paper can be seen both as an extension and as the discrete-time counterpart of that model, but above all as a first attempt to reverse the priority of retrial customers versus queued customers. As already indicated above, this reversed priority-scheme is mainly motivated by the priority of inbound calls over outbound calls in a call-center.

In the sections below we will study the joint steady-state distribution of the length of the queue of low-priority customers [outbound calls] and the size of the orbit with high-priority customers [inbound calls who lost their priority]. Not surprisingly, the mathematical analysis of our mixed retrial/delay model differs greatly from the analysis of the models discussed in the papers Choi and Kim [3] and Nobel and Moreno [8]. The analysis is also more involved than the analysis presented in Nobel [6].

As usual, we will derive the generating function of the joint steady-state distribution of the number of low-priority customers in the queue, the number of high-priority customers in the orbit and the residual service time of the (batch of) customer(s) in service. Notice that we do not keep track of the type of the ongoing service in the analysis. This generating function will be used to calculate several performance measures, e.g. the mean queue length and the mean orbit size. In Sect. 2 we describe the model in detail. In Sects. 3 and 4 we discuss the steady-state distributions and the first moment of the orbit size and the queue length. In Sect. 5 we will present some numerical results.

2   Description of the Model

For a detailed description of the discrete-time setup with late arrivals and delayed access [LAS/DA] we refer to Nobel and Moreno [9]. Recall that due to this LAS/DA setup in this classical retrial model the time slot after a departure the server always stays idle for at least one slot, because arrivals have precedence over departures. For the mixed retrial/delay model to be discussed in this paper we make the technical assumption that the slot following the completion of a (batch-)service the server always stays idle, *also in case low-priority customers are waiting in the queue*. Imposing this idle slot guarantees the priority of the (primary) retrial customers over the queued customers, by triggering the start of the batch-service of any incoming batch of high-priority customers in this idle slot at the start of the next slot, and so automatically blocking the possible start of the service of a (queued) low-priority customer, or a secondary arrival from the orbit. Only in case no primary batch of high-priority customers arrives during the idle slot, the service of the longest waiting low-priority customer will start his individual service the next slot. If no low-priority customers are present in the queue or no batch of low-priority customers will have arrived during the idle slot, then possibly a secondary arrival will start his individual service, and in case there are no secondary arrivals, the server stays idle also the next slot.

We will now give the precise description of the discrete-time mixed retrial/delay queueing model with one server and priorities for the primary retrial customers. In each time slot primary high-priority customers [inbound calls] arrive in batches. The batch sizes are mutually independent and follow a general probability distribution $\left\{ a_k^{(H)} \right\}_{k=0}^{\infty}$ with probability generating function (p.g.f.)

$$\mathcal{A}_H(z) = \sum_{k=0}^{\infty} a_k^{(H)} z^k.$$

In every time slot also low-priority customers [outbound calls] arrive in batches. These batch sizes follow a general probability distribution $\left\{ a_i^{(L)} \right\}_{i=0}^{\infty}$ with p.g.f.

$$\mathcal{A}_L(y) = \sum_{i=0}^{\infty} a_i^{(L)} y^i.$$

These batch sizes are again mutually independent and they are also independent of the batch sizes of the high-priority customers. Each individual high-priority customer requires a service time, measured as a number of time slots, which follows the discrete probability distribution $\left\{ b_j^{(H)} \right\}_{j=1}^{\infty}$ with p.g.f.

$$\mathcal{B}_H(w) = \sum_{j=1}^{\infty} b_j^{(H)} w^j.$$

Similarly, every low-priority customer requires a generally distributed service time with distribution $\left\{ b_j^{(L)} \right\}_{j=1}^{\infty}$ and p.g.f.

$$\mathcal{B}_L(w) = \sum_{j=1}^{\infty} b_j^{(L)} w^j.$$

All service times are mutually independent and they are also independent of the batch sizes of the arriving customers. A service time requires at least one time slot, so $b_0^{(H)} = b_0^{(L)} = 0$. As said before, the low-priority customers are placed in a queue, and are served individually on a first-come-first-served basis. Also primary high-priority customers are served individually, but *uninterruptedly as a batch-service*, i.e. after every individual service completion, the next customer of the batch starts his service *immediately* in the next slot. Only at the service completion of the last customer of the batch the server stays idle the next slot, even if low-priority customers are present in the queue, to enable the start of a batch-service in case a new batch of high-priority customers arrives in this idle slot. Rejected high-priority customers behave as the customers in the classical retrial queue, with the only difference that all incoming customers [inbound calls] from the orbit have lost their high-priority status. They even have lower-priority than the queued customers [outbound calls]. In each time slot retrial customers in the orbit [inbound calls who have lost their high-priority] try to reenter the system individually and independently with the so-called retrial probability r $[0 < r \leq 1]$.

We are interested in the joint steady-state distribution of the number of low-priority customers in the queue, the number of high-priority customers [strictly speaking a misnomer, because customers once in the orbit have lost their high-priority status] in orbit, and the residual service time of the (batch of) customer(s) currently in service. To analyze the mixed retrial/delay queueing model, we define a discrete-time Markov chain (DTMC) by observing the system at the epochs $k-$, that is at the start of the time slots k just after, possibly, a service of a (low- or high-priority) customer has started, but before the arrivals during time slot k have occurred. We define the following random variables,

R_k = the residual service time of the ongoing (batch-)service at time $k-$,

L_k = the number of low-priority customers present in the queue at time $k-$,

Q_k = the number of high-priority customers in orbit at time $k-$.

We define $R_k = 0$ when at epoch $k-$ the server is idle. Notice that the type of the residual service time is not part of the state description. Introduce the offered load

$$\varrho := \mathcal{A}'_L(1)\mathcal{B}'_L(1) + \mathcal{A}'_H(1)\mathcal{B}'_H(1).$$

Then, the stochastic process $\{(R_k, L_k, Q_k) : k = 0, 1, 2, \ldots\}$ is an irreducible aperiodic DTMC which is positive recurrent under the stability condition

$$a_0^{(H)}[1 - \varrho] - \mathcal{A}'_L(1) - \varrho\mathcal{A}'_H(1) > 0.$$

This complicated stability condition is due to the modeling assumption that a batch of primary high-priority customers is served uninterruptedly, imposing only one forced idle slot after the completion of the last customer of the batch, whereas all the other customers [low-priority and secondary customers arriving from the orbit] force the server to stay idle after each [individual] service. So the *total used capacity*, i.e. the fraction of time that the server is busy or waiting for a possible arrival of a batch of high-priority customers, say σ, should be smaller than 1, i.e.

$$\sigma := \varrho + \mathcal{A}'_L(1) + \varrho\mathcal{A}'_H(1) + \left(1 - a_0^{(H)}\right)(1 - \varrho) < 1.$$

A formal proof of this stability condition can be given using Foster's criterion [see Nobel and Moreno [9] for the details].

3   The Joint Distribution of Queue Length and Orbit Size

In this section we will derive the joint probability generating function [p.g.f.] of the steady-state distribution of the DTMC $\{(R_k, L_k, Q_k) : k = 0, 1, 2, \ldots\}$. Under the stability condition we can define the following limiting joint distribution of this DTMC

$$\pi(j, m, n) = \lim_{k \to \infty} \mathbf{P}(R_k = j; L_k = m; Q_k = n), \quad j, m, n = 0, 1, 2, \ldots,$$

with its associated three-dimensional generating function

$$\Pi(w, y, z) = \sum_{j=0}^{\infty} \sum_{m=0}^{\infty} \sum_{n=0}^{\infty} \pi(j, m, n) w^j y^m z^n.$$

In the following it is convenient to introduce also the partial generating functions,

$$\Pi_{jm}(z) = \sum_{n=0}^{\infty} \pi(j, m, n) z^n \quad \text{and}$$

$$\Pi_j(y, z) = \sum_{m=0}^{\infty} \sum_{n=0}^{\infty} \pi(j, m, n) y^m z^n = \sum_{m=0}^{\infty} \Pi_{jm}(z) y^m.$$

To find the p.g.f. $\Pi(w, y, z)$ we write down the system of balance equations,

$$
\pi(0, m, n) = \mathbf{I}_{\{m=0\}} a_0^{(L)} a_0^{(H)} (1 - r)^n \pi(0, 0, n)
$$
$$
+ \sum_{i=0}^{m} a_i^{(L)} \sum_{k=0}^{n} a_k^{(H)} \pi(1, m - i, n - k),
$$
$$
m, n = 0, 1, \ldots, \tag{1}
$$

$$
\pi(j, m, n) = \sum_{i=0}^{m} a_i^{(L)} \sum_{k=0}^{n} a_k^{(H)} \pi(j + 1, m - i, n - k)
$$
$$
+ a_0^{(H)} \sum_{i=0}^{m+1} a_i^{(L)} \pi(0, m + 1 - i, n) b_j^{(L)}
$$
$$
+ \sum_{i=0}^{m} a_i^{(L)} \sum_{k=1}^{j} a_k^{(H)} \pi(0, m - i, n) b_j^{(H)(*k)}
$$
$$
+ \mathbf{I}_{\{m=0\}} a_0^{(L)} a_0^{(H)} \left(1 - (1 - r)^{n+1} \right) \pi(0, 0, n + 1) b_j^{(H)},
$$
$$
j = 1, 2, \ldots; \quad m, n = 0, 1, 2, \ldots. \tag{2}
$$

Notice how our technical assumption that after the completion of a [batch-] service the server stays idle for at least one time slot plays its role in these balance equations.

From Eq. (1) we get by multiplying both sides with z^n and summing over $n = 0, 1, \ldots$, and subsequently multiplying both sides of the result by y^m and summing over $m = 0, 1, \ldots$,

$$\Pi_0(y, z) = a_0^{(L)} a_0^{(H)} \Pi_{00}((1 - r)z) + \mathcal{A}_L(y) \mathcal{A}_H(z) \Pi_1(y, z). \tag{3}$$

From Eq. (2) we get, acting similarly,

$$
\Pi_j(y, z) = \mathcal{A}_L(y) \mathcal{A}_H(z) \Pi_{j+1}(y, z)
$$
$$
+ \frac{a_0^{(H)} b_j^{(L)}}{y} \left[\mathcal{A}_L(y) \Pi_0(y, z) - a_0^{(L)} \Pi_{00}(z) \right]
$$
$$
+ \mathcal{A}_L(y) \sum_{k=1}^{j} a_k^{(H)} b_j^{(H)(*k)} \Pi_0(y, z)
$$
$$
+ \frac{a_0^{(L)} a_0^{(H)} b_j^{(H)}}{z} \left[\Pi_{00}(z) - \Pi_{00}((1 - r)z) \right]. \tag{4}
$$

Next, multiplying Eq. (4) by w^j and summing over $j = 1, 2, \ldots$ gives after some simple algebra, using Eq. (3) to get rid of $\Pi_1(y, z)$,

$$
\begin{aligned}
yz(w - \mathcal{A}_L(y)\mathcal{A}_H(z))\Pi(w, y, z) \\
= \mathcal{A}_L(y)z \left[a_0^{(H)} w(\mathcal{B}_L(w) - y) + y(w\mathcal{A}_H(\mathcal{B}_H(w)) - \mathcal{A}_H(z)) \right] \Pi_0(y, z) \\
+ a_0^{(L)} a_0^{(H)} w \left[y\mathcal{B}_H(w) - z\mathcal{B}_L(w) \right] \Pi_{00}(z) \\
+ a_0^{(L)} a_0^{(H)} wy \left[z - \mathcal{B}_H(w) \right] \Pi_{00}((1 - r)z).
\end{aligned}
\tag{5}
$$

So, the problem is to find the unknown partial generating functions $\Pi_0(y, z)$ and $\Pi_{00}(z)$. Firstly, take $w = \mathcal{A}_L(y)\mathcal{A}_H(z)$ in (5) to make the left-hand side zero. This gives

$$
\mathcal{A}_L(y)z \left[\begin{array}{c} a_0^{(H)} \omega(y, z)[\mathcal{B}_L(\omega(y, z)) - y] \\ +y[\omega(y, z)\mathcal{A}_H(\mathcal{B}_H(\omega(y, z))) - \mathcal{A}_H(z)] \end{array} \right] \Pi_0(y, z)
$$

$$
= a_0^{(L)} a_0^{(H)} \omega(y, z) \left[z\mathcal{B}_L(\omega(y, z)) - y\mathcal{B}_H(\omega(y, z)) \right] \Pi_{00}(z) +
$$

$$
- a_0^{(L)} a_0^{(H)} \omega(y, z) y \left[z - \mathcal{B}_H(\omega(y, z)) \right] \Pi_{00}((1 - r)z).
\tag{6}
$$

where $\omega(y, z) := \mathcal{A}_L(y)\mathcal{A}_H(z)$. Now consider the coefficient of $\Pi_0(y, z)$. Let

$$
\psi(y, z) := \begin{array}{c} a_0^{(H)} \omega(y, z)[\mathcal{B}_L(\omega(y, z)) - y] \\ +y[\omega(y, z)\mathcal{A}_H(\mathcal{B}_H(\omega(y, z))) - \mathcal{A}_H(z)] \end{array}
$$

Then we have

$$
\forall z \exists! y : \ \psi(y, z) = 0.
$$

For real $z \in (0, 1)$ this follows immediately

$$
\psi(0, z) = a_0^{(H)} a_0^{(L)} \mathcal{A}_H(z)\mathcal{B}_L \left(a_0^{(L)} \mathcal{A}_H(z) \right) > 0.
$$

$$
\psi(1, z) = \mathcal{A}_H(z) \left[a_0^{(H)} \mathcal{B}_L(\mathcal{A}_H(z)) + \mathcal{A}_H(\mathcal{B}_H(\mathcal{A}_H(z))) - \left(1 + a_0^{(H)} \right) \right] \leq 0
$$

with equality only for $z = 1$. Notice that $\psi(1, 1) = 0$.

Let $y^*(z)$ be the unique solution, i.e. $\psi(y^*(z), z) = 0$ and introduce

$$
\phi(z) := \omega(y^*(z), z) = \mathcal{A}_L(y^*(z))\mathcal{A}_H(z).
$$

Notice that from $\psi(y^*(z), z) = 0$ we get

$$
y^*(z) = \frac{a_0^{(H)} \phi(z)\mathcal{B}_L(\phi(z))}{a_0^{(H)} \phi(z) + \mathcal{A}_H(z) - \phi(z)\mathcal{A}_H(\mathcal{B}_H(\phi(z)))}.
\tag{7}
$$

It is easy to see that $y^*(1) = 1$ and so also $\phi(1) = 1$. Now from (6) and using (7) we find the recursion

$$\Pi_{00}(z) = \frac{y^*(z)[\mathcal{B}_H(\phi(z)) - z]}{y^*(z)\mathcal{B}_H(\phi(z)) - z\mathcal{B}_L(\phi(z))} \Pi_{00}((1-r)z)$$

$$= \frac{\frac{a_0^{(H)}\phi(z)\mathcal{B}_L(\phi(z))}{a_0^{(H)}\phi(z)+\mathcal{A}_H(z)-\phi(z)\mathcal{A}_H(\mathcal{B}_H(\phi(z)))}[\mathcal{B}_H(\phi(z)) - z]}{\frac{a_0^{(H)}\phi(z)\mathcal{B}_L(\phi(z))}{a_0^{(H)}\phi(z)+\mathcal{A}_H(z)-\phi(z)\mathcal{A}_H(\mathcal{B}_H(\phi(z)))}\mathcal{B}_H(\phi(z)) - z\mathcal{B}_L(\phi(z))} \Pi_{00}((1-r)z).$$

Some algebra leads to a simple recursion,

$$\Pi_{00}(z) = \frac{a_0^{(H)}\phi(z)[z - \mathcal{B}_H(\phi(z))]}{a_0^{(H)}\phi(z)[z - \mathcal{B}_H(\phi(z))] + z[\mathcal{A}_H(z) - \phi(z)\mathcal{A}_H(\mathcal{B}_H(\phi(z)))]} \Pi_{00}((1-r)z).$$

Now introduce the so-called *retrial function* for the primary batch-service model

$$\mathcal{R}_b(z) = \frac{a_0^{(H)}\phi(z)[z - \mathcal{B}_H(\phi(z))]}{a_0^{(H)}\phi(z)[z - \mathcal{B}_H(\phi(z))] + z[\mathcal{A}_H(z) - \phi(z)\mathcal{A}_H(\mathcal{B}_H(\phi(z)))]}.$$

Notice that $\mathcal{R}_b(0) = 1$ and after using L'Hôpital we find that

$$\mathcal{R}_b(1) = \frac{a_0^{(H)}[1 - \mathcal{B}'_H(1)\phi'(1)]}{a_0^{(H)}[1 - \mathcal{B}'_H(1)\phi'(1)] + \mathcal{A}'_H(1) - \phi'(1)[\mathcal{A}'_H(1)\mathcal{B}'_H(1) + 1]}$$

$$= \frac{a_0^{(H)}[1 - \mathcal{A}'_L(1)\mathcal{B}'_L(1) - \mathcal{A}'_H(1)\mathcal{B}'_H(1)] - \mathcal{A}'_L(1)}{a_0^{(H)}[1 - \mathcal{A}'_L(1)\mathcal{B}'_L(1) - \mathcal{A}'_H(1)\mathcal{B}'_H(1)] - \mathcal{A}'_L(1) - [\mathcal{A}'_L(1)\mathcal{B}'_L(1) + \mathcal{A}'_H(1)\mathcal{B}'_H(1)]\mathcal{A}'_H(1)}.$$

In the denominator we recognize the stability condition!

Now we get by iteration

$$\Pi_{00}(z) = \prod_{k=0}^{n-1} \mathcal{R}_b\left((1-r)^k z\right) \Pi_{00}\left((1-r)^n z\right).$$

Next, sending n to infinity we find

$$\Pi_{00}(z) = \prod_{k=0}^{\infty} \mathcal{R}_b\left((1-r)^k z\right) \Pi_{00}(0). \tag{8}$$

The problem is to calculate $\Pi_{00}(0)$, the steady-state probability that the system is empty!

From (6) we find

$$\Pi_0(y,z) = \frac{a_0^{(L)}a_0^{(H)}\omega(y,z)\left\{\begin{array}{l}[z\mathcal{B}_L(\omega(y,z)) - y\mathcal{B}_H(\omega(y,z))]\,\Pi_{00}(z) \\ -y\,[z - \mathcal{B}_H(\omega(y,z))]\,\Pi_{00}((1-r)z)\end{array}\right\}}{\mathcal{A}_L(y)z\left[\begin{array}{l}a_0^{(H)}\omega(y,z)[\mathcal{B}_L(\omega(y,z)) - y] \\ +y[\omega(y,z)\mathcal{A}_H(\mathcal{B}_H(\omega(y,z))) - \mathcal{A}_H(z)]\end{array}\right]}.$$

and using the recursion $\Pi_{00}(z) = \mathcal{R}_b(z)\Pi_{00}((1-r)z)$ we get

$$\Pi_0(y,z) = \frac{a_0^{(L)} a_0^{(H)} \mathcal{A}_H(z) \left\{ \begin{array}{c} [z\mathcal{B}_L(\omega(y,z)) - y\mathcal{B}_H(\omega(y,z))]\,\mathcal{R}_b(z) \\ -y\,[z - \mathcal{B}_H(\omega(y,z))] \end{array} \right\}}{z\left[\begin{array}{c} a_0^{(H)}\omega(y,z)[\mathcal{B}_L(\omega(y,z)) - y] \\ +y[\omega(y,z)\mathcal{A}_H(\mathcal{B}_H(\omega(y,z))) - \mathcal{A}_H(z)] \end{array} \right]}$$
$$\times \Pi_{00}((1-r)z). \tag{9}$$

We know that $\Pi_0(1,1) = 1 - \mathcal{A}'_L(1)\mathcal{B}'_L(1) - \mathcal{A}'_H(1)\mathcal{B}'_H(1)$. So we can find $\Pi_{00}(1-r)$, again using L'Hôpital, from (9),

$$1 - \mathcal{A}'_L(1)\mathcal{B}'_L(1) - \mathcal{A}'_H(1)\mathcal{B}'_H(1) = \Pi_0(1,1) = \lim_{y \to 1} \Pi_0(y,1)$$

$$= \lim_{y \to 1} \frac{a_0^{(L)} a_0^{(H)} \left\{ \begin{array}{c} [\mathcal{B}_L(\mathcal{A}_L(y)) - y\mathcal{B}_H(\mathcal{A}_L(y))]\,\mathcal{R}_b(1) + \\ -y\,[1 - \mathcal{B}_H(\mathcal{A}_L(y))] \end{array} \right\}}{a_0^{(H)}\mathcal{A}_L(y)[\mathcal{B}_L(\mathcal{A}_L(y)) - y] + y[\mathcal{A}_L(y)\mathcal{A}_H(\mathcal{B}_H(\mathcal{A}_L(y))) - 1]} \times \Pi_{00}(1-r)$$

$$= \frac{a_0^{(L)} a_0^{(H)} \{\mathcal{R}_b(1)[\mathcal{B}'_L(1)\mathcal{A}'_L(1) - 1 - \mathcal{B}'_H(1)\mathcal{A}'_L(1)] + \mathcal{B}'_H(1)\mathcal{A}'_L(1)\}}{a_0^{(H)}[\mathcal{B}'_L(1)\mathcal{A}'_L(1) - 1] + \mathcal{A}'_L(1) + \mathcal{A}'_H(1)\mathcal{B}'_H(1)\mathcal{A}'_L(1)} \Pi_{00}(1-r)$$

$$= \frac{a_0^{(L)} a_0^{(H)}[1 - \mathcal{A}'_L(1)\mathcal{B}'_L(1) - \mathcal{A}'_H(1)\mathcal{B}'_H(1)]}{a_0^{(H)}[1 - \mathcal{A}'_L(1)\mathcal{B}'_L(1) - \mathcal{A}'_H(1)\mathcal{B}'_H(1)] - \mathcal{A}'_L(1) - [\mathcal{A}'_L(1)\mathcal{B}'_L(1) + \mathcal{A}'_H(1)\mathcal{B}'_H(1)]\mathcal{A}'_H(1)}$$
$$\Pi_{00}(1-r).$$

So, using the *offered load* $\varrho = \mathcal{A}'_L(1)\mathcal{B}'_L(1) + \mathcal{A}'_H(1)\mathcal{B}'_H(1)$ we find

$$\Pi_{00}(1-r) = \frac{a_0^{(H)}(1-\varrho) - \mathcal{A}'_L(1) - \varrho\mathcal{A}'_H(1)}{a_0^{(L)} a_0^{(H)}}$$

and this leads to

$$\Pi_{00}(z) = \prod_{k=0}^{\infty} \mathcal{R}_b\left((1-r)^k z\right) \Pi_{00}(0)$$

$$= \frac{a_0^{(H)}(1-\varrho) - \mathcal{A}'_L(1) - \varrho\mathcal{A}'_H(1)}{a_0^{(L)} a_0^{(H)}} \mathcal{R}_b(1) \prod_{k=0}^{\infty} \frac{\mathcal{R}_b\left((1-r)^k z\right)}{\mathcal{R}_b\left((1-r)^k\right)}$$

$$= \frac{a_0^{(H)}[1-\varrho] - \mathcal{A}'_L(1)}{a_0^{(L)} a_0^{(H)}} \prod_{k=0}^{\infty} \frac{\mathcal{R}_b\left((1-r)^k z\right)}{\mathcal{R}_b\left((1-r)^k\right)}.$$

Now we can move to the next step in our search for a 'closed form formula' for $\Pi(w, y, z)$. Recall from (9) and the definition of $\mathcal{R}_b(z)$

$$\Pi_0(y, z) = \frac{a_0^{(L)} a_0^{(H)} \mathcal{A}_H(z) \left\{ \begin{array}{c} [z \mathcal{B}_L(\omega(y, z)) - y \mathcal{B}_H(\omega(y, z))] \mathcal{R}_b(z) \\ -y [z - \mathcal{B}_H(\omega(y, z))] \end{array} \right\}}{z \left[\begin{array}{c} a_0^{(H)} \omega(y, z)[\mathcal{B}_L(\omega(y, z)) - y] \\ +y[\omega(y, z)\mathcal{A}_H(\mathcal{B}_H(\omega(y, z))) - \mathcal{A}_H(z)] \end{array} \right]}$$
$$\times \Pi_{00}((1 - r)z) \tag{10}$$

$$= \left(\frac{a_0^{(L)} a_0^{(H)} \mathcal{A}_H(z) \left\{ \begin{array}{c} a_0^{(H)} \phi(z) \left[z \mathcal{B}_L(\omega(y, z)) - y \mathcal{B}_H(\omega(y, z)) \right] [z - \mathcal{B}_H(\phi(z))] + \\ -y [z - \mathcal{B}_H(\omega(y, z))] \\ \times \{a_0^{(H)} \phi(z)[z - \mathcal{B}_H(\phi(z))] + z[\mathcal{A}_H(z) - \phi(z)\mathcal{A}_H(\mathcal{B}_H(\phi(z)))]\} \end{array} \right\}}{z \left[\begin{array}{c} a_0^{(H)} \omega(y, z)[\mathcal{B}_L(\omega(y, z)) - y] \\ +y[\omega(y, z)\mathcal{A}_H(\mathcal{B}_H(\omega(y, z))) - \mathcal{A}_H(z)] \end{array} \right]}{\times \{a_0^{(H)} \phi(z)[z - \mathcal{B}_H(\phi(z))] + z[\mathcal{A}_H(z) - \phi(z)\mathcal{A}_H(\mathcal{B}_H(\phi(z)))]\}} \right)$$
$$\times \frac{a_0^{(H)}[1 - \varrho] - \mathcal{A}_L'(1) - \varrho \mathcal{A}_H'(1)}{a_0^{(L)} a_0^{(H)}} \prod_{k=1}^{\infty} \frac{\mathcal{R}_b\left((1 - r)^k z\right)}{\mathcal{R}_b\left((1 - r)^k\right)}.$$

Finally, we can find the full p.g.f. $\Pi(w, y, z)$! Recall (5) and use the result (10) for $\Pi_0(y, z)$ and $\Pi_{00}((1 - r)z)$,

$$\Pi(w, y, z)$$

$$= \frac{\mathcal{A}_L(y)z \left[a_0^{(H)} w(\mathcal{B}_L(w) - y) + y(w \mathcal{A}_H(\mathcal{B}_H(w)) - \mathcal{A}_H(z)) \right] \Pi_0(y, z)}{yz(w - \mathcal{A}_L(y)\mathcal{A}_H(z))}$$
$$\quad + a_0^{(L)} a_0^{(H)} w [y \mathcal{B}_H(w) - z \mathcal{B}_L(w)] \Pi_{00}(z)$$
$$\quad + a_0^{(L)} a_0^{(H)} wy [z - \mathcal{B}_H(w)] \Pi_{00}((1 - r)z)$$

Substitution of our previous results gives

$$\Pi(w, y, z) = \left(a_0^{(H)}[1 - \varrho] - \mathcal{A}_L'(1) - \varrho \mathcal{A}_H'(1) \right) \prod_{k=1}^{\infty} \frac{\mathcal{R}_b\left((1 - r)^k z\right)}{\mathcal{R}_b\left((1 - r)^k\right)}$$

$$\mathcal{A}_L(y)z \left[a_0^{(H)} w(\mathcal{B}_L(w) - y) + y(w \mathcal{A}_H(\mathcal{B}_H(w)) - \mathcal{A}_H(z)) \right]$$

$$\times \frac{\left[\mathcal{A}_H(z) \frac{\left\{ \begin{array}{c} [z \mathcal{B}_L(\omega(y, z)) - y \mathcal{B}_H(\omega(y, z))] \mathcal{R}_b(z) \\ -y [z - \mathcal{B}_H(\omega(y, z))] \end{array} \right\}}{z \left[\begin{array}{c} a_0^{(H)} \omega(y, z)[\mathcal{B}_L(\omega(y, z)) - y] \\ +y[\omega(y, z)\mathcal{A}_H(\mathcal{B}_H(\omega(y, z))) - \mathcal{A}_H(z)] \end{array} \right]} \right]}{}$$
$$\times \frac{+w [y \mathcal{B}_H(w) - z \mathcal{B}_L(w)] \mathcal{R}_b(z)}{yz(w - \omega(y, z))} . \tag{11}$$
$$\quad + wy [z - \mathcal{B}_H(w)]$$

4   The Queue Size and the Orbit Size

From expression (11) we find the marginal p.g.f.'s $\mathcal{L}(y) := \Pi(1, y, 1)$ and $\mathcal{Q}(z) := \Pi(1, 1, z)$ of the limiting distribution of the queue length and the orbit size, respectively. After some simplifications we find

$$\mathcal{L}(y) = \left(a_0^{(H)}[1 - \varrho] - \mathcal{A}_L'(1) - \varrho \mathcal{A}_H'(1) \right) \left(\frac{1 - y}{1 - \mathcal{A}_L(y)} \right)$$

$$\times \frac{a_0^{(H)} \mathcal{A}_L(y)(\mathcal{R}_b(1) - 1)[1 - \mathcal{B}_H(\mathcal{A}_L(y))] + \mathcal{R}_b(1)[1 - \mathcal{A}_L(y)\mathcal{A}_H(\mathcal{B}_H(\mathcal{A}_L(y)))]}{a_0^{(H)} \mathcal{A}_L(y)[\mathcal{B}_L(\mathcal{A}_L(y)) - y] + y[\mathcal{A}_L(y)\mathcal{A}_H(\mathcal{B}_H(\mathcal{A}_L(y))) - 1]}$$

$$\mathcal{Q}(z) = \left(a_0^{(H)}[1 - \varrho] - \mathcal{A}_L'(1) - \varrho \mathcal{A}_H'(1) \right) \prod_{k=1}^{\infty} \frac{\mathcal{R}_b\left((1 - r)^k z\right)}{\mathcal{R}_b\left((1 - r)^k\right)}$$

$$\times \left[\frac{\mathcal{R}_b(z)[z\mathcal{B}_L(\mathcal{A}_H(z)) - \mathcal{B}_H(\mathcal{A}_H(z))] + \mathcal{B}_H(\mathcal{A}_H(z)) - z}{z \left\{ a_0^{(H)}[\mathcal{B}_L(\mathcal{A}_H(z)) - 1] + \mathcal{A}_H(\mathcal{B}_H(\mathcal{A}_H(z))) - 1 \right\}} \right.$$

$$\left. + \frac{(1 - z)(\mathcal{R}_b(z) - 1)}{z(1 - \mathcal{A}_H(z))} \right].$$

To find the *mean queue length* $\overline{\mathcal{L}} = \mathcal{L}'(1)$ we write

$$\mathcal{L}(y) = (1 - \sigma) \times F(y) \times \frac{N(z)}{D(z)}$$

where σ is again the total used capacity

$$\sigma = \varrho + \mathcal{A}_L'(1) + \varrho \mathcal{A}_H'(1) + \left(1 - a_0^{(H)} \right)(1 - \varrho)$$

and

$$F(y) = \frac{1 - y}{1 - \mathcal{A}_L(y)}$$

$$N(y) = a_0^{(H)} \mathcal{A}_L(y)(\mathcal{R}_b(1) - 1)[1 - \mathcal{B}_H(\mathcal{A}_L(y))] + \mathcal{R}_b(1)[1 - \mathcal{A}_L(y)\mathcal{A}_H(\mathcal{B}_H(\mathcal{A}_L(y)))]$$

$$D(y) = a_0^{(H)} \mathcal{A}_L(y)[\mathcal{B}_L(\mathcal{A}_L(y)) - y] + y[\mathcal{A}_L(y)\mathcal{A}_H(\mathcal{B}_H(\mathcal{A}_L(y))) - 1].$$

Differentiating $\mathcal{L}(y)$ gives

$$\mathcal{L}'(y) = (1 - \sigma)\left(F(y) \cdot \frac{N'(y)D(y) - N(y)D'(y)}{[D(y)]^2} + F'(y) \cdot \frac{N(y)}{D(y)} \right).$$

So we want to calculate

$$\mathcal{L}'(1) = (1 - \sigma) \lim_{y \to 1} \left(F(y) \cdot \frac{N'(y)D(y) - N(y)D'(y)}{[D(y)]^2} + F'(y) \cdot \frac{N(y)}{D(y)} \right).$$

After tedious calculations using L'Hôpital we find for the mean queue length

$$\overline{\mathcal{L}} = \mathcal{L}'(1) = \frac{1-\sigma}{\mathcal{A}'_L(1)} \cdot \frac{N''(1)}{2D'(1)} - \frac{D''(1)}{2D'(1)} - \frac{\mathcal{A}''_L(1)}{2\mathcal{A}'_L(1)}$$

where

$$
\begin{aligned}
N''(1) &= a_0^{(H)}\left(1 - \mathcal{R}_b(1)\right)\left(2\mathcal{A}'_L(1)^2\mathcal{B}'_H(1) + \mathcal{A}'_L(1)^2\mathcal{B}''_H(1) + \mathcal{A}''_L(1)\mathcal{B}'_H(1)\right) + \\
&\quad - \mathcal{R}_b(1)\left(2\mathcal{A}'_L(1)^2\mathcal{B}'_H(1)\mathcal{A}'_H(1) + \mathcal{A}'_L(1)^2\mathcal{B}''_H(1)\mathcal{A}'_H(1) + \mathcal{A}''_L(1)\mathcal{B}'_H(1)\mathcal{A}'_H(1)\right. \\
&\quad \left. + \mathcal{A}''_L(1) + \mathcal{A}'_L(1)^2\mathcal{A}''_H(1)\mathcal{B}'_H(1)^2\right) \\
D'(1) &= a_0^{(H)}\left(\mathcal{B}'_L(1)\mathcal{A}'_L(1) - 1\right) + \mathcal{A}'_L(1) + \mathcal{A}'_H(1)\mathcal{B}'_H(1)\mathcal{A}'_L(1) \\
D''(1) &= 2a_0^{(H)}\mathcal{A}'_L(1)\left(\mathcal{A}'_L(1)\mathcal{B}'_L(1) - 1\right) + a_0^{(H)}\mathcal{A}''_L(1)\mathcal{B}'_L(1) + a_0^{(H)}\mathcal{A}'_L(1)^2\mathcal{B}''_L(1) \\
&\quad + 2\mathcal{A}'_L(1)^2\mathcal{A}'_H(1)\mathcal{B}'_H(1) + \mathcal{A}'_L(1)^2\mathcal{B}''_H(1)\mathcal{A}'_H(1) + \mathcal{A}''_L(1)\mathcal{A}'_H(1)\mathcal{B}'_H(1) + \mathcal{A}''_L(1) \\
&\quad + \mathcal{A}'_L(1)^2\mathcal{A}''_H(1)\mathcal{B}'_H(1)^2 + 2\mathcal{A}'_L(1) + 2\mathcal{A}'_L(1)\mathcal{A}'_H(1)\mathcal{B}'_H(1).
\end{aligned}
$$

To calculate $\overline{\mathcal{Q}} = \mathcal{Q}'(1)$ first rewrite $\mathcal{Q}(z)$ as

$$\mathcal{Q}(z) = (1-\sigma)\prod_{k=1}^{\infty}\frac{\mathcal{R}_b((1-r)^kz)}{\mathcal{R}_b((1-r)^k)}\left(\frac{N_1(z)}{D_1(z)} + \frac{N_2(z)}{D_2(z)}\right)$$

with

$$
\begin{aligned}
N_1(z) &= \left[z\mathcal{B}_L(\mathcal{A}_H(z)) - \mathcal{B}_H(\mathcal{A}_H(z))\right]\mathcal{R}_b(z) + \mathcal{B}_H(\mathcal{A}_H(z)) - z \\
D_1(z) &= z\left[a_0^{(H)}(\mathcal{B}_L(\mathcal{A}_H(z)) - 1) + \mathcal{A}_H(\mathcal{B}_H(\mathcal{A}_H(z))) - 1\right] \\
N_2(z) &= (1-z)\left[\mathcal{R}_b(z) - 1\right] \\
D_2(z) &= z\left[1 - \mathcal{A}_H(z)\right].
\end{aligned}
$$

Differentiating $\mathcal{Q}(z)$ gives

$$
\begin{aligned}
\frac{\mathcal{Q}'(z)}{1-\sigma} &= \prod_{k=1}^{\infty}\frac{\mathcal{R}_b((1-r)^kz)}{\mathcal{R}_b((1-r)^k)}\left(\frac{N_1'(z)D_1(z) - N_1(z)D_1'(z)}{D_1(z)^2} + \frac{N_2'(z)D_2(z) - N_1(z)D_2'(z)}{D_2(z)^2}\right) \\
&\quad + \sum_{k=1}^{\infty}\frac{(1-r)^k\mathcal{R}_b'((1-r)^kz)}{\mathcal{R}_b((1-r)^k)}\prod_{i\neq k}\frac{\mathcal{R}_b((1-r)^iz)}{\mathcal{R}_b((1-r)^i)}\left(\frac{N_1(z)}{D_1(z)} + \frac{N_2(z)}{D_2(z)}\right)
\end{aligned}
$$

and we need to calculate

$$
\begin{aligned}
\frac{\mathcal{Q}'(1)}{1-\sigma} &= \lim_{z\to1}\left(\frac{N_1'(z)D_1(z) - N_1(z)D_1'(z)}{D_1(z)^2} + \frac{N_2'(z)D_2(z) - N_1(z)D_2'(z)}{D_2(z)^2}\right) \\
&\quad + \sum_{k=1}^{\infty}\frac{(1-r)^k\mathcal{R}_b'((1-r)^k)}{\mathcal{R}_b((1-r)^k)}\lim_{z\to1}\left(\frac{N_1(z)}{D_1(z)} + \frac{N_2(z)}{D_2(z)}\right).
\end{aligned}
$$

Again after tedious calculations we find

$$Q'(1) = \frac{(1-\sigma)N_1''(1)}{2D_1'(1)} - \frac{(1-\varrho)D_1''(1)}{2D_1'(1)} + \frac{(1-\sigma)N_2''(1)}{2D_2'(1)} - \frac{\varrho D_2''(1)}{2D_2'(1)}$$
$$+ \sum_{k=1}^{\infty} \frac{(1-r)^k \mathcal{R}_b'((1-r)^k)}{\mathcal{R}_b((1-r)^k)} \tag{12}$$

with

$$N_1''(1) = \mathcal{R}_b(1)\left(\mathcal{B}_L'(1)\mathcal{A}_H''(1) + \mathcal{A}_H'(1)^2\mathcal{B}_L''(1) + 2\mathcal{B}_L'(1)\mathcal{A}_H'(1) + \right.$$
$$\left. -\mathcal{B}_H'(1)\mathcal{A}_H''(1) - \mathcal{A}_H'(1)^2\mathcal{B}_H''(1)\right)$$
$$+ \mathcal{R}_b'(1)\left(\mathcal{B}_L'(1)\mathcal{A}_H'(1) - \mathcal{B}_H'(1)\mathcal{A}_H'(1) + 1\right) + \mathcal{B}_H'(1)\mathcal{A}_H''(1) + \mathcal{B}_H''(1)\mathcal{A}_H'(1)^2$$
$$D_1'(1) = a_0^{(H)}\mathcal{B}_L'(1)\mathcal{A}_H'(1) + \mathcal{A}_H'(1)^2\mathcal{B}_H'(1)$$
$$D_1''(1) = 2a_0^{(H)}\mathcal{B}_L'(1)\mathcal{A}_H'(1) + 2\mathcal{A}_H'(1)^2\mathcal{B}_H'(1) + a_0^{(H)}\mathcal{B}_L'(1)\mathcal{A}_H''(1) + a_0^{(H)}\mathcal{B}_L''(1)\mathcal{A}_H'(1)^2$$
$$+ \mathcal{A}_H'(1)\mathcal{B}_H'(1)\mathcal{A}_H''(1) + \mathcal{A}_H'(1)^3\mathcal{B}_H''(1) + \mathcal{A}_H'(1)^2\mathcal{B}_H'(1)^2\mathcal{A}_H''(1)$$

and

$$N_2''(1) = -2\mathcal{R}_b'(1)$$
$$D_2'(1) = -\mathcal{A}_H'(1), \quad D_2''(1) = -\mathcal{A}_H''(1) - 2\mathcal{A}_H'(1).$$

Of course, we also have to calculate the derivative $\mathcal{R}_b'(1)$. Recall that

$$\mathcal{R}_b(z) = \frac{a_0^{(H)}\phi(z)[z - \mathcal{B}_H(\phi(z))]}{a_0^{(H)}\phi(z)[z - \mathcal{B}_H(\phi(z))] + z[\mathcal{A}_H(z) - \phi(z)\mathcal{A}_H(\mathcal{B}_H(\phi(z)))]}$$
$$= \frac{y^*(z)[\mathcal{B}_H(\phi(z)) - z]}{y^*(z)\mathcal{B}_H(\phi(z)) - z\mathcal{B}_L(\phi(z))}$$

where

$$\phi(z) = \omega(y^*(z), z), \quad \omega(y, z) = \mathcal{A}_L(y)\mathcal{A}_H(z)$$

and $y = y^*(z)$ is the *unique solution* of the equation

$$a_0^{(H)}\omega(y, z)[\mathcal{B}_L(\omega(y, z)) - y] + y[\omega(y, z)\mathcal{A}_H(\mathcal{B}_H(\omega(y, z))) - \mathcal{A}_H(z)] = 0.$$

Now, introduce

$$N(z) = y^*(z)[\mathcal{B}_H(\phi(z)) - z]$$
$$D(z) = y^*(z)\mathcal{B}_H(\phi(z)) - z\mathcal{B}_L(\phi(z)).$$

Then differentiation gives

$$\mathcal{R}_b'(z) = \frac{D(z)N'(z) - N(z)D'(z)}{D(z)^2}$$
$$\mathcal{R}_b'(1) = \lim_{z \to 1} \mathcal{R}_b(z) = \frac{N''(1) - \mathcal{R}_b(1)D''(1)}{2D'(1)}$$

where we have

$$
\begin{aligned}
N''(1) &= 2y^{*\prime}(1)\left(1 - \mathcal{B}'_H(1)\phi'(1)\right) - \mathcal{B}'_H(1)\phi''(1) - \mathcal{B}''_H(1)\phi'(1)^2 \\
D'(1) &= 1 + \mathcal{B}'_L(1)\phi'(1) - \mathcal{B}'_H(1)\phi'(1) - y^{*\prime}(1) \\
D''(1) &= \phi''(1)\left(\mathcal{B}'_L(1) - \mathcal{B}'_H(1)\right)\phi'(1)^2\left(\mathcal{B}''_L(1) - \mathcal{B}''_H(1)\right) \\
&\quad + 2\mathcal{B}'_L(1)\phi'(1) - 2\mathcal{B}'_H(1)y^{*\prime}(1)\phi'(1) - y^{*\prime\prime}(1),
\end{aligned}
$$

and we recall that

$$
\mathcal{R}_b(1) = \frac{a_0^{(H)}[1 - \mathcal{B}'_H(1)\phi'(1)]}{a_0^{(H)}[1 - \mathcal{B}'_H(1)\phi'(1)] + \mathcal{A}'_H(1) - \phi'(1)[\mathcal{A}'_H(1)\mathcal{B}'_H(1) + 1]}.
$$

The expressions for $y^{*\prime}(1)$ and $\phi'(1)$ are given by

$$
y^{*\prime}(1) = \frac{\mathcal{A}'_H(1)\left[a_0^{(H)}\mathcal{B}'_L(1) + \mathcal{A}'_H(1)\mathcal{B}'_H(1)\right]}{a_0^{(H)}\left[1 - \mathcal{A}'_L(1)\mathcal{B}'_L(1)\right] - \mathcal{A}'_L(1)\left[1 + \mathcal{A}'_H(1)\mathcal{B}'_H(1)\right]}
$$

$$
\phi'(1) = \frac{\mathcal{A}'_H(1)\left[a_0^{(H)} - \mathcal{A}'_L(1)\right]}{a_0^{(H)}\left[1 - \mathcal{A}'_L(1)\mathcal{B}'_L(1)\right] - \mathcal{A}'_L(1)\left[1 + \mathcal{A}'_H(1)\mathcal{B}'_H(1)\right]}
$$

and after many further calculations we find

$$
y^{*\prime\prime}(1) = \frac{
\begin{aligned}
&\left(a_0^{(H)}\mathcal{B}'_L(1) + \mathcal{A}'_H(1)\mathcal{B}'_H(1) + 1\right) \\
&\times \left(2\mathcal{A}'_L(1)\mathcal{A}'_H(1)y^{*\prime}(1) + \mathcal{A}''_H(1) + \mathcal{A}''_L(1)y^{*\prime}(1)^2\right) \\
&+\phi'(1)^2\left(2a_0^{(H)}\mathcal{B}'_L(1) + a_0^{(H)}\mathcal{B}''_L(1) + \mathcal{A}'_H(1)\mathcal{B}''_H(1)\right. \\
&\quad\left. + \mathcal{A}''_H(1)\mathcal{B}'_H(1)^2 + 2\mathcal{A}'_H(1)\mathcal{B}'_H(1)\right) + \\
&-\mathcal{A}''_H(1) - 2y^{*\prime}(1)\left(\mathcal{A}'_H(1) + \phi'(1)\left(a_0^{(H)} - \mathcal{A}'_H(1)\mathcal{B}'_H(1) - 1\right)\right)
\end{aligned}
}{
a_0^{(H)} - a_0^{(H)}\mathcal{A}'_L(1)\mathcal{B}'_L(1) - \mathcal{A}'_L(1)\mathcal{A}'_H(1)\mathcal{B}'_H(1) - \mathcal{A}'_L(1)
}
$$

$$
\phi''(1) = \mathcal{A}''_H(1) + 2\mathcal{A}'_L(1)\mathcal{A}'_H(1)y^{*\prime}(1) + \mathcal{A}''_L(1)y^{*\prime}(1)^2 + \mathcal{A}'_L(1)y^{*\prime\prime}(1).
$$

Plugging in all these results in (12) gives a closed form expression for $\overline{\mathcal{Q}} = \mathcal{Q}'(1)$.

5   Numerical Results

The starting position for our numerical results is

– All distributions geometric [batch size shifted to 0]

$$
\mathcal{B}'_H(1) = \mathcal{B}'_L(1) = 2, \quad \mathcal{A}'_H(1) = 0.21, \quad \mathcal{A}'_L(1) = 0.09,
$$

– So $a_0^{(H)} = 0.826$,
– The offered load of L-customers is $\varrho_L := \mathcal{A}'_L(1)\mathcal{B}'_L(1) = 0.18$,

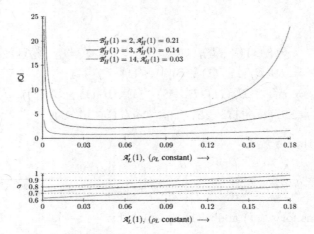

Fig. 1. The mean size of the orbit $\bar{Q}$ as a function of $\mathcal{A}'_L(1)$, with $r = 0.5$ and $\rho_L = 0.18$ constant.

- The offered loa4d of H-customers is $\varrho_H := \mathcal{A}'_H(1)\mathcal{B}'_H(1) = 0.42$,
- The total offered load is $\varrho = \varrho_L + \varrho_H = 0.60$,
- The total used capacity is

$$\sigma = \varrho + \mathcal{A}'_L(1) + \varrho\mathcal{A}'_H(1) + \left(1 - a_0^{(H)}\right)(1 - \varrho) = 0.885.$$

Firstly, we take the offered load of L-customers $\varrho_L := \mathcal{A}'_L(1)\mathcal{B}'_L(1)$ constant and $\mathcal{A}'_L(1)$ increasing. The numerical results are presented in Fig. 1.

Next, we keep the total offered load $\varrho = 0.6$ constant, $\varrho_H := \mathcal{A}'_H(1)\mathcal{B}'_H(1)$ increasing. The results are presented in Fig. 2.

Finally, we keep the total offered load $\varrho = 0.65$ constant, and again ϱ_H increasing. The results are presented in Fig. 3.

From the Figs. 1, 2 and 3 we can draw the following conclusions.

- Keeping both the offered load of inbound calls $\varrho_H = \mathcal{A}'_H(1)\mathcal{B}'_H(1)$ and the offered load of outbound calls $\varrho_L = \mathcal{A}'_L(1)\mathcal{B}'_L(1)$ constant we have seen that
  - Increasing the arrival intensity $\mathcal{A}'_H(1)$ of inbound calls [and simultaneously decreasing the mean service time $\mathcal{B}'_H(1)$] *decreases the mean queue length $\overline{\mathcal{L}}$ of outbound calls* and *increases the mean orbit size $\overline{Q}$ of inbound calls.*
  - Increasing the arrival intensity $\mathcal{A}'_L(1)$ of outbound calls [and simultaneously decreasing the mean service time $\mathcal{B}'_L(1)$] *increases the mean queue length $\overline{\mathcal{L}}$ of outbound calls* and *first decreases and then increases the mean orbit size $\overline{Q}$ of inbound calls.*

Keeping the total offered load $\varrho = \varrho_H + \varrho_L$ constant we have seen that *increasing the offered load ϱ_H of inbound calls* [and simultaneously decreasing the offered load ϱ_L of outbound calls]

- *decreases the mean queue length $\overline{\mathcal{L}}$ of outbound calls* and *increases the mean orbit size $\overline{Q}$ of inbound calls.*

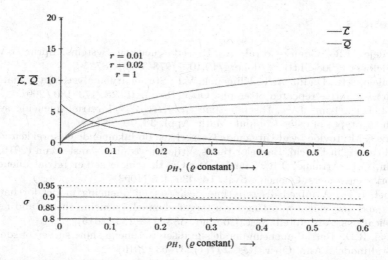

Fig. 2. The mean length of the queue $\bar{\mathcal{L}}$ and mean size of the orbit $\bar{\mathcal{Q}}$ as a function of ρ_H, with $\varrho = 0.6$ constant.

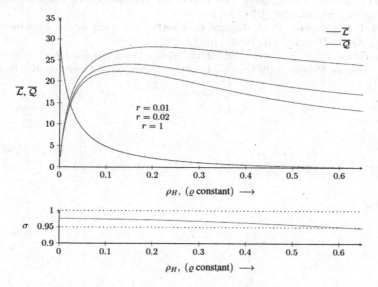

Fig. 3. The mean length of the queue $\bar{\mathcal{L}}$ and mean size of the orbit $\bar{\mathcal{Q}}$ as a function of ρ_H, with offered load $\varrho = 0.65$ constant.

- increases the mean orbit size $\overline{\mathcal{Q}}$ of inbound calls *for a moderate total offered load, say $\varrho = 0.6$,*
- first increases and then decreases the mean orbit size $\overline{\mathcal{Q}}$ of inbound calls *for a high total offered load, say $\varrho = 0.65$.*

For more numerical results we refer to Dekker [4].

References

1. Artalejo, J.R., Gómez-Corral, A.: Retrial Queueing Systems. Springer-Verlag, Heidelberg (2008). https://doi.org/10.1007/978-3-540-78725-9
2. Artalejo, J.R., Dudin, A.N., Klimenok, V.I.: Stationary analysis of a retrial queue with preemptive repeated attempts. Oper. Res. Lett. **28**, 173–180 (2001)
3. Choi, B.D., Kim, J.W.: Discrete-time $Geo_1, Geo_2/G/1$ retrial queueing systems with two types of calls. Comput. Math. Appl. **33**, 79–88 (1997)
4. Dekker, M.: Invloed van Uitgaande Gesprekken op Inkomende Gesprekken Bij Een Call-center [in Dutch], Bachelor thesis, Vrije Universiteit Amsterdam (2016)
5. Falin, G.I., Artalejo, J.R., Martin, M.: On the single server retrial queue with priority customers. Queueing Syst. **14**, 439–455 (1993)
6. Nobel, R.D.: A mixed discrete-time delay/retrial queueing model for handover calls and new calls competing for a targat channel, queueing theory and network applications. Adv. Intell. Syst. Comput. **383**, 173–185 (2015)
7. Nobel, R.D.: Retrial queueing models in discrete time: a short survey of some late arrival models. Ann. Oper. Res. **247**(1), 37–63 (2016)
8. Nobel, R.D., Moreno, P.: A discrete-time priority loss/retrial queueing model with two types of traffic. In: Choi, B.D. (ed.) Proceedings of the Korea-Netherlands Joint Conference on Queueing Theory and its Applications to Telecommunication Systems, Seoul, pp. 189–207 (2005)
9. Nobel, R.D., Moreno, P.: A discrete-time retrial queueing model with one server. Eur. J. Oper. Res. **189**(3), 1088–1103 (2008)
10. el Sharkawy, S.: Een Discrete-tijd Retrial/delay Wachtrijmodel Met Klanten Van Lage en Hoge Prioriteit [in Dutch], Bachelor thesis, Vrije Universiteit Amsterdam (2014)

Fluid and Diffusion Limits for Bike Sharing Systems

Quan-Lin Li[✉], Zhi-Yong Qian, and Rui-Na Fan

School of Economics and Management Sciences, Yanshan University,
Qinhuangdao 066004, China
liquanlin@tsinghua.edu.cn, {qianzhiyong0926,fanruina}@stumail.ysu.edu.cn

Abstract. Bike sharing systems have rapidly developed around the world, and they are served as a promising strategy to improve urban traffic congestion and to decrease polluting gas emissions. So far performance analysis of bike sharing systems always exists many difficulties and challenges under some more general factors. In this paper, a more general large-scale bike sharing system is discussed by means of heavy traffic approximation of multiclass closed queueing networks with non-exponential factors. Based on this, the fluid scaled equations and the diffusion scaled equations are established by means of the numbers of bikes both at the stations and on the roads, respectively. Furthermore, the scaling processes for the numbers of bikes both at the stations and on the roads are proved to converge in distribution to a semimartingale reflecting Brownian motion (SRBM) in a N^2-dimensional box, and also the fluid and diffusion limit theorems are obtained. Furthermore, performance analysis of the bike sharing system is provided. Thus the results and methodology of this paper provide new highlight in the study of more general large-scale bike sharing systems.

Keywords: Bike sharing systems · Fluid limit · Diffusion limit Semimartingale reflecting Brownian motion

1   Introduction

Bike sharing systems have become an important way of urban transportation due to its accessibility and affordability, and they are widely deployed in more than 600 major cities around the world. Bike sharing systems are regarded as promising solutions to reduce congestion of traffic and parking, automobile exhaust pollution, transportation noise, and so on. For some survey and development of bike sharing systems, readers may refer to, DeMaio [10], Shaheen et al. [33], Shu et al. [35], Labadi et al. [23], and Meddin and DeMaio [28].

Two major operational issues of bike sharing systems are to care for (i) the non-empty: sufficient bikes parked at each station in order to be able to rent a bike at any time; and (ii) the non-full: suitable bike parking capacity designed for each station in order to be able to return a bike in real time. Thus the empty

© Springer International Publishing AG 2017
W. Yue et al. (Eds.): QTNA 2017, LNCS 10591, pp. 217–245, 2017.
https://doi.org/10.1007/978-3-319-68520-5_14

or full stations are called problematic stations. Up till now, efficient measures are developed in the study of problematic stations, including time-nonhomogeneous demand forecasting, average bike inventory level, real-time bike repositioning, and probability analysis of problematic stations.

So far queueing models and Markov processes have been applied to characterizing some key steady-state performance of the bike sharing systems. Important prior works on the bike sharing models include the $M/M/1/C$ queue by Leurent [22] and Schuijbroek et al. [34]; the time-inhomogeneous $M(t)/M(t)/1/C$ model by Raviv et al. [31] and Raviv and Kolka [30]; the queueing networks by Kochel et al. [20], Savin et al. [32], Adelman [1], George and Xia [14,15] and Li et al. [26]; the fluid models combining with Markov decision processes by Waserhole and Jost [36,37]; the mean-field theory by Fricker et al. [11], Fricker and Gast [12] and Fricker and Tibi [13]; the time-inhomogeneous $M(t)/M(t)/1/K$ and $MAP(t)/MAP(t)/1/K + 2L + 1$ queues combining with mean-field theory by Li et al. [24] and Li and Fan [25].

An important and realistic feature of bike sharing systems is the time-varying arrivals of bike users and their random travel times. In general, analysis of bike sharing systems with non-Poisson user arrivals and general travel times are always very difficult and challenging because more complicated multiclass closed queueing networks are established to deal with bike sharing systems. See Li et al. [26] for more interpretations. For this, fluid and diffusion approximations may be an effective and better method in the study of more general bike sharing systems. This motivates us in this paper to develop fluid and diffusion limits for more general large-scale bike sharing systems.

Fluid and diffusion approximations are usually applied to analysis of more general large-scale complicated queueing networks, which possibly originate in some practical systems including communication networks, manufacturing systems, transportation networks and so forth. See excellent monographs by, for example, Harrison [16], Chen and Yao [4], Whitt [38]. For the bike sharing system, further useful information is introduced as follows. (a) For heavy traffic approximation of closed queueing networks, readers may refer to, such as, Harrison et al. [19] for a closed queueing network with homogeneous customer population and infinite buffer. Chen and Mandelbaum [3] for a closed Jackson network, Harrison and Williams [18] for a multiclass closed network with two single-server stations and a fixed customer population. Kumar [21] for a two-server closed networks in heavy traffic. (b) For heavy traffic approximation of queueing networks with finite buffers, important examples include, Dai and Dai [6] obtained the SRBM of queue-length process relying on a uniform oscillation result for solutions to a family of Skorohod problems. Dai [8] modeled the queueing networks with finite buffers under a communication blocking scheme, showed that the properly normalized queue length process converges weakly to a reflected Brownian motion in a rectangular box, and presented a general implementation via finite element method to compute the stationary distribution of SRBM. Furthermore, Dai [9] analyzed a multiclass queueing networks with finite buffers and a feedforward routing structure under a blocking scheme, and showed a pseudo-heavy-traffic limit theorem which stated that the limit process of queue length is a reflecting Brownian motion. (c) There are some available results on heavy traffic approximation of multiclass queueing networks, readers may refer to, for instance, Harrison and Nguyen [17], Dai [5], Bramson [2], Meyn [29] and Majewski [27].

Contributions of this Paper: The main contributions of this paper are three-fold. The first contribution is to propose a more general large-scale bike sharing system having renewal arrival processes of bike users and general travel times, and to establish a multiclass closed queueing network from the practical factors of the bike-sharing system where bikes are abstracted as virtual customers, while both stations and roads are regarded as virtual nodes or servers. Note that the virtual customers (i.e. bikes) at stations are of single class; while the virtual customers (i.e. bikes) on roads are of two different classes due to two classes of different bike travel or return times. The second contribution is to set up the queue-length processes of the multiclass closed queueing network through observing both some bikes parked at stations and the other bikes ridded on roads. Such analysis gives the fluid scaled equations and the diffusion scaled equations by means of the numbers of bikes both at the stations and on the roads. The third contribution is to prove that the scaling processes, corresponding to the numbers of bikes both at the stations (having one class of virtual customers) and on the roads (having two classes of virtual customers), converge in distribution to a semimartingale reflecting Brownian motion, and the fluid and diffusion limit theorems are obtained in some simple versions. Based on this, performance analysis of the bike sharing system is also given. Therefore, the results and methodology given in this paper provide new highlight on the study of more general large-scale bike sharing systems.

Organization of this Paper: The structure of this paper is organized as follows. In Sect. 2, we describe a more general large-scale bike sharing system with N different stations and with $N(N-1)$ different roads, while this system has renewal arrival processes of bike users and general travel times on the roads. In Sect. 3, we establish a multiclass closed queueing network from practical factors of the bike-sharing system where bikes are abstracted as virtual customers, while both stations and roads are regarded as virtual nodes or servers. In Sect. 4, we set up the queue-length processes of the multiclass closed queueing network by means of the numbers of bikes both at the stations and on the roads, and establish the fluid scaled equations and the diffusion scaled equations. In Sects. 5 and 6, we prove that the scaling processes of the bike sharing system converge in distribution to a semimartingale reflecting Brownian motion under heavy traffic conditions, and obtain the fluid limit theorem and the diffusion limit theorem, respectively. In Sect. 7, we give performance analysis of the bike sharing system by means of the fluid and diffusion limits. Finally, some concluding remarks are described in Sect. 8.

Useful Notation: We now introduce the notation used in the paper. For positive integer n, the n-dimensional Euclidean space is denoted by $\mathcal{R}^n$ and the

n-dimensional positive orthant is denoted by $\mathcal{R}_+^n = \{x \in \mathcal{R}^n : x_i \geq 0\}$. We definite $D_{\mathcal{R}^n}[0,T]$ as the path space of all functions $f : [0,T] \to \mathcal{R}^n$ which are right continuous and have left limits. Define $\delta_{j,k} = 1$ if $j = k$, else, $\delta_{j,k} = 0$. For a set $\mathcal{K}$, let $|\mathcal{K}|$ denote its cardinality. u.o.c. means that the convergence is uniformly on compact set. A triple $(\Omega, \mathcal{F}, \{\mathcal{F}_t\})$ is called a filtered space if Ω is a set, $\mathcal{F}$ is a σ-field of subsets of Ω, and $\{\mathcal{F}_t, t \geq 0\}$ is an increasing family of sub-σ-fields of $\mathcal{F}$, i.e., a filtration. If, in addition, P is a probability measure on $(\Omega, \mathcal{F})$, then $(\Omega, \mathcal{F}, \{\mathcal{F}_t\}, P)$ is called a filtered probability space. Let P_x denote the unique family of probability measures on $(\Omega, \mathcal{F})$, and E_x be the expectation operator under P_x.

2   Model Description

In this section, we describe a more general large-scale bike sharing system with N different stations and with $N(N-1)$ different roads, which has renewal arrival processes of bike users and general travel times.

In the large-scale bike sharing system, a customer arrives at a nonempty station, rents a bike, and uses it for a while, then he returns the bike to a destination station and immediately leaves this system. If a customer arrives at a empty station, then he immediately leaves this system.

Now, we describe the bike sharing system including operations mechanism, system parameters and mathematical notation as follows:

(1) Stations and roads: We assume that the bike sharing system contains N different stations and at most $N(N-1)$ different roads, where a pair of directed roads may be designed from any station to another station. Also, we assume that at the initial time $t = 0$, each station has C_i bikes and K_i parking positions, where $1 \leq C_i \leq K_i < \infty$ for $i = 1, \ldots, N$ and $\sum_{i=1}^N C_i > K_j$ for $j = 1, \ldots, N$. Note that these conditions make that some bikes can result in at least a full station.

(2) Arrival processes: The arrivals of outside bike users (or customers) at each station is a general renewal process. For station j, let $u_j = \{u_j(n), n \geq 1\}$ be an i.i.d. random sequence of exogenous interarrival times, where $u_j(n) \geq 0$ is the interarrival time between the $(n-1)$st customer and the nth customer. We assume that $u_j(n)$ has the mean $1/\lambda_j$ and the coefficient of variation $c_{a,j}$.

(3) The bike return times:

(3.1) The first return: Once an outside customer successfully rents a bike from station i, then he rides on a road directed to station j with probability $p_{i \to j}$ for $\sum_{j \neq i}^N p_{i \to j} = 1$, and his riding-bike time $v_{i \to j}^{(1)}$ on the road $i \to j$ is a general distribution with the mean $1/\mu_{i \to j}^{(1)}$ and the coefficient of variation $c_{s, i \to j}^{(1)}$. If there is at least one available parking position at station j, then the customer directly returns his bike to station j, and immediately leaves this system. Let $r^i = \{r_j^i(n), n \geq 1\}$ be a sequence of routing selections for $i, j = 1, \ldots, N$ with $i \neq j$, where $r_j^i(n) = 1$ means that the nth customer rents a bike from station i and rides on a road directed to station j (i.e., the customer rides on road $i \to j$), hence $\Pr\{r_j^i(n) = 1\} = p_{i \to j}$.

(3.2) The second return: From (3.1), if no parking position is available at station j, then the customer has to ride the bike to another station l_1 with probability $\alpha_{j \to l_1}$ for $\sum_{l_1 \neq j}^{N} \alpha_{j \to l_1} = 1$, and his riding-bike time $v_{j \to l_1}^{(2)}$ on road $j \to l_1$ is also a general distribution with the mean $1/\mu_{j \to l_1}^{(2)}$ and the coefficient of variation $c_{s,j \to l_1}^{(2)}$. If there is at least one available parking position at station l_1, then the customer directly returns his bike and immediately leaves this bike sharing system.

　(3.3) The $(k+1)$st return for $k \geq 2$: From (3.2) and more, we assume that this bike has not been returned at any station yet through k consecutive returns. In this case, the customer has to try his $(k+1)$st lucky return, he will ride bike from the l_{k-1}th full station to the l_kth station with probability $\alpha_{l_{k-1} \to l_k}$ for $\sum_{l_k \neq l_{k-1}}^{N} \alpha_{l_{k-1} \to l_k} = 1$, and his riding-bike time $v_{l_{k-1} \to l_k}^{(2)}$ on road $l_{k-1} \to l_k$ is also a general distribution with the mean $1/\mu_{l_{k-1} \to l_k}^{(2)}$ and the coefficient of variation $c_{s,l_{k-1} \to l_k}^{(2)}$. If there is at least one available parking station, then the customer directly returns his bike and immediately leaves this bike sharing system; otherwise he has to continuously try another station again. In the next section, those bikes ridden under their first return are called the first class of virtual customers; while those bikes ridden under the k $(k \geq 2)$ returns are called the second class of virtual customers. Let $\bar{r}^j = \{\bar{r}_i^j(n), n \geq 1\}$ be a sequence of routing selections for $i, j = 1, \ldots, N$ with $i \neq j$, where $\bar{r}_i^j(n) = 1$ means that the nth customer who can not return the bike to the full station j will deflect into road $j \to i$, thus $\Pr\{\bar{r}_i^j(n) = 1\} = \alpha_{j \to i}$. Similarly, let $r^{j \to i,(d)} = \{r^{j \to i,(d)}(n), n \geq 1\}$ be a sequence of routing selections for $i, j = 1, \ldots, N$ with $i \neq j, d = 1, 2$, where $r^{j \to i,(d)}(n) = 1$ means the nth customer of class d who completes his short trip on road $j \to i$ will return the bike to station i, hence $\Pr\{r^{j \to i,(d)}(n) = 1\} = p_{j \to i, i} = 1$.

　(4) Two classes of riding-bike times: In (3), there are two classes of riding-bike times, who have two general distributions, that is, there are two classes of virtual customers riding on each road. Let $v_{j \to i}^{(d)} = \{v_{j \to i}^{(d)}(n), n \geq 1\}$ be a random sequence of riding-bike times of class d for $i, j = 1, \ldots, N$ with $i \neq j, d = 1, 2$, where $v_{j \to i}^{(d)}(n)$ is the riding-bike time for the nth customer of class d riding on the road $j \to i$. We assume that $v_{j \to i}^{(d)}$ has the mean $1/\mu_{j \to i}^{(d)}$ and the coefficient of variation $c_{s,j \to i}^{(d)}$. To care for the expected riding-bike times, we set that $\mu_{j \to i}^{(d)} = 1/m_{j \to i}$ for $d = 1$ and $\mu_{j \to i}^{(d)} = 1/\xi_{j \to i}$ for $d = 2$.

　(5) The departure disciplines: The customer departure has two different cases: (a) an outside customer directly leaves the bike sharing system if he arrives at an empty station; (b) if one customer rents and uses a bike, and he finally returns the bike to a station, then the customer completes his trip and immediately leaves the bike sharing system.

　For such a bike sharing system, Fig. 1 outlines its physical structure and associated operations.

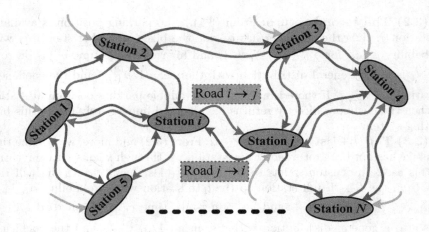

Fig. 1. The physical structure of the bike sharing system.

3   The Closed Queueing Network

In this section, we establish a multiclass closed queueing network from the bike-sharing system where bikes are abstracted as virtual customers, and both stations and roads are regarded as virtual nodes or servers. Specifically, the stations contain only one class of virtual customers; while the roads can contain two classes of virtual customers.

In the bike sharing system, there are N stations and $N(N-1)$ roads, and each bike can not leave this system, hence, the total number of bikes in this system is fixed as $\sum_{i=1}^{N} C_i$. Base on this, such a system can be regarded as a closed queueing network with multiclass customers due to two types of different travel or return times.

Let S_i and $R_{i \to j}$ denote station i and road $i \to j$, respectively. Let SN denote the set of nodes abstracted by the stations, and RN the set of nodes abstracted by the roads. Clearly SN $= \{S_i, i = 1, \ldots, N\}$ and RN $= \{R_{i \to j} : i, j = 1, \ldots, N \text{ with } i \neq j\}$. Let n_j and $n_{i \to j}^{(d)}$ denote the numbers of bikes parking in the jth station node and of bikes of class d riding on the road $i \to j$ node, respectively.

(1) Virtual nodes: Although the stations and the roads have different physical attributes, they are all regarded as abstract nodes in the closed queueing network.

(2) Virtual customers: The virtual customers are abstracted by the bikes, which are either parked in the stations or ridden on the roads. It is seen that only one class of virtual customers are packed in the station nodes; while two classes of different virtual customers are ridden on the road nodes due to their different return times.

(3) **The routing matrix P:** To express the routing matrix, we first define a mapping $\sigma(\cdot)$ as follow,

$$\begin{cases} \sigma(S_i) = i & \text{for } i = 1, \ldots, N, \\ \sigma(R_{i \to j}) = i\langle j \rangle & \text{for } i, j = 1, \ldots, N, \text{ with } i \neq j. \end{cases}$$

It is necessary to understand the mapping $\sigma(\cdot)$. For example, $N = 2, \sigma(S_1) = 1$, $\sigma(S_2) = 2, \sigma(R_{1 \to 2}) = 1\langle 2 \rangle$, $\sigma(R_{2 \to 1}) = 2\langle 1 \rangle$, thus the routing matrix is written as

$$P = \begin{array}{c} 1 \\ 2 \\ 1\langle 2 \rangle \\ 2\langle 1 \rangle \end{array} \overset{\displaystyle 1 \;\; 2 \;\; 1\langle 2 \rangle \;\; 2\langle 1 \rangle}{\left[\begin{array}{cccc} & & & \\ & & & \\ & & & \\ & & & \end{array} \right]}.$$

In this case, the component $p_{\tilde{i}, \tilde{j}}$ of the routing matrix P denotes the probability that a customer leaves node $\tilde{i}$ to node $\tilde{j}$, where

$$p_{\tilde{i}, \tilde{j}} = \begin{cases} 1 & \text{if } \tilde{i} = \sigma(R_{i \to j}), \tilde{j} = \sigma(S_j), \\ p_{i \to j} & \text{if } \tilde{i} = \sigma(S_i), \tilde{j} = \sigma(R_{i \to j}), \\ \alpha_{j \to k} & \text{if } \tilde{i} = \sigma(R_{i \to j}), \tilde{j} = \sigma(R_{j \to k}), \\ 0 & \text{otherwise.} \end{cases}$$

(4) **The service processes in the station nodes:** For $j \in \text{SN}$, the service process $S_j = \{S_j(t), t \geq 0\}$ of station node j, associated with the interarrival time sequence $u_j = \{u_j(n), n \geq 1\}$ of the outside customers who arrive at station j, is given by

$$S_j(t) = \sup\{n : U_j(n) \leq t\},$$

where $U_j(n) = \sum_{l=1}^{n} u_j(l), n \geq 1$ and $U_j(0) = 0$. Let $b_j = \lambda_j 1_{\{1 \leq n_j \leq K_j\}}$.

(5) **The service processes in the road nodes:** For $i, j = 1, \ldots, N$ with $i \neq j$ and $d = 1, 2$, the service process $S_{j \to i}^{(d)} = \{S_{j \to i}^{(d)}(t), t \geq 0\}$ of road node $j \to i$, associated with the riding-bike time sequence $v_{j \to i}^{(d)} = \{v_{j \to i}^{(d)}(n), n \geq 1\}$ of the customers of class d ridden on road $j \to i$, is given by

$$S_{j \to i}^{(d)}(t) = \sup\{n : V_{j \to i}^{(d)}(n) \leq t\},$$

where $V_{j \to i}^{(d)}(n) = \sum_{l=1}^{n} v_{j \to i}^{(d)}(l), \ n \geq 1$ and $V_{j \to i}^{(d)}(0) = 0$. We write

$$b_{j \to i}^{(d)} = n_{j \to i}^{(d)} \mu_{j \to i}^{(d)} = \begin{cases} n_{j \to i}^{(1)} \frac{1}{m_{j \to i}} & d = 1, \\ n_{j \to i}^{(2)} \frac{1}{\xi_{j \to i}} & d = 2. \end{cases}$$

(6) **The routing processes in the station nodes:**

Case one: For $j \in \text{SN}$, the routing process $R^j = \{R_i^j, i \neq j, i = 1, \ldots, N\}$ and $R_i^j = \{R_i^j(n), n \geq 1\}$, associated with the routing selecting sequence $r^i = \{r_j^i(n), n \geq 1\}$ of station j, is given by

$$R^j(n) = \sum_{l=1}^{n} r^j(l) \text{ or } R_i^j(n) = \sum_{l=1}^{n} r_i^j(l), n \geq 1,$$

and the ith component of $R^j(n)$ is $R_i^j(n)$ associated with probability $p_{j \to i}$.

Case two: For $j \in \text{SN}$, the routing process $\bar{R}^j = \{\bar{R}^j_i, i \neq j, i = 1, \ldots, N\}$ and $\bar{R}^j_i = \{\bar{R}^j_i(n), n \geq 1\}$, associated with the routing deflecting sequence $\bar{r}^j = \{\bar{r}^j_i(n), n \geq 1\}$ of station j, is given by

$$\bar{R}^j(n) = \sum_{l=1}^{n} \bar{r}^j(l) \text{ or } \bar{R}^j_i(n) = \sum_{l=1}^{n} \bar{r}^j_i(l), n \geq 1,$$

and the ith component of $\bar{R}^j(n)$ is $\bar{R}^j_i(n)$ associated with probability $\alpha_{j \to i}$.

(7) The routing processes in the road nodes: For $i, j = 1, \ldots, N$ with $i \neq j$ and $d = 1, 2$, the routing process $R^{j \to i, (d)} = \{R^{j \to i, (d)}(n), n \geq 1\}$, associated with the routing transferring sequence $r^{j \to i, (d)} = \{r^{j \to i, (d)}(n), n \geq 1\}$ of road $j \to i$, is given by

$$R^{j \to i, (d)}(n) = \sum_{l=1}^{n} r^{j \to i, (d)}(l), n \geq 1,$$

and the $R^{j \to i, (d)}(n)$ is associated with probability $p_{j \to i, i} = 1$.

(8) Service disciplines: The first come first served (FCFS) discipline is assumed for all station nodes. A new processor sharing (PS) is used for all the road nodes, where each customer of either class one or class two is served by a general service time distribution, as described in (4) and (5).

4   The Joint Queueing Process

In this section, we set up the queue-length processes of the multiclass closed queueing network by means of the numbers of bikes both at the stations and on the roads, and establish the fluid scaled equations and the diffusion scaled equations.

(1) $Q(t) = \{(Q_j(t), Q^{(d)}_{j \to i}(t)), i \neq j, i, j = 1, \ldots, N; d = 1, 2; t \geq 0\}$, where $Q_j(t)$ and $Q^{(d)}_{j \to i}(t)$ are the number of virtual customers at station node j and the numbers of virtual customers of class d at the road $j \to i$ at time t, respectively. Specifically, $Q_j(0)$ and $Q^{(d)}_{j \to i}(0)$ are the number of virtual customers at station node j and the number of virtual customers of class d on the road node $j \to i$ at time $t = 0$, respectively.

(2) $Y^K(t) = \{(Y^K_j(t)), j = 1, \ldots, N; t \geq 0\}$, where $Y^K_j(t)$ is the cumulative number of virtual customers deflecting from station node j whose parking positions are full in the time interval $[0, t]$.

(3) $Y^0(t) = \{(Y^0_j(t), Y^{0, (d)}_{j \to i}(t)), i \neq j, i, j = 1, \ldots, N; d = 1, 2; t \geq 0\}$, where $Y^0_j(t)$ and $Y^{0, (d)}_{j \to i}(t)$ are the cumulative amount of time that station node j and the road node $j \to i$ are idle (no available bike, i.e., empty) in the time interval $[0, t]$, respectively.

$$Y_j^0(t) = \int_0^t 1\{Q_j(s) = 0\}ds = t - B_j(t),$$

$$Y_{j\to i}^{0,(d)}(t) = \int_0^t 1\{Q_{j\to i}^{(d)}(s) = 0\}ds = t - B_{j\to i}^{(d)}(t).$$

(4) $B(t) = \{(B_j(t), B_{j\to i}^{(d)}(t)), i \neq j, i, j = 1, \ldots, N; d = 1, 2; t \geq 0\}$, where $B_j(t)$ and $B_{j\to i}^{(d)}(t)$ are the cumulative amount of time that the station node j and the road node $j \to i$ are busy (available bike, non-empty) in the time interval $[0, t]$, respectively.

$$B_j(t) = \int_0^t 1\{0 < Q_j(s) \leq K_j\}ds,$$

$$B_{j\to i}^{(d)}(t) = \int_0^t 1\{Q_{j\to i}^{(d)}(s) > 0\}ds.$$

(5) $B^F(t) = \{(B_j^F(t)), j = 1, \ldots, N; t \geq 0\}$, where $B_j^F(t)$ is the cumulative amount of time that station node j is full (no available parking position) in the time interval $[0, t]$,

$$B_j^F(t) = \int_0^t 1\{Q_j(s) = K_j\}ds.$$

(6) $S_j(B_j(t))$ denotes the number of virtual customers that have completed service at station node j during the time interval $[0, t]$; $S_{j\to i}^{(d)}(B_{j\to i}^{(d)}(t))$ denotes the number of virtual customers of class d that have completed service at road node $j \to i$ during the time interval $[0, t]$.

(7) $R_i^j(S_j(B_j(t)))$ denotes the number of virtual customers that enter station node i (i.e., riding on road $j \to i$) from station node j during the time interval $[0, t]$; $\bar{R}_i^j(Y_j^K(t))$ denotes the number of virtual customers that enter station node i from station j whose parking positions are full during the time interval $[0, t]$; and $R^{j\to i,(d)}(S_{j\to i}^{(d)}(B_{j\to i}^{(d)}(t)))$ denotes the number of virtual customers of class d that enter station node i from road node $j \to i$ during the time interval $[0, t]$.

Now, we have the following flow balance relations for the station nodes and the road nodes. For station node $j = 1, \ldots, N$,

$$Q_j(t) = Q_j(0) + \sum_{d=1}^2 \sum_{i \neq j}^N \left[R^{i\to j,(d)}(S_{i\to j}^{(d)}(B_{i\to j}^{(d)}(t))) - R^{i\to j,(d)}(S_{i\to j}^{(d)}(B_j^F(t))) \right]$$

$$- S_j(B_j(t)). \tag{1}$$

Note that $Y_j^K(t) = \sum_{d=1}^2 \sum_{i\neq j}^N R^{i\to j,(d)}(S_{i\to j}^{(d)}(B_j^F(t)))$, we have

$$Q_j(t) = Q_j(0) + \sum_{d=1}^2 \sum_{i \neq j}^N R^{i\to j,(d)}(S_{i\to j}^{(d)}(B_{i\to j}^{(d)}(t))) - S_j(B_j(t)) - Y_j^K(t). \tag{2}$$

For road node $j \to i$ for $i, j = 1, \ldots, N$ with $i \neq j$ and $d = 1, 2$, we have

$$Q_{j\to i}^{(1)}(t) = Q_{j\to i}^{(1)}(0) + R_i^j(S_j(B_j(t))) - S_{j\to i}^{(1)}(B_{j\to i}^{(1)}(t)), \tag{3}$$

$$Q_{j\to i}^{(2)}(t) = Q_{j\to i}^{(2)}(0) + \bar{R}_i^j(Y_j^K(t)) - S_{j\to i}^{(2)}(B_{j\to i}^{(2)}(t)). \tag{4}$$

Because the total number of bikes in this bike sharing system is fixed as $\sum_{i=1}^{N} C_i$, we get that for $t \geq 0$

$$\sum_{i=1}^{N} Q_i(t) + \sum_{d=1}^{2} \sum_{i \neq j}^{N} Q_{i \to j}^{(d)}(t) = \sum_{i=1}^{N} C_i. \tag{5}$$

We now elaborate to apply a centering operation to the queue-length representations of the station nodes and of the road nodes, and rewrite (2), (3) and (4) as follows:

$$Q(t) = X(t) + R^0 Y^0(t) + R^K Y^K(t), \tag{6}$$

where $X(t) = (X_1(t), X_2(t), \ldots, X_N(t))$, and $X_j(t)$ is given by

$$X_j(t) = Q_j(0) + \sum_{d=1}^{2} \sum_{i \neq j}^{N} \left[R^{i \to j,(d)}(S_{i \to j}^{(d)}(B_{i \to j}^{(d)}(t))) - S_{i \to j}^{(d)}(B_{i \to j}^{(d)}(t)) \right]$$

$$+ \sum_{d=1}^{2} \sum_{i \neq j}^{N} \left[S_{i \to j}^{(d)}(B_{i \to j}^{(d)}(t)) - b_{i \to j}^{(d)} B_{i \to j}^{(d)}(t) \right] - [S_j(B_j(t)) - b_j B_j(t)]$$

$$- Y_j^K(t) + \theta_j t, \tag{7}$$

note that $R^{i \to j,(d)}(S_{i \to j}^{(d)}(B_{i \to j}^{(d)}(t))) = S_{i \to j}^{(d)}(B_{i \to j}^{(d)}(t))$, $X_j(t)$ is simplified as

$$X_j(t) = Q_j(0) + \sum_{d=1}^{2} \sum_{i \neq j}^{N} \left[S_{i \to j}^{(d)}(B_{i \to j}^{(d)}(t)) - b_{i \to j}^{(d)} B_{i \to j}^{(d)}(t) \right]$$

$$- [S_j(B_j(t)) - b_j B_j(t)] - Y_j^K(t) + \theta_j t, \tag{8}$$

$$\theta_j = \sum_{d=1}^{2} \sum_{i \neq j}^{N} b_{i \to j}^{(d)} - b_j, \tag{9}$$

$$\left(R^0 Y^0(t) \right)_{\tilde{i},\tilde{j}} = \begin{cases} b_j Y_j^0(t), & \text{if } \tilde{i} = \sigma(S_i), \text{ and } \tilde{j} = \tilde{i}, \\ -\sum_{d=1}^{2} b_{i \to j}^{(d)} Y_{i \to j}^{0,(d)}(t), & \text{if } \tilde{i} = \sigma(S_i), \text{ and } \tilde{j} = \sigma(R_{i \to j}), \\ 0, & \text{otherwise}, \end{cases} \tag{10}$$

$$\left(R^K Y^K(t) \right)_{\tilde{i},\tilde{j}} = \begin{cases} -Y_j^K(t), & \text{if } \tilde{i} = \sigma(S_i), \text{ and } \tilde{i} = \tilde{j}, \\ 0, & \text{otherwise}. \end{cases} \tag{11}$$

For road node $j \to i$ $(i, j = 1, \ldots, N$ with $i \neq j$ and $d = 1, 2)$, $X_{j \to i}^{(d)}(t)$ is given by,

$$X_{j \to i}^{(1)}(t) = Q_{j \to i}^{(1)}(0) + \left[R_i^j(S_j(B_j(t))) - p_{j \to i} S_j(B_j(t)) \right]$$

$$+ [p_{j \to i}(S_j(B_j(t)) - b_j B_j(t))]$$

$$- \left[S_{j \to i}^{(1)}(B_{j \to i}^{(1)}(t)) - b_{j \to i}^{(1)} B_{j \to i}^{(1)}(t) \right] + \theta_{j \to i}^{(1)} t, \tag{12}$$

$$\theta_{j\to i}^{(1)} = p_{j\to i}b_j - b_{j\to i}^{(1)}, \tag{13}$$

$$\left(R^0 Y^0(t)\right)_{\tilde{i},\tilde{j}} = \begin{cases} b_{j\to i}^{(1)} Y_{j\to i}^{0,(1)}(t), & \text{if } \tilde{i} = \sigma(R_{j\to i}) \text{ and } \tilde{j} = \tilde{i}, \\ -p_{j\to i}b_j Y_j^0(t), & \text{if } \tilde{i} = \sigma(R_{j\to i}) \text{ and } \tilde{j} = \sigma(S_j), \\ 0, & \text{otherwise}, \end{cases} \tag{14}$$

$$\left(R^K Y^K(t)\right)_{\tilde{i},\tilde{j}} = 0. \tag{15}$$

$$X_{j\to i}^{(2)}(t) = Q_{j\to i}^{(2)}(0) + \left[\bar{R}_i^j(Y_j^K(t)) - \alpha_{j\to i}Y_j^K(t)\right]$$
$$- \left[S_{j\to i}^{(2)}(B_{j\to i}^{(2)}(t)) - b_{j\to i}^{(2)}B_{j\to i}^{(2)}(t)\right] + \theta_{j\to i}^{(2)}t, \tag{16}$$

$$\theta_{j\to i}^{(2)} = -b_{j\to i}^{(2)}, \tag{17}$$

$$\left(R^0 Y^0(t)\right)_{\tilde{i},\tilde{j}} = \begin{cases} b_{j\to i}^{(2)} Y_{j\to i}^{0,(2)}(t), & \text{if } \tilde{i} = \sigma(R_{j\to i}) \text{ and } \tilde{j} = \tilde{i}, \\ 0, & \text{otherwise}, \end{cases} \tag{18}$$

$$\left(R^K Y^K(t)\right)_{\tilde{i},\tilde{j}} = \begin{cases} \alpha_{j\to i} Y_j^K(t), & \text{if } \tilde{i} = \sigma(R_{j\to i}) \text{ and } \tilde{j} = \sigma(S_j), \\ 0, & \text{otherwise}, \end{cases} \tag{19}$$

For $i, j = 1, \ldots, N$ with $i \neq j$, and $d = 1, 2$, $Q_j(t), Q_{j\to i}^{(d)}(t), Y_j^0(t)$, $Y_j^K(t), Y_{j\to i}^{0,(d)}(t)$ have some important properties as follows:

$$0 \leq Q_j(t) \leq K_j; \; 0 \leq Q_{j\to i}^{(d)}(t) \leq \sum_{i=1}^N C_i; \; t \geq 0, \tag{20}$$

$$Y_j^0(0) = 0, \; Y_j^0(t) \text{ is continuous and nondecreasing}, \tag{21}$$

$$Y_j^K(0) = 0, \; Y_j^K(t) \text{ is continuous and nondecreasing}, \tag{22}$$

$$Y_{j\to i}^{0,(d)}(0) = 0, \; Y_{j\to i}^{0,(d)}(t) \text{ is continuous and nondecreasing}, \tag{23}$$

$$Y_j^0(t) \text{ increases at times } t \text{ only when } Q_j(t) = 0, \tag{24}$$

$$Y_j^K(t) \text{ increases at times } t \text{ only when } Q_j(t) = K_j, \tag{25}$$

$$Y_{j\to i}^{0,(d)}(t) \text{ increases at times } t \text{only when } Q_{j\to i}^{(d)}(t) = 0. \tag{26}$$

In the remainder of this section, we provide a lemma to prove that the matrix $R = (R^0, R^K)$ is an $\mathcal{S}$ - matrix, which plays a key role in discussing existence and uniqueness of the SRBM through the box polyhedron for the closed queueing network. Note that R^0 and R^K are defined in (14) and (15) for $d = 1$, and in (18) and (19) for $d = 2$. Also, the ith column of R is denoted as the vector v_i. To analyze the matrix R, readers may refer to Theorem 1.3 in Dai and Williams [7] for more details.

The following definition comes from Dai and Williams [7], here we restate it for convenience of readers.

Definition 1. *A square matrix A is called an $\mathcal{S}$ - matrix if there is a vector $x \geq 0$ such that $Ax > 0$. The matrix A is completely - $\mathcal{S}$ if and only if each principal submatrix of A is an $\mathcal{S}$ - matrix.*

Notice that the capacity of station nodes is finite and the total number of bikes in this bike sharing system is a fixed constant. Without loss of generality, we assume that the capacity of each road node is also finite, and the maximal capacity of each road is $\sum_{i=1}^{N} C_i$ due to the fact that the total number of bikes in this bike sharing system is $\sum_{i=1}^{N} C_i$. Therefore, the state space S of this close queueing network is a N^2-dimensional box space with $2N^2$ boundary faces F_i, given by

$$S \equiv \{x = (x_1, \ldots, x_{N^2})' \in \mathcal{R}_+^{N^2} : 0 \leq x_i \leq \sum_{i=1}^{N} C_i\}. \tag{27}$$

We write

$$F_i \equiv \{x \in S : x_i = 0\}, F_{i+N^2} \equiv \{x \in S : x_i = K_i\} \text{ for } i \in \text{SN}, \tag{28}$$

$$F_j \equiv \{x \in S : x_j = 0\}, F_{j+N^2} \equiv \{x \in S : x_j = \sum_{i=1}^{N} C_i\} \text{ for } j \in \text{RN}. \tag{29}$$

Let $J \equiv \{1, 2, \ldots, 2N^2\}$ be the index set of the faces, and for each $\emptyset \neq \mathcal{K} \subset J$, define $F_{\mathcal{K}} = \cap_{i \in \mathcal{K}} F_i$. We indicate that the set $\mathcal{K} \subset J$ is maximal if $\mathcal{K} \neq \emptyset, F_{\mathcal{K}} \neq \emptyset$, and $F_{\mathcal{K}} \neq F_{\tilde{\mathcal{K}}}$ for any $\mathcal{K} \subset \tilde{\mathcal{K}}$ such that $\mathcal{K} \neq \tilde{\mathcal{K}}$. Thus, we can obtain that the maximal set $\mathcal{K}$ is precisely the set of indexes of N^2 distinct faces meeting at any vertex of S. Let $\mathcal{N}$ be a $2N^2 \times N^2$ matrix whose ith row is given by the unit normal of face F_i, which directs to the interior of S. We obtain,

$$\mathcal{N} = \begin{bmatrix} 1 & 0 & \cdots & 0 \\ 0 & 1 & \cdots & 0 \\ \cdot & \cdot & \cdots & \cdot \\ 0 & 0 & \cdots & 1 \\ \hline -1 & 0 & \cdots & 0 \\ 0 & -1 & \cdots & 0 \\ \cdot & \cdot & \cdots & \cdot \\ 0 & 0 & \cdots & -1 \end{bmatrix}.$$

The state space S has 2^{N^2} vertexes due to its box space and each vertex given by $(\cap_{i \in \alpha} F_i) \cap (\cap_{i \in \beta} F_{i+N^2})$ for a unique index set $\alpha \subset \{1, \ldots, N^2\}$ with $\beta = \{1, \ldots, N^2\} \backslash \alpha$. Before we provide a lemma to prove the $(NR)_{\mathcal{K}}$ (exactly $|\mathcal{K}|$ distinct faces contain $F_{\mathcal{K}}$) is a special $\mathcal{S}$-matrix, we give a geometric interpretation for a $|\mathcal{K}| \times |\mathcal{K}|$ $\mathcal{S}$-matrix $(NR)_{\mathcal{K}}$. At the each vertex of the box, we should make sure that there is a positive linear combination $x_i v_i + x_j v_{j+N^2}, x_i > 0$ for $i \in \alpha$ and $x_j > 0$ for $j \in \beta$ such that $x_i v_i + x_j v_{j+N^2}$ directs to the interior of the state space S.

Now, we provide a lemma to indicate the matrix $(NR)_{\mathcal{K}}$ is an $\mathcal{S}$-matrix.

Lemma 1. *The matrix $(NR)_{\mathcal{K}}$ is an S-matrix for each maximal $\mathcal{K} \subset J$.*

Proof: It is easy to check that

$$\mathcal{N}R = \begin{pmatrix} R^0 & R^K \\ -R^0 & -R^K \end{pmatrix}.$$

Because the state space of the closed queueing network is a N^2-dimensional box space, it has $2N^2$ faces. Now, let us make a classify of those vertexes in this box space as follows:

Type-1: the vertexes are given by $(\cap_{i \in A_S} F_i) \cap (\cap_{j \in A_R} F_j)$;
Type-2: the vertexes are given by $(\cap_{i \in A_S} F_i) \cap (\cap_{k \in B_R} F_k)$;
Type-3: the vertexes are given by $(\cap_{l \in B_S} F_l) \cap (\cap_{j \in A_R} F_j)$;
Type-4: the vertexes are given by $(\cap_{l \in B_S} F_l) \cap (\cap_{k \in B_R} F_k)$;
Type-5: the vertexes are given by $(\cap_{i \in A_S} F_i) \cap (\cap_{j \in A_R} F_j) \cap (\cap_{k \in B_R \backslash A_R} F_k)$;
Type-6: the vertexes are given by $(\cap_{l \in B_S} F_l) \cap (\cap_{j \in A_R} F_j) \cap (\cap_{k \in B_R \backslash A_R} F_k)$;
Type-7: the vertexes are given by $(\cap_{j \in A_R} F_j) \cap (\cap_{i \in A_S} F_i) \cap (\cap_{l \in B_S \backslash A_S} F_l)$;
Type-8: the vertexes are given by $(\cap_{k \in B_R} F_k) \cap (\cap_{i \in A_S} F_i) \cap (\cap_{l \in B_S \backslash A_S} F_l)$;
Type-9: the vertexes are given by $(\cap_{i \in A_S} F_i) \cap (\cap_{l \in B_S \backslash A_S} F_l) \cap (\cap_{j \in A_R} F_j) \cap (\cap_{k \in B_R \backslash A_R} F_k)$;

where A_S and A_R denote the set of index of face $F_i = \{x_i = 0\}$ for $i \in$ SN and $F_j = \{x_j = 0\}$ for $j \in$ RN, respectively; B_S and B_R denote the set of index of face $F_l = \{x_l = K_l\}$ for $l \in$ SN and $F_k = \{x_k = \sum_{i=1}^{N} C_i + 1\}$ for $k \in$ RN, respectively. According to the model description in Sect. 2, it is seen that the following two cases can not be established:

Case 1: All the station nodes are saturated when $1 \leq C_i < K_i < \infty$, namely, the reflection direction vector v_i on face $F_i (i \in B_S)$ can not simultaneously exist in the box state space S due to $\sum_{i=1}^{N} K_i > \sum_{i=1}^{N} C_i$. Therefore, at the vertexes of type-3, there must be a positive linear combination $x_i v_i + x_j v_j > 0$ to direct to the interior of state space S, where $x_i \geq 0$ for $i \in A_R$ and $x_j \geq 0$ for $j \in B_S$.

Case 2: Any road node is full, namely, the faces $F_i (i \in B_R)$ does not have the reflection direction vector v_i in the box state space S. In other word, the reflection direction vector v_i on face F_i ($i \in B_R$) is zero vector. Therefore, at the vertexes of type-2, type-4, type-5, type-6, type-8 and type-9, there must be a positive linear combination who directs to the interior of state space S.

Now, we should only prove that at these vertexes of type-1, type-7 and type-3, where $C_i = K_i$, there also is a positive linear combination who directs to the interior of the state space S.

At the vertexes of type-1, we only should prove that the matrix R^0 in the matrix $\mathcal{N}R$ is an S-matrix for $d = 1, 2$. It is clear that the matrix R^0 is an S-matrix due to the fact that all the diagonal elements of R^0 are positive.

At the vertexes of type-7 and of type-3, for $C_i = K_i$ and $d = 1, 2$, we can rewrite the $(\mathcal{N}R)_{\mathcal{K}}$ as the following form:

$$M = (\mathcal{N}R)_{\mathcal{K}} = \begin{pmatrix} M_1 & M_2 \\ M_1 & M_4 \end{pmatrix} = \begin{pmatrix} M_1 & 0 \\ 0 & M_4 \end{pmatrix} + \begin{pmatrix} 0 & M_2 \\ M_3 & 0 \end{pmatrix}.$$

where M_1 is a submatrix of R^0, which contains ith row (column) and ith column (row) of R^0 simultaneously with $i \in \alpha \subset \{1, \ldots, N^2\}$. Because the R^0 is a complete $\mathcal{S}$-matrix, M_1 is an $\mathcal{S}$-matrix. M_4 is also a submatrix of $-R^K$, which also contains $i + N^2$th row (column) and $i + N^2$th column (row) of $-R^K$ simultaneously with $i \in \beta = \{1, \ldots, N^2\} \backslash \alpha$. At the same time, M_4 is a diagonal matrix whose diagonal element is unit one, hence M_4 is also an $\mathcal{S}$-matrix. M_2 is a submatrix of R^K and M_3 is a submatrix of $-R^0$. Because M_2 and M_3 do not contain any diagonal elements of R^K and $-R^0$, M_2 and M_3 are both nonnegative matrices. Therefore, there must be a positive linear combination who directs to the interior of the state space S at the vertexes of type-7 and type-3, for $C_i = K_i$ and $d = 1, 2$. This completes the proof. ∎

5   Fluid Limits

In this section, we provide a fluid limit theorem for the queueing processes of the closed queueing network corresponding to the bike sharing system.

It follows from the functional strong law of large numbers (FSLLN) that as $t \to \infty$

$$(\frac{1}{t}S_j(t), \frac{1}{t}S_{j \to i}^{(d)}(t)) \to (b_j, b_{j \to i}^{(d)}), \quad d = 1, 2, \tag{30}$$

and as $n \to \infty$

$$(\frac{1}{n}R_i^j(n), \frac{1}{n}\bar{R}_i^j(n), \frac{1}{n}R^{i \to j,(d)}(n)) \to (p_{j \to i}, \alpha_{j \to i}, 1), \quad d = 1, 2. \tag{31}$$

We consider a sequence of closed queueing networks, indexed by $n = 1, 2, \ldots$, as described in Sect. 3. Let $(\Omega^n, \mathcal{F}^n, P^n)$ be the probability space on which the nth closed queueing network is defined for the bike sharing system. All the processes and parameters associated with the nth network are appended with a superscript n.

For the nth network, the renewal service processes of the station nodes and of the road nodes are expressed by $S_j^n = \{S_j^n(t), t \geq 0\}$ and $S_{j \to i}^{(d),n} = \{S_{j \to i}^{(d),n}(t), t \geq 0\}$, respectively. Let b_j^n and $b_{j \to i}^{(d),n}$ be the long run average service rates of $S_i^n(t)$ and $S_{j \to i}^{(d),n}(t)$, respectively. The vectors of the N station capacities and of their initial bike numbers are denoted as $K^n = (K_1^n, \ldots, K_N^n)'$ and $C^n = (C_1^n, \ldots, C_N^n)'$, respectively, where $1 \leq C_1^n \leq K_N^n < \infty$. For simplicity of description, we write $R^{j,n}$ as R^j, $\bar{R}^{j,n}$ as $\bar{R}^j$ and $R^{j \to i,(d),n}$ as $R^{j \to i,(d)}$ for all $n \geq 1$, i.e., the routing processes of the station nodes and of the road nodes are compressed the number n. We append a superscript n to the performance indexes such as $Y_j^{0,n}(t), Y_{j \to i}^{0,(d),n}(t), B_j^n(t)$ and $B_{j \to i}^n(t)$, and the interesting processes $Q^n = ((Q_j^n(t), Q_{j \to i}^{(d),n}(t))'$ and $Y_j^{K,n}(t)$.

The Heavy Traffic Conditions: We assume that as $n \to \infty$

$$(b_j^n, b_{j \to i}^{(d),n}, \sqrt{n}\theta_j^n, \sqrt{n}\theta_{j \to i}^{(d),n}, \frac{1}{\sqrt{n}}C_i^n, \frac{1}{\sqrt{n}}K_i^n) \to (b_j, b_{j \to i}^{(d)}, \theta_j, \theta_{j \to i}^{(d)}, C_i, K_i), \tag{32}$$

where $\theta_j^n = \sum_{d=1}^{2} \sum_{j \neq i}^{N} b_{j \to i}^{(d),n} - b_j^n$; $\theta_{j \to i}^{(1),n} = p_{j \to i} b_j^n - b_{j \to i}^{(1),n}$ and $\theta_{j \to i}^{(2),n} = -b_{j \to i}^{(2),n}$. At the same time, we assume that for $i, j = 1, \ldots, N$ with $i \neq j, d = 1, 2$, all these limits are finite.

For the initial queue lengths $Q_j^n(0)$ and $Q_{j \to i}^{(d),n}(0)$, we assume that as $n \to \infty$

$$\bar{Q}_j^n(0) \equiv \frac{1}{n} Q_j^n(0) \to 0 \text{ and } \bar{Q}_{j \to i}^{(d),n}(0) \equiv \frac{1}{n} Q_{j \to i}^{(d),n}(0) \to 0. \tag{33}$$

It follows from the functional strong law of large numbers that for $d = 1, 2$, as $n \to \infty$

$$(\bar{S}_j^n(t), \bar{S}_{j \to i}^{(d),n}(t), \bar{R}_i^{j,n}(t), \bar{\bar{R}}_i^{j,n}(t), \bar{R}^{j \to i,(d),n}(t))$$

$$\to (b_j t, b_{j \to i}^{(d)} t, p_{j \to i} t, \alpha_{j \to i} t, t), \text{ u.o.c.,} \tag{34}$$

where

$$\bar{S}_j^n(t) = \frac{1}{n} S_j^n(nt), \bar{S}_{j \to i}^{(d),n}(t) = \frac{1}{n} S_{j \to i}^{(d),n}(nt), \ \bar{R}_i^{j,n}(t) = \frac{1}{n} R_i^j(\lfloor nt \rfloor),$$

$$\bar{\bar{R}}_i^{j,n}(t) = \frac{1}{n} \bar{R}_i^j(\lfloor nt \rfloor), \ \bar{R}^{j \to i,(d),n}(t) = \frac{1}{n} R^{j \to i,(d)}(\lfloor nt \rfloor),$$

and $\lfloor x \rfloor$ is the maximal integer part of the real number x.

We give a notation: for any process $W^n = \{W^n(t), t \geq 0\}$, we define its centered processes $\hat{W}^n = \{\hat{W}^n(t), t \geq 0\}$ by

$$\hat{W}^n(nt) = W^n(nt) - w^n nt,$$

where w^n is the mean of the process W^n.

For the station nodes and road nodes, we write some centered processes as

$$\hat{S}_j^n(nt) = S_j^n(nt) - b_j^n nt, \ \hat{S}_{j \to i}^{(d),n}(t) = S_{j \to i}^{(d),n}(nt) - b_{j \to i}^{(d),n} nt, \tag{35}$$

$$\hat{R}_i^{j,n}(t) = R_i^{j,n}(\lfloor nt \rfloor) - p_{j \to i} \lfloor nt \rfloor), \ \hat{\bar{R}}_i^{j,n}(t) = \bar{R}_i^{j,n}(\lfloor nt \rfloor) - \alpha_{j \to i} \lfloor nt \rfloor). \tag{36}$$

For convenience of readers, we restate a lemma for the oscillation result of a sequence of (S^n, R^n)-regulation problems in convex polyhedrons, which is a summary restatement of Lemma 4.3 of Dai and Williams [7] and the Theorem 3.1 of Dai [8], whose proof is omitted here and can easily be referred to Dai and Williams [7] and Dai [8] for more details.

This lemma prevails due to the fact that the state space of the box polyhedron of this bike sharing system belongs to a simple convex polyhedrons as analyzed in the last of Sect. 4. For a function f defined from $[t_1, t_2] \subset [0, \infty]$ into $\mathcal{R}^k$ for some $k \geq 1$, let

$$Osc(f, [t_1, t_2]) = \sup_{t_1 \leq s \leq t \leq t_2} |f(t) - f(s)|.$$

Lemma 2. *For any $T > 0$, given a sequence of $\{x^n\}_{n=1}^{\infty} \in D_{\mathcal{R}^{N^2}}[0, T]$ with the initial values $x^n(0) \in S^n$. Let (z^n, y^n) be an (S^n, R^n)-regulation of x^n over*

$[0, T]$, *where* $(z^n, y^n) \in D_{\mathcal{R}^{N^2}}[0, T] \times D_{\mathcal{R}_+^{2N^2}}[0, T]$. *Assuming that all* S^n *have the same shape, i.e., the only difference is the corresponding boundary size* K_i^n. *Assuming that* $\{K_i^n\}$ *belongs to some bounded set, and the jump sizes of* y^n *are bounded by* Γ^n *for each* n. *Then if* $(\mathcal{N}R)_\mathcal{K}$ *is an* S *- matrix and* $R^n \to R$ *as* $n \to \infty$, *we have*

$$Osc(z^n, [t_1, t_2]) \leq C \ \max\{Osc(x^n, [t_1, t_2]), \Gamma^n\},$$

$$Osc(y^n, [t_1, t_2]) \leq C \ \max\{Osc(x^n, [t_1, t_2]), \Gamma^n\},$$

where C *depends only on* $(\mathcal{N}, R, |\mathcal{K}|)$ *for all* $\mathcal{K} \subset \Xi$, *where* Ξ *denotes the collection of subsets of* $J \equiv \{1, 2, \ldots, 2N^2\}$ *consisting of all maximal sets in* J *together with the empty set.*

Theorem 1 *(Fluid Limit Theorem). Under Assumptions (32) to (34), as* $n \to \infty$, *we have*

$$\left(\bar{B}_j^n(t), \bar{B}_{j \to i}^{(d),n}(t), \bar{Y}_j^{0,n}(t), \bar{Y}_{j \to i}^{0,(d),n}(t)\right) \to \left(\bar{\tau}_j(t), \bar{\tau}_{j \to i}^{(d)}(t), \bar{Y}_j^0(t), \bar{Y}_{j \to i}^{0,(d)}(t)\right) \ u.o.c,$$

where $\bar{\tau}_j(t) \equiv et, \bar{\tau}_{j \to i}^{(d)}(t) \equiv et, \bar{Y}_j(t) \equiv 0$ *and* $\bar{Y}_{j \to i}^{(d)}(t) \equiv 0$; $\bar{Y}_j^{0,n}(t) = \frac{1}{n}Y_j^{0,n}(nt), \bar{Y}_{j \to i}^{0,(d),n}(t) = \frac{1}{n}Y_{j \to i}^{0,(d),n}(nt)$, $\bar{B}_j^n(t) = \frac{1}{n}B_j^n(nt)$ *and* $\bar{B}_{j \to i}^{(d),n}(t) = \frac{1}{n}B_{j \to i}^{(d),n}(nt)$ *for* $i, j = 1, \ldots, N$ *with* $i \neq j, d = 1, 2$.

Proof: Recall the queue length process: $Q(t) = X(t) + R^0 Y^0(t) + R^K Y^K(t)$, where $X(t)$ is given by (8), (12) and (16) in Sect. 4. It follows from (2) to (4) that the scaling queueing processes for the station nodes and the road nodes are given by

$$\bar{Q}^n(t) = \bar{Q}^n(0) + \bar{X}^n(t) + R^{0,n}\bar{Y}^{0,n}(t) + R^{K,n}\bar{Y}^{K,n}(t),$$

where $\bar{Q}^n(t) = \frac{1}{n}Q^n(nt), \bar{Q}^n(t) = \{(\bar{Q}_j^n(t), \bar{Q}_{j \to i}^{(d),n}(t)), i \neq j, i, j = 1, \ldots, N; d = 1, 2; t \geq 0\}$; $\bar{X}^n(t) = \frac{1}{n}X^n(nt), \ \bar{X}^n(t) = \{(\bar{X}_j^n(t), \ \bar{X}_{j \to i}^{(d),n}(t)), i \neq j, i, j = 1, \ldots, N; d = 1, 2; t \geq 0\}$; $\bar{Y}^{0,n}(t) = \frac{1}{n}Y^{0,n}(nt), \bar{Y}^{0,n}(t) = \{(\bar{Y}_j^{0,n}(t), \bar{Y}_{j \to i}^{0,(d),n}(t)), i \neq j, i, j = 1, \ldots, N; d = 1, 2; t \geq 0\}$; $\bar{Y}^{K,n}(t) = \frac{1}{n}Y^{K,n}(nt), \ \bar{Y}^{K,n}(t) = \{(\bar{Y}_j^{K,n}(t)), \ j = 1, \ldots, N\}$. For each $n, \bar{Q}^n(t), \bar{Y}^n(t)$ and $\bar{Y}^{K,n}(t)$ satisfy the properties (20) to (26) with the state space S^n, given by

$$S^n \equiv \left\{x = (x_1, \ldots, x_{N^2})' \in R_+^{N^2} : x_i \leq \bar{K}_i^n = \frac{K_i^n}{n} \text{ for } i \in \text{SN}; \right.$$

$$\left. \text{and } x_i \leq \frac{\sum_{i=1}^N C_i^n}{n} + 1 \text{ for } i \in \text{RN}\right\}.$$

For station node $j = 1, \ldots, N$, by using (2), (8), (35) and (36), we have

$$\bar{X}_j^n(t) \equiv \frac{1}{n}Q_j^n(0) + \frac{1}{n}\sum_{d=1}^2 \sum_{i \neq j}^N \hat{S}_{i \to j}^{(d),n}(n\bar{B}_{i \to j}^{(d),n}(t)) - \frac{1}{n}\hat{S}_j^n(n\bar{B}_j^n(t)) + \frac{1}{n}\theta_j^n nt. \quad (37)$$

For road node $j \to i$ $(i, j = 1, \ldots, N$ with $i \neq j)$, by using (12), (16), (35) and (36), we have,

$$\bar{X}_{j\to i}^{(1),n}(t) \equiv \frac{1}{n} Q_{j\to i}^{(1),n}(0) + \frac{1}{n} \hat{R}_i^{j,n}(n\bar{S}_j^n(\bar{B}_j^n(t)))$$
$$+ \frac{1}{n} p_{j\to i} \hat{S}_j^n(n\bar{B}_j^n(t)) - \frac{1}{n} \hat{S}_{j\to i}^{(1),n}(n\bar{B}_{j\to i}^{(1),n}(t)) + \frac{1}{n} \theta_{j\to i}^{(1),n} nt, \quad (38)$$

$$\bar{X}_{j\to i}^{(2),n}(t) \equiv \frac{1}{n} Q_{j\to i}^{(2),n}(0) + \frac{1}{n} \hat{R}_i^{j,n}(n\bar{Y}_j^{K,n}(t)) - \frac{1}{n} \hat{S}_{j\to i}^{(2),n}(n\bar{B}_{j\to i}^{(2),n}(t))$$
$$+ \frac{1}{n} \theta_{j\to i}^{(2),n} nt. \quad (39)$$

Note that $\bar{B}_{j\to i}^{(1),n}(t) \leq t, \bar{B}_j^n(t) \leq t, \bar{Y}_j^{K,n}(t) \leq \sum_{i=1}^N C_i^n - K_j^n$, by using (32) to (34) and the Skorohod Representation Theorem, as $n \to \infty$, we have

$$\bar{X}^n(t) = (\bar{X}_j^n(t), \bar{X}_{j\to i}^{(d),n}(t)) \to 0, \quad \text{u.o.c.}$$

Since the state space S^n of this bike sharing system are the boxes of the same shape in the N^2-dimensional space, $(\mathcal{N}R)_\mathcal{K}$ is an S-matrix and $R^n \to R$ as $n \to \infty$. Then by Lemma 2 we have

$$Osc(\bar{Y}^{0,n}, [s,t] \subseteq [0,T]) \leq C \ Osc(\bar{X}^n, [s,t] \subseteq [0,T]),$$

for any $T \geq 0$, where C depends only on R and $\mathcal{N}$ for n large enough.

$$0 \leq \lim_{n\to\infty} \inf Osc(\bar{Y}^{0,n}, [s,t] \subseteq [0,T])$$
$$\leq \lim_{n\to\infty} \sup Osc(\bar{Y}^{0,n}, [s,t] \subseteq [0,T])$$
$$\leq C \lim_{n\to\infty} Osc(\bar{X}^n, [s,t] \subseteq [0,T])$$
$$= 0, \quad \text{a.s.}$$

where $\bar{Y}^n(t) = (\bar{Y}_j^{0,n}(t)', \bar{Y}_{j\to i}^{(d),0,n}(t)')'$. Notice that $Y^n(0) = 0$ for all n, we have

$$\lim_{n\to\infty} \bar{Y}^n(t) = 0, \quad \text{u.o.c.} \quad (40)$$

Since $\bar{B}_j^n(t) = t - \bar{Y}_j^{0,n}(t)$ and $\bar{B}_{j\to i}^{(d),n}(t) = t - \bar{Y}_{j\to i}^{(d),0,n}(t)$, we obtain the convergence of $\bar{B}_j^n(t)$ and $\bar{B}_{j\to i}^{(d),n}(t)$ for $i, j = 1, \ldots, N$ with $i \neq j, d = 1, 2$. This competes the proof. ∎

6 Diffusion Limits

In this section, we set up the diffusion scaled processes of the queueing processes, and give their weak convergence results for the multiclass closed queueing network corresponding to the bike sharing system.

We introduce the diffusion scaling process for the process $\hat{W}^n = \{\hat{W}^n(nt), t \geq 0\}$, given by

$$\tilde{W}^n(t) \equiv \frac{1}{\sqrt{n}}\hat{W}^n(nt) = \frac{1}{\sqrt{n}}(W^n(nt) - w^n nt).$$

For the station nodes and the road nodes, we write

$$\tilde{S}_j^n(t) = \sqrt{n}\left(\frac{S_j^n(nt)}{n} - b_j^n t\right), \quad \tilde{S}_{j\to i}^{(d),n}(t) = \sqrt{n}\left(\frac{S_{j\to i}^{(d),n}(nt)}{n} - b_{j\to i}^{(d),n}t\right), \quad (41)$$

$$\tilde{R}_i^{j,n}(t) = \sqrt{n}\left(\frac{R_i^{j,n}(nt)}{n} - p_{j\to i}t\right), \quad \tilde{\bar{R}}_i^{j,n}(t) = \sqrt{n}\left(\frac{\bar{R}_i^{j,n}(nt)}{n} - \alpha_{j\to i}t\right), \quad (42)$$

$$\tilde{R}^{j\to i,(d),n}(t) = \sqrt{n}\left(\frac{R^{j\to i,(d),n}(nt)}{n} - t\right). \quad (43)$$

For the initial queueing processes $Q_j^n(0)$ and $Q_{j\to i}^{n,(d)}(0)$ for $i, j = 1, \ldots, N$ with $i \neq j$, $d = 1, 2$, we assume that as $n \to \infty$

$$\tilde{Q}_j^n(0) \equiv \frac{1}{\sqrt{n}}Q_j^n(0) \Rightarrow \tilde{Q}(0), \quad (44)$$

$$\tilde{Q}_{j\to i}^{(d),n}(0) \equiv \frac{1}{\sqrt{n}}Q_{j\to i}^{(d),n}(0) \Rightarrow \tilde{Q}_{j\to i}^{(d)}(0). \quad (45)$$

It follows from the Skorohod Representation Theorem and the Donsker's Theorem that

$$(\tilde{S}_j^n(t), \tilde{S}_{j\to i}^{(d),n}(t), \tilde{R}_i^{j,n}(t), \tilde{\bar{R}}_i^{j,n}(t), \tilde{R}^{j\to i,(d),n}(t))$$
$$\Rightarrow (\tilde{S}_j(t), \tilde{S}_{j\to i}^{(d)}(t), \tilde{R}_i^j(t), \tilde{\bar{R}}_i^j(t), \tilde{R}^{j\to i,(d)}(t)), \quad (46)$$

where $\Rightarrow$ denotes weak convergence, and $\tilde{S}_j(t), \tilde{S}_{j\to i}^{(d)}(t), \tilde{R}_i^j(t), \tilde{\bar{R}}_i^j(t)$ and $\tilde{R}^{j\to i,(d)}(t)$ are all the Brownian motions with drift zero and covariance matrices Γ^S, $\Gamma^{R,S,l}$, $\Gamma^{\bar{R},S,l}$ and $\Gamma^{R,S,j\to i}$, which are given by

(1) The covariance matrix of $\tilde{S}(t) = (\tilde{S}_j(t), \tilde{S}_{j\to i}^{(d)}(t))$ for $i, j = 1, \ldots, N$ with $i \neq j, d = 1, 2$, is given by

$$\Gamma^S = \begin{pmatrix} (\Gamma^{S,S})_{N\times N} & 0 \\ 0 & (\Gamma^{S,R,(d)})_{(N^2-N)\times(N^2-N)} \end{pmatrix}_{N^2\times N^2},$$

where

$$(\Gamma^{S,S})_{\tilde{i},\tilde{j}} = \begin{cases} b_i c_{a,i}^2 \delta_{\tilde{i},\tilde{j}}, & \sigma(S_i) = \tilde{i}, \\ 0, & \text{otherwise}, \end{cases}$$

$$\left(\Gamma^{S,R,(d)}\right)_{\tilde{i},\tilde{j}} = \begin{cases} b_{i\to j}^{(d)}(c_{s,i\to j}^{(d)})^2 \delta_{\tilde{i},\tilde{j}}, & \sigma(R_{i\to j}) = \tilde{i}, \\ 0, & \text{otherwise}. \end{cases}$$

(2) The covariance matrix of $\tilde{R}(t) = (\tilde{R}^l(t))$ for $l = 1, \ldots, N$, is given by

$$\Gamma^{R,S,l} = \begin{pmatrix} 0 & 0 \\ 0 & \left(\Gamma^{R,S,l}\right)_{(N-1)\times(N-1)} \end{pmatrix}_{N^2 \times N^2},$$

where

$$\left(\Gamma^{R,S,l}\right)_{\tilde{i},\tilde{j}} = \begin{cases} p_{l \to k_1}(\delta_{\tilde{i},\tilde{j}} - p_{l \to k_2}), & \sigma(R_{l \to k_1}) = \tilde{i}, \sigma(R_{l \to k_2}) = \tilde{j}, \\ 0, & \text{otherwise.} \end{cases}$$

(3) The covariance matrix of $\bar{\tilde{R}}(t) = (\bar{\tilde{R}}^l(t))$ for $l = 1, \ldots, N$, is given by

$$\Gamma^{\bar{R},S,l} = \begin{pmatrix} 0 & 0 \\ 0 & \left(\Gamma^{\bar{R},S,l}\right)_{(N-1)\times(N-1)} \end{pmatrix}_{N^2 \times N^2},$$

where

$$\left(\Gamma^{\bar{R},S,l}\right)_{\tilde{i},\tilde{j}} = \begin{cases} \alpha_{l \to k_1}(\delta_{\tilde{i},\tilde{j}} - \alpha_{l \to k_2}), & \sigma(R_{l \to k_1}) = \tilde{i}, \sigma(R_{l \to k_2}) = \tilde{j}, \\ 0, & \text{otherwise.} \end{cases}$$

(4) The covariance matrix of $\tilde{R}(t) = (\tilde{R}^{j \to i,(d)}(t))$ for $i, j = 1, \ldots, N$ with $i \neq j, d = 1, 2$, is given by

$$\Gamma^{R,S,j \to i} = \begin{pmatrix} \left(\Gamma^{R,R,j \to i}\right)_{N \times N} & 0 \\ 0 & 0 \end{pmatrix}_{N^2 \times N^2},$$

where

$$\left(\Gamma^{R,R,j \to i}\right)_{\tilde{l},\tilde{k}} = \begin{cases} p_{j \to i,l}(\delta_{\tilde{l},\tilde{k}} - p_{j \to i,k}) = 0, & \sigma(S_l) = \tilde{l}, \sigma(S_k) = \tilde{k}, \\ 0, & \text{otherwise.} \end{cases}$$

Now, we prove adaptedness properties of the diffusion scaling processes $(\tilde{Q}^n(t), \tilde{X}^n(t), \tilde{Y}^n(t))$, where $\tilde{Q}^n(t) = \frac{1}{\sqrt{n}}Q^n(nt), \tilde{Q}^n(t) = (\tilde{Q}_j^n(t), \tilde{Q}_{j \to i}^{(d),n}(t))$; $\tilde{X}^n(t) = \frac{1}{\sqrt{n}}X^n(nt)$, $\tilde{X}^n(t) = (\tilde{X}_j^n(t), \tilde{X}_{j \to i}^{(d),n}(t))$; $\tilde{Y}^{0,n}(t) = \frac{1}{\sqrt{n}}Y^{0,n}(nt)$, $\tilde{Y}^{0,n}(t) = (\tilde{Y}_j^{0,n}(t)', \tilde{Y}_{j \to i}^{0,(d),n}(t)')$, $\tilde{Y}_j^{K,n}(t) = \frac{1}{\sqrt{n}}Y_j^{K,n}(nt)$.

Define

$$\varsigma_t^n = \sigma\{\tilde{Q}^n(0), \tilde{S}^n(s), \tilde{Y}^{0,n}(s), \tilde{Y}^{K,n}(t), s \leq t\}, \tag{47}$$

where $\tilde{Q}^n(0), \tilde{S}^{(d),n}(s), \tilde{R}^{(d),n}(s)$ and $\bar{\tilde{R}}^n(s)$ are defined in (41) to (45). Define $T_k^n = (T_k^{j,n}, T_k^{j \to i,(d),n})$, where $T_k^{j,n}$ and $T_k^{j \to i,(d),n}$ denote the partial sum of the service time sequence at station node j and road node $j \to i$, respectively, for the nth network, that is,

$$T_k^{j,n} = \sum_{l=1}^k u_j^n(l), \quad T_k^{j \to i,(d),n} = \sum_{l=1}^k v_{j \to i}^{(d),n}(l),$$

with the initial condition $T_0^n \equiv 0$. Notice that $T_k^n = (T_k^{j,n}, T_k^{j \to i,(d),n})$ is a ς_t^n − *stopping time*, and, $0 = T_0^n < T_1^n < T_2^n < \cdots < T_k^n \to \infty$ a.s. as $k \to \infty$ for each n and $i, j = 1, \dots, N$ with $i \neq j, d = 1, 2$. Let $\varsigma_{T_k^{(n)}-}$ denote the strict past at time T_k^n. Then

$$\varsigma_{T_k^{(n)}-} = \sigma(A_t \cap \{t < T_k^n\}, A_t \in \varsigma_t^n, t \geq 0).$$

Because T_k^n is a ς_t^n-stopping time, $u_j^n(k+1)$ and $v_{j \to i}^{(d),n}(k+1)$ are independent of the history of the network before the time at which the kth customer is served at station node j and road node $j \to i$. Therefore, T_k^n is $\varsigma_{T_k^{(n)}-}$-measurable, $u_j^n(k+1)$ is independent of $\varsigma_{T_k^{(j,n)}-}$, and $v_{j \to i}^{(d),n}(k+1)$ is independent of $\varsigma_{T_k^{(j \to i,(d),n)}-}$.

Theorem 2. *Under Assumption (32), we have that*

$$\left(\tilde{Q}^n(t), \tilde{X}^n(t), \tilde{Y}^{0,n}(t), \tilde{Y}^{K,n}(t) \right) \Rightarrow \left(\tilde{Q}(t), \tilde{X}(t), \tilde{Y}^0(t), \tilde{Y}^K(t) \right), \text{ as } n \to \infty,$$

or, in component form,

$$\left(\tilde{Q}_j^n(t), \tilde{Q}_{j \to i}^{(d),n}(t), \tilde{X}_j^n(t), \tilde{X}_{j \to i}^{(d),n}(t), \tilde{Y}_j^{0,n}(t), \tilde{Y}_{j \to i}^{0,(d),n}(t), \tilde{Y}_j^{K,n}(t) \right)$$
$$\Rightarrow \left(\tilde{Q}_j(t), \tilde{Q}_{j \to i}^{(d)}(t), \tilde{X}_j(t), \tilde{X}_{j \to i}^{(d)}(t), \tilde{Y}_j^0(t), \tilde{Y}_{j \to i}^{0,(d)}(t), \tilde{Y}_j^K(t) \right), \quad \text{as } n \to \infty,$$

where $\tilde{X}(t)$ is a Brownian motion with covariance matrix Γ. Moreover, $\tilde{X}(t) - \theta t$ is a martingale with respect to the filtration $\mathcal{F}_t = \sigma(\tilde{Q}(s), \tilde{Y}^0(s), \tilde{Y}^K(s), s \leq t)$.

Proof: First, we define

$$\tau_+^n(t) = \min\{T_k^n : T_k^n > t\} \text{ and } \tau_-^n(t) = \max\{T_k^n : T_k^n \leq t\}. \qquad (48)$$

For the station node $j \in \text{SN}$, when $\tau_+^{j,n}(nt)$ approximates nt from its right side, we have

$$\lim_{n \to \infty} E\left[\left| \frac{1}{\sqrt{n}} (S_j^n(\tau_+^{j,n}(nt)) - b_j^n \tau_+^{j,n}(nt)) - \tilde{S}_j^n(t) \right| \right]$$

$$= \lim_{n \to \infty} E\left[\left| \frac{1}{\sqrt{n}} (1 - b_j^n(\tau_+^{jn}(nt)) - nt) \right| \right]$$

$$\leq \frac{1}{\sqrt{n}} \lim_{n \to \infty} b_j^n E\left[\tau_+^{j,n}(nt) - \tau_-^{j,n}(nt) \right]$$

$$= \lim_{n \to \infty} \frac{1}{\sqrt{n}} b_j^n E\left[u_j^n(1) \right] = 0. \qquad (49)$$

Similarly, when $\tau_-^{j,n}(nt)$ approximates nt from its left side, we have

$$\lim_{n \to \infty} E[|\frac{1}{\sqrt{n}} (S_j^n(\tau_-^{j,n}(nt)) - b_j^n \tau_-^{j,n}(nt)) - \tilde{S}_j^n(t)|] = 0. \qquad (50)$$

Moreover, we obtain

$$E[\tilde{S}_j^n(T_{k+1}^{j,n}) - \tilde{S}_j^n(T_k^{j,n})|\varsigma_{T_k^{j,n}}^n] = \frac{1}{\sqrt{n}}\{1 - b_j^n E[u_j^n(k+1)|\varsigma_{T_k^{j,n}}^n]\} = 0, \quad (51)$$

where the filtration $\{\varsigma_t^n\}$ is defined in (47). Notice that for any $\{\varsigma_t^n\}$-stopping time T and any random variable X with $E[|X|] < \infty$,

$$E[E[X|\varsigma_t^n]|\varsigma_t^n]I_{\{T>t\}} = E[X|\varsigma_t^n]I_{\{T>t\}} = E[XI_{\{T>t\}}|\varsigma_t^n]. \quad (52)$$

Also, for each $j \in SN$ and all $s, t \geq 0$,

$$E\left[\tilde{S}_j^n(t+s) - \tilde{S}_j^n(t)|\varsigma_t^n\right]$$

$$= E\left[\tilde{S}_j^n(t+s) - \frac{1}{\sqrt{n}}\left(S_j^n(\tau_-^{j,n}(n(t+s))) - b_j^n\tau_-^{j,n}(n(t+s))\right)|\varsigma_t^n\right]$$

$$+ E\left[\frac{1}{\sqrt{n}}\left(S_j^n(\tau_+^{j,n}(nt)) - b_j^n\tau_+^{j,n}(nt)\right) - \tilde{S}_j^n(t)|\varsigma_t^n\right]$$

$$- \sum_k E\left[E\left[\tilde{S}_j^n(T_{k+1}^{j,n}) - \tilde{S}_j^n(T_k^{j,n})\Big|\varsigma_{T_k^{j,n}}^n\right]I_{\{nt<T_k^{j,n}\leq n(t+s)\}}|\varsigma_t^n\right].$$

Hence, it follows from (49) to (52) that

$$\lim_{n\to\infty} E\left[\left|E\left[\tilde{S}_j^n(t+s) - \tilde{S}_j^n(t)|\varsigma_t^n\right]\right|\right] = 0. \quad (53)$$

For road node $j \to i$ $(i, j = 1, \ldots, N$ with $i \neq j)$. When we approximate nt from both sides, a similar analysis to the proof of (53) for station node j. For all $s, t \geq 0$, we have

$$\lim_{n\to\infty} E\left[\left|E\left[\tilde{S}_{j\to i}^{(d),n}(t+s) - \tilde{S}_{j\to i}^{(d),n}(t)|\varsigma_t^n\right]\right|\right] = 0. \quad (54)$$

Next, we can set up the scaling queueing processes by mean of (2) to (4) for the station nodes and of the road nodes through the scaling processes (41) to (45), given by:

$$\tilde{Q}^n(t) = \tilde{Q}^n(0) + \tilde{X}^n(t) + R^{0,n}\tilde{Y}^{0,n}(t) + R^{K,n}\tilde{Y}^{K,n}(t), \quad (55)$$

and for each n, $(\tilde{Q}^n(t), \tilde{Y}^{0,n}(t), \tilde{Y}^{K,n}(t))$ has the properties (20) to (26) with the state space S^n as follow:

$$S^n \equiv \left\{x = (x_1, \ldots, x_{N^2})' \in R_+^{N^2} : x_i \leq \tilde{K}_i^n = \frac{K_i^n}{\sqrt{n}} \text{ for } i \in SN,\right.$$

$$\left.\text{and } x_i \leq \frac{\sum_{i=1}^N C_i^n}{\sqrt{n}} + 1 \text{ for } i \in RN\right\}.$$

For station node $j = 1, \ldots, N$, by using (3), (12), (41) to (45) and $\tilde{X}_j^n(t) = \frac{1}{\sqrt{n}}X_j^n(nt) = \sqrt{n}\bar{X}_j^n(t)$, we have

$$\tilde{X}_j^n(t) = \tilde{Q}_j^n(0) + \frac{1}{\sqrt{n}} \sum_{d=1}^{2} \sum_{i \neq j}^{N} \hat{S}_{i \to j}^{(d),n}(n \bar{B}_{i \to j}^{(d),n}(t)) - \frac{1}{\sqrt{n}} \hat{S}_j^n(n \bar{B}_j^n(t)) + \sqrt{n} \theta_j^n t. \quad (56)$$

For road node $j \to i$ ($i, j = 1, \ldots, N$ with $i \neq j$), by using (12), (16), (41) to (45) and $\tilde{X}_{j \to i}^{(d),n}(t) = \frac{1}{\sqrt{n}} X_{j \to i}^{(d),n}(nt) = \sqrt{n} \bar{X}_{j \to i}^{(d),n}(t)$, we have,

$$\tilde{X}_{j \to i}^{(1),n}(t) = \tilde{Q}_{j \to i}^{(1),n}(0) + \frac{1}{\sqrt{n}} \hat{R}_i^{j,n}(n \bar{S}_j^n(\bar{B}_j(t)))$$

$$+ \frac{1}{\sqrt{n}} p_{j \to i} \hat{S}_j^n(n \bar{B}_j^n(t)) - \frac{1}{\sqrt{n}} \hat{S}_{j \to i}^{(1),n}(n \bar{B}_{j \to i}^{(1),n}(t)) + \frac{1}{\sqrt{n}} \theta_{j \to i}^{(1),n} nt \quad (57)$$

and

$$\tilde{X}_{j \to i}^{(2),n}(t) = \tilde{Q}_{j \to i}^{(2),n}(0) + \frac{1}{\sqrt{n}} \hat{R}_i^{j,n}(n \bar{Y}_j^{K,n}(t)) - \frac{1}{\sqrt{n}} \hat{S}_{j \to i}^{(2),n}(n \bar{B}_{j \to i}^{(2),n}(t))$$

$$+ \frac{1}{\sqrt{n}} \theta_{j \to i}^{(2),n} nt. \quad (58)$$

From Assumption (32), using the Continuous Mapping Theorem and Theorem 1 (Fluid Limit), we obtain that for station node j,

$$\tilde{X}_j^n(t) \Rightarrow \tilde{X}_j(t) = \tilde{Q}_j(0) + \sum_{d=1}^{2} \sum_{i \neq j}^{N} \tilde{S}_{i \to j}^{(d)}(t) - \tilde{S}_j(t) + \theta_j t, \quad (59)$$

where $\tilde{X}_j(t)$ is an Brownian motion with the initial queue length $\tilde{Q}_j(0)$ and the drift θ_j. For road station $j \to i$,

$$\tilde{X}_{j \to i}^{(1),n}(t) \Rightarrow \tilde{X}_{j \to i}^{(1)}(t) = \tilde{Q}_{j \to i}^{(1)}(0) + \tilde{R}_i^j(b_j t) + p_{j \to i} \tilde{S}_j(t) - \tilde{S}_{j \to i}^{(1)}(t) + \theta_{j \to i}^{(1)} t, \quad (60)$$

where $\tilde{X}_{j \to i}^{(1)}(t)$ is an Brownian motion with the initial queue length $\tilde{Q}_{j \to i}^{(1)}(0)$ and the drift $\theta_{j \to i}^{(1)}$. Similarly we have

$$\tilde{X}_{j \to i}^{(2),n}(t) \Rightarrow \tilde{X}_{j \to i}^{(2)}(t) = \tilde{Q}_{j \to i}^{(2)}(0) + \tilde{\tilde{R}}_i^{j,n}(\bar{Y}_j^{K,n}(t)) - \tilde{S}_{j \to i}^{(2)}(t) + \theta_{j \to i}^{(2)} t, \quad (61)$$

where $\tilde{X}_{j \to i}^{(2)}(t)$ is an Brownian motion with the initial queue length $\tilde{Q}_{j \to i}^{(2)}(0)$ and the drift $\theta_{j \to i}^{(2)}$. The covariance matrix $\Gamma = (\Gamma_{\tilde{k}, \tilde{l}})_{N^2 \times N^2}$ of $\tilde{X}(t) = (\tilde{X}_j(t), \tilde{X}_{j \to i}^{(d)}(t))$ is given by

$$\Gamma_{\tilde{k}, \tilde{l}} = \begin{cases} \sum_{d=1}^{2} \sum_{i \neq k}^{N} b_{i \to k}^{(d)} (c_{s,i \to k}^{(d)})^2 \delta_{\tilde{k}, \tilde{l}} \\ \quad + b_l c_{a,l}^2 \delta_{\tilde{k}, \tilde{l}}, & \text{if } \sigma(S_k) = \tilde{k}, \sigma(S_l) = \tilde{l}; \\ p_{k \to l} b_k c_{a,k}^2, & \text{if } \sigma(S_k) = \tilde{k}, \sigma(R_{k \to l}) = \tilde{l}, d = 1; \\ b_i p_{i \to k} (\delta_{\tilde{k}, \tilde{l}} - p_{i \to l}) \\ \quad + p_{k \to l} b_k c_{a,k}^2 + b_{k \to l}^{(d)} (c_{s,k \to l}^{(d)})^2, & \text{if } \sigma(R_{k \to l}) = \tilde{k}, \tilde{l} = \tilde{k}, d = 1; \\ b_k \alpha_{i \to k} (\delta_{\tilde{k}, \tilde{l}} - \alpha_{i \to l}) \\ \quad + b_{k \to l}^{(d)} (c_{s,k \to l}^{(d)})^2, & \text{if } \sigma(R_{k \to l}) = \tilde{k}, \tilde{l} = \tilde{k}, d = 2; \\ 0, & \text{otherwise.} \end{cases} \quad (62)$$

Now, let $h(t)$ be an arbitrary real, bounded and continuous function. For an arbitrary positive integer m, let $t_i \leq t \leq t + s, i \leq m$. Define

$$\tilde{H}^n(t) = \left(\tilde{Q}^n(t), \tilde{Y}^{0,n}(t), \tilde{Y}^{K,n}(t)\right), \quad \tilde{H}(t) = \left(\tilde{Q}(t), \tilde{Y}^0(t), \tilde{Y}^K(t)\right),$$

$$G^n(t,s) = \left(G_j^n(t,s), G_{j \to i}^{(d),n}(t,s)\right),$$

$$G_j^n(t,s) = \tilde{X}_j^n(t+s) - \tilde{X}_j^n(t), \; G_{j \to i}^{(d),n}(t,s) = \tilde{X}_{j \to i}^{(d),n}(t+s) - \tilde{X}_{j \to i}^{(d),n}(t).$$

Notice that

$$\tilde{S}_j^n(t) = \frac{1}{\sqrt{n}}\left(\sup\left\{k : \sum_{l=1}^{k} u_j^n(l) \leq b_j^n nt\right\} - b_j^n nt\right),$$

$$\tilde{S}_{j \to i}^{(d),n}(t) = \frac{1}{\sqrt{n}}\left(\sup\left\{k : \sum_{l=1}^{k} v_{j \to i}^{(d),n}(l) \leq b_{j \to i}^{(d),n} nt\right\} - b_{j \to i}^{(d),n} nt\right),$$

by using the Assumption (32), there exist some nonnegative constants C_1 and C_2 such that $b_j^n \leq C_1$ and $b_{j \to i}^{(d),n} \leq C_2$. From the convergences of (53) and (54), we have

$$\left|E\left[h\left(\tilde{H}(t_i), i \leq m\right)\left(\tilde{X}(t+s) - \tilde{X}(t) - \theta s\right)\right]\right|$$

$$= \left|\lim_{n \to \infty} E\left[h\left(\tilde{H}^n(t_i), i \leq m\right) G^n(t,s)\right]\right|$$

$$= \lim_{n \to \infty}\left|E\left[h\left(\tilde{H}^n(t_i), i \leq m\right) E\left[G^n(t,s) | \varsigma_t^n\right]\right]\right|$$

$$\leq M \lim_{n \to \infty} E\left[\left|E\left[G^n(t,s) | \varsigma_t^n\right]\right|\right]$$

$$= 0,$$

where M is some positive constant. The arbitrariness of $h(t)$, t_i, t and $t + s$ implies that

$$E\left[\tilde{X}(t+s) - \tilde{X}(t) - \theta s | \mathcal{F}_u, u \leq t\right] = 0.$$

This shows that $\tilde{X}(t) - \theta t$ is an $\{\mathcal{F}_t\}$-martingale. This completes the proof. ■

Remark 1. *Note that Dai [8] discussed the queueing networks with finite buffers, this paper is related well to fluid and diffusion limits in Dai [8] in order to deal with a two-class closed queueing network.*

Now, we give the diffusion limit for the bike sharing system. In Sect. 5, we set up a sequence of closed queueing networks corresponding to the bike sharing systems, and prove the limit theorems of the fluid scaled equations of the busy period processes and the idle period processes through the functional strong law of large numbers and the oscillation property of an (S^n, R^n)-regulation. This is summarized as the Fluid Limit Theorem 1. Furthermore, based on the

Fluid Limit Theorem, we prove the weak limit of the diffusion scaled processes of some performance measures and obtain a key martingale. Also see Theorem 2.

The following theorem provides a diffusion limit, and its proof is easy by means of some similar analysis to Theorems 3.2 and 3.3 in Dai [8] or Theorem 3.1 in Dai and Dai [6].

Theorem 3 *(Diffusion Limit Theorem). Under Assumption (32), we have*

$$
\left(\frac{1}{\sqrt{n}} Q^n(nt), \frac{1}{\sqrt{n}} Y^{0,n}(nt), \frac{1}{\sqrt{n}} Y^{K,n}(nt) \right) \Rightarrow \left(\tilde{Q}(t), \tilde{Y}^0(t), \tilde{Y}^K(t) \right),
$$

where $\tilde{Q}(t) = \left(\tilde{Q}_j(t), \tilde{Q}^{(d)}_{j \to i}(t) \right), \tilde{Y}^0(t) = \left(\tilde{Y}^0_j(t), \tilde{Y}^{0,(d)}_{j \to i}(t) \right); \tilde{Q}(t)$ *together with* $\tilde{Y}^0(t)$ *and* $\tilde{Y}^K(t)$ *are an* (S, θ, Γ, R)*-semimartingale reflecting Brownian motion with* $\tilde{Q}(t) = \tilde{Q}(0) + \tilde{X}(t) + R^0 \tilde{Y}^0(t) + R^K \tilde{Y}^K(t)$. *The state space* S *is given by (27) to (29). For station node* j, $\tilde{X}_j(t)$ *is given by (59),* R^0 *and* R^K *are given by (10), (11). For road node* $j \to i$, *when* $d = 1, \tilde{X}^{(1)}_{j \to i}(t)$ *is given by (60),* R^0 *and* R^K *are given by (14) and (15); when* $d = 2, \tilde{X}^{(2),n}_{j \to i}(t)$ *is given by (61),* R^0 *and* R^K *are given by (18) and (19), and the covariance matrix* $\Gamma = (\Gamma_{\tilde{k}, \tilde{l}})_{N^2 \times N^2}$ *of* $\tilde{X}(t) = (\tilde{X}_j(t), \tilde{X}^{(d)}_{j \to i}(t))$ *is given by (62).*

7   Performance Analysis

In this section, we first set up a basic adjoint relationship for the steady-state probabilities of N station nodes and of $N(N-1)$ road nodes in the multiclass closed queueing network. Then we analyze some key performance measures of the bike sharing system.

From Theorem 3, it is seen that the scaling queueing processes, for the numbers of bikes at the stations and on the roads, converge in distribution to a semi-martingale reflecting Brownian motion $\tilde{Q}(t) = \left(\tilde{Q}_{\tilde{i}}(t), \tilde{Q}^{(d)}_{\tilde{j}}(t) \right)$ for $\tilde{i} = \sigma(S_i)$ $(i = 1, \ldots, N)$ and $\tilde{j} = \sigma(R_{j \to i})$ $(i, j = 1, \ldots, N$ with $i \neq j, d = 1, 2)$, where the state space S, the drift vector $\theta = \left(\theta_{\tilde{i}}, \theta^{(d)}_{\tilde{j}} \right)$ for $\tilde{i} = \sigma(S_i), \tilde{j} = \sigma(R_{j \to l})$, the covariance matrix

$$
\Gamma = \begin{pmatrix} (\Gamma_{\tilde{i}, \tilde{k}}) & (\Gamma^{(d)}_{\tilde{i}, \tilde{j}}) \\ (\Gamma^{(d)}_{\tilde{l}, \tilde{k}}) & (\Gamma^{(d)}_{\tilde{l}, \tilde{j}}) \end{pmatrix}_{N^2 \times N^2}
$$

for $\tilde{i} = \sigma(S_i), \tilde{k} = \sigma(S_k), \tilde{j} = \sigma(R_{j \to h}), \tilde{l} = \sigma(R_{l \to g})$ and the reflecting matrix $R = \left(\left(R^0_{\tilde{i}} \right), \left(R^{K,(d)}_{\tilde{j}} \right) \right)$ for $\tilde{i} = \sigma(S_i), \tilde{j} = \sigma(R_{j \to l})$, as seen in those previous sections. Hence, it is natural to approximate the steady-state distribution of the queue-length process by means of the steady-state distribution of the semi-martingale reflecting Brownian motion.

From Lemma 1 and Theorem 1.3 in Dai and Williams [7], it is seen that there exists a unique stationary distribution $\pi = \left(\pi_{\tilde{i}}, \pi^{(d)}_{\tilde{j}} \right)$ on $(S, \mathcal{B}_S)$ for the

SRBM $\tilde{Q}(t) = \left(\tilde{Q}_{\tilde{i}}(t), \tilde{Q}_{\tilde{j}}^{(d)}(t)\right)$. Furthermore, $\pi = \left(\pi_{\tilde{i}}, \pi_{\tilde{j}}^{(d)}\right)$ is equivalent to the Lebesgue measure on the state space S, thus for every bounded Borel function f on S and for $t \geq 0$, we have

$$E_\pi \left[f\left(\tilde{Q}(t)\right) \right] \equiv \int_S \left(E_x \left[f\left(\tilde{Q}(t)\right) \right] \right) \pi(dx) = \int_S f(x)\pi(dx).$$

Then for each $\tilde{i} = 1, \ldots, N$ (i.e., $\tilde{i} = \sigma(S_i)$, $i = 1, \ldots, N$) and $\tilde{j} = 1, \ldots, N(N-1)$ (i.e., $\tilde{j} = \sigma(R_{j\to i})$, $i, j = 1, \ldots, N$ with $i \neq j$), let $\delta = \left(\delta_{\tilde{i}}, \delta_{\tilde{j}}^{(d)}\right)$ denote $(N^2 - 1)$-dimensional Lebesgue measure (surface measure) vector on face $(F, \mathcal{B}_F)$. Thus, there is a finite Borel measure vector $\beta^F = \left(\beta_{\tilde{i}}^F, \beta_{\tilde{j}}^{F,(d)}\right)$ on face $F = (F_{\tilde{i}}, F_{\tilde{j}})$ such that $\beta^F \approx \delta$ and

$$E_\pi \left\{ \int_0^t 1_A \left(\tilde{Q}(s)\right) d\tilde{Y}(s) \right\} = t\beta^F(A), \ t \geq 0, A \in \mathcal{B}_F,$$

where $\tilde{Y}(t) = \left(\tilde{Y}^0(t), \tilde{Y}^K(t)\right)$. Notice that the SRBM $\tilde{Q}(t) = (\tilde{Q}_{\tilde{i}}(t), \tilde{Q}_{\tilde{j}}^{(d)}(t))$ is a strong Markov process with continuous sample paths. Furthermore, let $p(x) = \left(p_{\tilde{i}}(x_{\tilde{i}}), p_{\tilde{j}}^{(d)}\left(x_{\tilde{j}}^{(d)}\right)\right)$, $p^F(x) = \left(p_{\tilde{i}}^F\left(\delta_{\tilde{i}}\right), p_{\tilde{j}}^{F,(d)}\left(\delta_{\tilde{j}}^{(d)}\right)\right)$, and define $d\pi = pdx$, i.e., $d\pi_{\tilde{i}} = p_{\tilde{i}}dx_{\tilde{i}}$ for $\tilde{i} = \sigma(S_i)$ ($i = 1, \ldots, N$) and $d\pi_{\tilde{j}}^{(d)} = p_{\tilde{j}}^{(d)}dx_{\tilde{j}}^{(d)}$ for $\tilde{j} = \sigma(R_{j\to i})$ ($i, j = 1, \ldots, N$ with $i \neq j, d = 1, 2$). Further, we define $d\beta^F = p^F d\delta$, i.e., $d\beta_{\tilde{i}}^F = p_{\tilde{i}}^F d\delta_{\tilde{i}}$ for $\tilde{i} = \sigma(S_i)$ ($i = 1, \ldots, N$) and $d\beta_{\tilde{j}}^{F,(d)} = p_{\tilde{j}}^{F,(d)}d\delta_{\tilde{j}}^{(d)}$ for $\tilde{j} = \sigma(R_{j\to i})$ ($i, j = 1, \ldots, N$ with $i \neq j, d = 1, 2$). Let $\nabla f(x)$ be the gradient of f, and $C_b^2(S)$ the space of twice differentiable functions whose first and second order partial derivative are continuous and bounded on the state space S. Base on this, it follows from the Ito's formula that the probability measures $p(x)$ and $p^F(x)$ have a basic adjoint relationship as follows: for $\forall f \in C_b^2(S)$,

$$\int_S (\mathcal{L}f(x)p(x))\,dx + \sum_{\tilde{i}=1}^N \int_{F_{\tilde{i}}^{\blacktriangledown}} (\mathcal{D}_{\tilde{i}}f(\delta_{\tilde{i}})p_{\tilde{i}}^F(\delta_{\tilde{i}}))d\delta_{\tilde{i}} + \sum_{\tilde{i}=1}^N \int_{F_{\tilde{i}}^{\blacktriangle}} (\mathcal{D}_{\tilde{i}}f(\delta_{\tilde{i}})p_{\tilde{i}}^F(\delta_{\tilde{i}}))d\delta_{\tilde{i}}$$

$$+ \sum_{d=1}^2 \sum_{\tilde{j}=1}^{N^2-N} \int_{F_{\tilde{j}}^{\blacktriangledown}} (\mathcal{D}_{\tilde{j}}f(\delta_{\tilde{j}}^{(d)})p_{\tilde{j}}^{F,(d)}(\delta_{\tilde{j}}^{(d)}))d\delta_{\tilde{j}}^{(d)}$$

$$+ \sum_{d=1}^2 \sum_{\tilde{j}=1}^{N^2-N} \int_{F_{\tilde{j}}^{\blacktriangle}} (\mathcal{D}_{\tilde{j}}f(\delta_{\tilde{j}}^{(d)})p_{\tilde{j}}^{F,(d)}(\delta_{\tilde{j}}^{(d)}))d\delta_{\tilde{j}}^{(d)} = 0, \tag{63}$$

where

$$\mathcal{L}f = \sum_{\tilde{i}=1}^N \mathcal{L}f(x_{\tilde{i}}) + \sum_{d=1}^2 \sum_{\tilde{j}}^{N^2-N} \mathcal{L}f(x_{\tilde{j}}^{(d)}),$$

for $i, k, j = 1, \ldots, N$ with $i \neq j, d = 1, 2$, and $\tilde{k} = \sigma(S_k), \tilde{j} = \sigma(R_{j \to i}), \tilde{i} = \sigma(S_i) \in \{1, 2, \ldots, N\}$,

$$\mathcal{L}f(x_{\tilde{i}}) = \frac{1}{2} \sum_{\tilde{k}=1}^{N} \Gamma_{\tilde{i}, \tilde{k}} \frac{\partial^2 f(x_{\tilde{i}})}{\partial x_{\tilde{i}} \partial x_{\tilde{k}}} + \frac{1}{2} \sum_{d=1}^{2} \sum_{\tilde{j}=1}^{N^2-N} \Gamma_{\tilde{i}, \tilde{j}}^{(d)} \frac{\partial^2 f(x_{\tilde{i}})}{\partial x_{\tilde{i}} \partial x_{\tilde{j}}^{(d)}} + \theta_{\tilde{i}} \frac{\partial f(x_{\tilde{i}})}{\partial x_{\tilde{i}}},$$

$$\mathcal{D}_{\tilde{i}} f(\delta_{\tilde{i}}) \equiv v_{\tilde{i}}' \nabla f(\delta_{\tilde{i}}) = \sum_{\tilde{k}=1}^{N} v_{\tilde{k}, \tilde{i}} \frac{\partial}{\partial \delta_{\tilde{k}}} f(\delta_{\tilde{i}}) + \sum_{d=1}^{2} \sum_{\tilde{j}=1}^{N^2-N} v_{\tilde{j}, \tilde{i}} \frac{\partial}{\partial \delta_{\tilde{j}}^{(d)}} f(\delta_{\tilde{i}}),$$

for $l, k, i, j, h = 1, \ldots, N$ with $j \neq i, l \neq k$ and $d = 1, 2$, and $\tilde{l} = \sigma(R_{l \to k}), \tilde{h} = \sigma(S_h), \tilde{j} = \sigma(R_{j \to i}) \in \{1, 2, \ldots, N^2 - N\}$,

$$\mathcal{L}f(x_{\tilde{j}}^{(d)}) = \frac{1}{2} \sum_{\tilde{h}=1}^{N} \Gamma_{\tilde{j}, \tilde{h}}^{(d)} \frac{\partial^2 f(x_{\tilde{j}}^{(d)})}{\partial x_{\tilde{j}}^{(d)} \partial x_{\tilde{h}}} + \sum_{\tilde{l}=1}^{N^2-N} \Gamma_{\tilde{j}, \tilde{l}}^{(d)} \frac{\partial^2 f(x_{\tilde{j}}^{(d)})}{\partial x_{\tilde{j}}^{(d)} \partial x_{\tilde{l}}^{(d)}} + \theta_{\tilde{j}}^{(d)} \frac{\partial f(x_{\tilde{j}}^{(d)})}{\partial x_{\tilde{j}}^{(d)}},$$

$$\mathcal{D}_{\tilde{j}} f(\delta_{\tilde{j}}^{(d)}) \equiv v_{\tilde{j}}' \nabla f(\delta_{\tilde{j}}^{(d)}) = \sum_{\tilde{l}=1}^{N(N-1)} v_{\tilde{l}, \tilde{j}} \frac{\partial}{\partial \delta_{\tilde{l}}^{(d)}} f(\delta_{\tilde{j}}^{(d)}) + \sum_{\tilde{h}=1}^{N} v_{\tilde{h}, \tilde{j}} \frac{\partial}{\partial \delta_{\tilde{h}}} f(\delta_{\tilde{j}}^{(d)}),$$

$F_{\tilde{i}}^{\blacktriangledown}$ and $F_{\tilde{i}}^{\blacktriangle}$ denote the "bottom face" and the "top face" in this box state space S corresponding to empty station i and full station i, respectively. As a similar expression, it is clear that $F_{\tilde{j}}^{\blacktriangledown}$ and $F_{\tilde{j}}^{\blacktriangle}$ are related to road $j \to i$; $v_{\tilde{k}}$ is the $\tilde{k}$th column of the reflection matrix $R = \left(\left(R_{\tilde{i}}^0 \right), \left(R_{\tilde{j}}^{K,(d)} \right) \right)$.

Now, we consider some key performance measures of the bike sharing system in terms of the steady-state probability density function p on $(S, \mathcal{B}_S)$ and an nonnegative integrable Borel function p^F on $(F, \mathcal{B}_F)$. Here, it is easy to see that for $\tilde{i} = 1, \ldots, N$ and $\tilde{j} = 1, \ldots, N(N-1)$, the "bottom face" $F_{\tilde{i}}^{\blacktriangledown}$ $(F_{\tilde{j}}^{\blacktriangledown})$ and the "top face" $F_{\tilde{i}}^{\blacktriangle}$ $(F_{\tilde{j}}^{\blacktriangle})$ are precisely parallel in this box state space S.

(1) The steady-state probability that station i is empty is given by

$$\int_S p_{\tilde{i}}^F 1_{\{x_{\tilde{i}} \in F_{\tilde{i}}^{\blacktriangledown}\}} dx_{\tilde{i}}, \qquad \text{for } \tilde{i} = \sigma(S_i).$$

(2) The steady-state probability that station i is full is given by

$$\int_S p_{\tilde{i}}^F 1_{\{x_{\tilde{i}} \in F_{\tilde{i}}^{\blacktriangle}\}} dx_{\tilde{i}}, \qquad \text{for } \tilde{i} = \sigma(S_i).$$

(3) The steady-state probability that road $j \to i$ is empty for bikes of class d is given by

$$\int_S p_{\tilde{j}}^{F,(d)} 1_{\{x_{\tilde{j}}^{(d)} \in F_{\tilde{j}}^{\blacktriangledown}\}} dx_{\tilde{j}}^{(d)}, \qquad \text{for } \tilde{j} = \sigma(R_{j \to i}), d = 1, 2.$$

(4) The steady-state probability that road $j \to i$ is full for bikes of class d is given by

$$\int_S p_{\tilde{j}}^{F,(d)} 1_{\{x_{\tilde{j}}^{(d)} \in F_{\tilde{j}}^{\blacktriangle}\}} dx_{\tilde{j}}^{(d)}, \quad \text{for } \tilde{j} = \sigma(R_{j \to i}), d = 1, 2.$$

(5) The steady-state means of the number of bikes parked at the station i and the number of bikes of class d ridden on road $j \to i$ are respectively given by

$$\mathcal{Q}_{\tilde{i}} = \int_S x_{\tilde{i}} p_{\tilde{i}}(x_{\tilde{i}}) dx_{\tilde{i}}, \text{ for } \tilde{i} = \sigma(S_i),$$

$$\mathcal{Q}_{\tilde{j}}^{(d)} = \int_S x_{\tilde{j}}^{(d)} p_{\tilde{j}}^{(d)} \left(x_{\tilde{j}}^{(d)}\right) dx_{\tilde{j}}^{(d)}, \text{ for } \tilde{j} = \sigma(R_{j \to i}), d = 1, 2.$$

(6) The steady-state mean of the number of bikes of class d deflecting from the full station i is given by

$$\mathcal{E}_{\tilde{i}}^{(d)} = \int_{F_{\tilde{i}}^{\blacktriangle}} x_{\tilde{i}}^{(d)} p_{\tilde{i}}^{F,(d)} \left(x_{\tilde{i}}^{(d)}\right) dx_{\tilde{i}}^{(d)}, \text{ for } \tilde{i} = \sigma(S_i), d = 1, 2.$$

8 Concluding Remarks

In this paper, we describe a more general large-scale bike sharing system having renewal arrival processes and general travel times, and develop fluid and diffusion approximation of a multiclass closed queuing network which is established from the bike sharing system where bikes are regarded as virtual customers, and stations and roads are viewed as virtual nodes or servers. From the multiclass closed queuing network, we show that the scaling queue-length processes, which are set up by means of the number of bikes both at stations and on roads, converge in distribution to a semimartingale reflecting Brownian motion. Also, we obtain the Fluid Limit Theorem and the Diffusion Limit Theorem. Based on this, we provide performance analysis of the bike sharing system. Therefore, the results of this paper give new highlight in the study of more general large-scale bike sharing systems. The methodology developed here can be applicable to deal with more general bike sharing systems by means of the fluid and diffusion approximation. Along such a line, there are some interesting directions in our future research, for example,

- analyzing bike repositioning policies through several fleets of trucks under information technologies;
- making price regulation of bike sharing systems through Brownian approximation of multiclass closed queuing network;
- developing heavy traffic approximation for time-varying or periodic bike sharing systems; and
- developing heavy traffic approximation for new ride sharing (bike or car) systems with scheduling, matching and control.

Acknowledgments. This work was supported in part by the National Natural Science Foundation of China under grant Nos. 71671158 and 71471160, and Natural Science Foundation of Hebei under grant No. G2017203277.

References

1. Adelman, D.: Price-directed control of a closed logistics queueing network. Oper. Res. **55**, 1022–1038 (2007)
2. Bramson, M.: State space collapse with application to heavy traffic limits for multiclass queueing networks. Queueing Syst. **30**, 89–140 (1998)
3. Chen, H., Mandelbaum, A.: Stochastic discrete flow networks: diffusion approximations and bottlenecks. Ann. Probab. **19**, 1463–1519 (1991)
4. Chen, H., Yao, D.D.: Fundamentals of Queueing Networks: Performance, Asymptotics, and Optimization. Springer, New York (2001). https://doi.org/10.1007/978-1-4757-5301-1
5. Dai, J.G.: On positive Harris recurrence of multiclass queueing networks: a unified approach via fluid limit models. Ann. Appl. Probab. **5**, 49–77 (1995)
6. Dai, J.G., Dai, W.: A heavy traffic limit theorem for a class of open queueing networks with finite buffers. Queueing Syst. **32**, 5–40 (1999)
7. Dai, J.G., Williams, R.J.: Existence and uniqueness of semimaringale reflecting Brownian motions in convex polyhedrons. Theory Probab. Appl. **40**, 1–40 (1995)
8. Dai, W.: Brownian approximations for queueing networks with finite buffers: modeling, heavy traffic analysis and numerical implementations. Ph.D. thesis, School of Mathematics, Georgia Institute of Technology (1996)
9. Dai, W.: A Brownian model for multiclass queueing networks with finite buffers. J. Comput. Appl. Math. **144**, 145–160 (2002)
10. DeMaio, P.: Bike-sharing: history, impacts, models of provision, and future. J. Public Transp. **12**, 41–56 (2009)
11. Fricker, C., Gast, N., Mohamed, A.: Mean field analysis for inhomogeneous bike-sharing systems. DMTCS Proc. **1**, 365–376 (2012)
12. Fricker, C., Gast, N.: Incentives and regulations in bike-sharing systems with stations of finite capacity. EURO J. Transp. Logist. **3**, 1–31 (2014)
13. Fricker, C., Tibi, D.: Equivalence of ensembles for large vehicle-sharing models. Ann. Appl. Probab. **27**, 883–916 (2017)
14. George, D.K., Xia, C.H.: Asymptotic analysis of closed queueing networks and its implications to achievable service levels. ACM Sigmetrics Perform. Eval. Rev. **38**, 3–5 (2010)
15. George, D.K., Xia, C.H.: Fleet-sizing and service availability for a vehicle rental system via closed queueing networks. Eur. J. Oper. Res. **211**, 198–207 (2011)
16. Harrison, J.M.: Brownian Motion and Stochastic Flow Systems. Wiley, New York (1985)
17. Harrison, J.M., Nguyen, V.: Brownian models of multiclass queueing networks: current status and open problems. Queueing Syst. **13**, 5–40 (1993)
18. Harrison, J.M., Williams, R.J.: A multiclass closed queueing network with unconventional heavy traffic behavior. Ann. Appl. Probab. **6**, 1–47 (1996)
19. Harrison, J.M., Williams, R.J., Chen, H.: Brownian models of closed queueing networks with homogeneous customer populations. Stochast.: Int. J. Probab. Stochast. Process. **29**, 37–74 (1990)
20. Kochel, P., Kunze, S., Nielander, U.: Optimal control of a distributed service system with moving resources: application to the fleet sizing and allocation problem. Int. J. Prod. Econ. **81**, 443–459 (2003)
21. Kumar, S.: Two-server closed networks in heavy traffic: diffusion limits and asymptotic optimality. Ann. Appl. Probab. **10**, 930–961 (2000)

22. Leurent, F.: Modelling a vehicle-sharing station as a dual waiting system: stochastic framework and stationary analysis. HAL Id: hal-00757228 (2012)
23. Labadi, K., Benarbia, T., Barbot, J.P., Hamaci, S., Omari, A.: Stochastic Petri net modeling, simulation and analysis of public bicycle sharing systems. IEEE Trans. Autom. Sci. Eng. **12**, 1380–1395 (2015)
24. Li, Q.L., Chen, C., Fan, R.N., Xu, L., Ma, J.Y.: Queueing analysis of a large-scale bike sharing systems through mean-field theory. arXiv Preprint: arXiv:1603.09560, pp. 1–51 (2016)
25. Li, Q.L., Fan, R.N.: Bike-sharing systems under Markovian environment. arXiv Preprint: arXiv:1610.01302, pp. 1–44 (2016)
26. Li, Q.-L., Fan, R.-N., Ma, J.-Y.: A unified framework for analyzing closed queueing networks in bike sharing systems. In: Dudin, A., Gortsev, A., Nazarov, A., Yakupov, R. (eds.) ITMM 2016. CCIS, vol. 638, pp. 177–191. Springer, Cham (2016). https://doi.org/10.1007/978-3-319-44615-8_16
27. Majewski, K.: Fractional Brownian heavy traffic approximations of multiclass feedforward queueing networks. Queueing Syst. **50**, 199–230 (2005)
28. Meddin, R., DeMaio, P.: The Bike-Sharing World Map. http://www.metrobike.net
29. Meyn, S.P.: Sequencing and routing in multiclass queueing networks part I: feedback regulation. SIAM J. Control Optim. **40**, 741–776 (2001)
30. Raviv, T., Kolka, O.: Optimal inventory management of a bike-sharing station. IIE Trans. **45**, 1077–1093 (2013)
31. Raviv, T., Tzur, M., Forma, I.A.: Static repositioning in a bike-sharing system: models and solution approaches. EURO J. Transp. Logist. **2**, 187–229 (2013)
32. Savin, S., Cohen, M., Gans, N., Katala, Z.: Capacity management in rental businesses with two customer bases. Oper. Res. **53**, 617–631 (2005)
33. Shaheen, S.A., Guzman, S.Y., Zhang, H.: Bike-sharing in Europe, the American and Asia: past, present and future. In: 89th Transportation Research Board Annual Meeting, Washington, D.C. (2010)
34. Schuijbroek, J., Hampshire, R., van Hoeve, W.J.: Inventory rebalancing and vehicle routing in bike-sharing systems. Eur. J. Oper. Res. **257**, 992–1004 (2017)
35. Shu, J., Chou, M.C., Liu, Q., Teo, C.P., Wang, I.L.: Models for effective deployment and redistribution of bicycles within public bicycle-sharing systems. Oper. Res. **61**, 1346–1359 (2013)
36. Waserhole, A., Jost, V.: Vehicle sharing system pricing regulation: transit optimization of intractable queueing network. HAL Id: hal-00751744, pp. 1–20 (2012)
37. Waserhole, A., Jost, V.: Pricing in vehicle sharing systems: optimization in queuing networks with product forms. EURO J. Transp. Logisti. **5**, 1–28 (2016)
38. Whitt, W.: Stochastic-Process Limits: An Introduction to Stochastic-Process Limits and Their Application to Queues. Springer, New York (2002). https://doi.org/10.1007/b97479

Model and Algorithm for Rescue Resource Assignment Problem in Disaster Response Based on Demand-Ability-Equipment Matching

Zhiwen Xiao, Guoqing Wang[(✉)], and Jianming Zhu

School of Engineering Science, University of Chinese Academy of Science,
Beijing, People's Republic of China
Xiaozhiwen15@mails.ucas.ac.cn,
{wgq94,jmzhu}@ucas.ac.cn

Abstract. The rapid and accurate response in post-disaster is of vital importance of emergency management. This article mainly focuses on the optimizing assignment process and aims to allocate rescue resources to diverse disaster points in an attacked area. The proposed model contains the allocation of rescue teams and equipment, and a corresponding transportation strategy was provided when the rescue team and equipment are assigned. A multi-objective optimization problem was established on the basis of cost and time consideration. Moreover, an improved epsilon-constraint algorithm was developed to find Pareto fronts of the multi-objective optimization problem. Some numerical examples are analyzed and the computational results confirm the feasibility of the assignment method.

Keywords: Assignment method · DAEM · Multi-objective · ε-constraint

1 Introduction

A disaster is the result of a vast ecological breakdown in the relations between man and his environment [1], so that timely response and effective disposal in order to restore the original situation after a disaster occurred are crucial aspects in emergency management., there are not only casualties, but also the destruction of infrastructures in an area of post-disaster, frequent natural disaster sets unprecedented challenges for emergency decision maker especially in the process of distributing rescue resources which always expose the shortcomings when reevaluating the rescue activities in post-disaster. Thus, taking the earthquake as a background, providing optimization strategy for distribution problem in restore infrastructures has the profound realistic significance. This article will mainly focus on the response phase of which optimizing the distribution problem from aid center includes the warehouse of engineering equipment and the location of rescue teams which assigned to restore the infrastructures in disaster points. Aiming at designing a distribution approach which on the premise of Demand-Ability-Equipment Matching (DAEM), and the concept of DAEM will be described in detail as follow.

W. Yue et al. (Eds.): QTNA 2017, LNCS 10591, pp. 246–261, 2017.
https://doi.org/10.1007/978-3-319-68520-5_15

The limited district in an affected area in post-disaster is described as disaster point. Because the types and the degrees of damages within each disaster point is not the same, the demand of each point to rescue and restore infrastructure is different. Since the Wenchuan earthquake happened in 2008, Chinese government had been pay more attention to emergency response ability. Standardization Administration of the People's Republic of China (SAC) tries to establish a series of comprehensive national standard for the process of post-earthquake rescue. One of these standards is Classification of Earthquake Damage to Lifeline Engineering (CEDLE), it defines all kinds of damage to lifeline engineering in post-earthquake [5], definitions are showed as Table 1.

Table 1. Related infrastructure of different classification of earthquake damage to lifeline engineering

Classification	Related infrastructure
Traffic system	Road; Bridge; Tunnel; Railway line
Water supply	Treatment pool; Treatment plant; Supply pumping station: Distribution network
Oil system	Refinery; Oil pumping station; Oil depot; Oil pipeline
Gas system	Gas station; Gas storage; Gas pipeline
Electric system	Generating station; Transformer substation; Transmission line
Communication system	Center control room; Telecommunication line
Water conservancy	Earth and rockfill dam

For a long time, rescue teams are the key of response phase in the disaster management. The rescue technology owned by rescue teams determines whether the objective of rescue will be achieved. These standards attend to categorize the different kind of damage in disaster points and designate related rescue skill for each classification. As any kind of damage has a corresponding rescue skill owned by rescue teams to restore, it form a mechanism for matching rescue skill between demand of disaster points and ability of rescue team. When a natural or an anthropogenic disaster occurs in a region, there are some incipient disaster points appearing. Decision makers estimate what rescue skills do each disaster point needs by damage characteristics, and assign rescue teams to disaster points. For example, a disruptive earthquake damages the electric power system and the transportation system in a disaster point, this disaster point needs those two corresponding rescue skills to carry out rescue, And a team or teams will go to the disaster point to complete the rescue mission. In addition, it is understandably that large engineering equipments are used by rescue teams in rescue activities. For example, excavators are widely used in transportation system rescue, especially in a road destroyed. Hence, as previous research [2], all elements of the objective function including the commodities are represented in an equivalent common unit of number of persons not served. So, in this paper, D-A-E Matching is a direct and universal approach of decision-making which not only matches rescue teams based on disaster point demand, but also ensure the assigned rescue teams to have the ability to complete the mission of disaster points in response phase. Considering the condition of constrained resources and the demand of disaster point, D-A Mating proves that dispatching the closet rescue team to the accident point is not always the best strategy.

Moreover, *CEDLE* divides post-earthquake damage into different levels, the feature of each level are shown in Table 2. As with in-person training, levels of rescue skill can be uneven in all rescue teams, and problems can be sloved from the same or a lower level. It indicates that the best match inters to when a team be assigned to a disaster point, all the abilities of this rescue team are used. For example, the rescue teams which have the abilities of *fourth-level* power system rescue skill and *third-level* transportation system rescue skill save a disaster point with only *second-level* power system damage is a waste of rescue resources. So, D-A Matching further means to assign teams of a disaster point by a fit conjunction with less redundancy should be concerned.

Table 2. Feature of each level

Level	Feature
I	Basically intact
II	Slight damage
III	Moderate damage
IV	Serious damage
V	Destroy

However, only chasing after the little redundancy in the process of assignment can not make the assignment scheme get a high performance in practice in terms of response and rescue. Other constraints based on practical issues need to be considered.

The remainder of the paper is organized as follows. A literature review is given in the next section. In Sect. 3 the proposed multi-objective mathematical formulation is described. In Sect. 4, an improved exact solution method is proposed. Numerical examples are illustrated in Sect. 5, and the concluding remarks are given in the final section.

2 Literature Review

In recent years, the response phase has been the main focus of emergency management researching in previous studies. Meanwhile, the study of rescue resource distribution problem has become one of the most popular topics within the response phase. Tofighi et al. [3] developed a novel two-stage scenario-based possibilistic-stochastic programming approach to formulate the problem under a mixture of probabilistic and possibilistic uncertainties for solve the problem of two-echelon humanitarian logistics network design involving multiple central warehouses and local distribution centers. Torabi et al. [4] accounted for epistemic uncertainty of critical data and proposed a bi-objective model for the supplier selection and order allocation problem to build the resilient supply base under operational and disruption risks. Wallace et al. [5] has developed a formulation of mixed-integer programming to minimize the cost terms from traditional network flow models; Campbell et al. [6] developed some methodologies of traveling salesman problem (TSP) and the vehicle routing problem (VRP) for disaster relief. Minimized the maximum arrival time and the average arrival time to bounded the worst-case performance of optimal TSP solutions. And presented solution approaches for these two variants of the TSP and VRP, which are based on well-known insertion and local search techniques. Nurre et al. [7] considered the integrated network design and scheduling (INDS) problems applicate in infrastructure restoration after an extreme event and building humanitarian logistics networks. And they proposed a novel heuristic dispatching rule algorithm to solve the NP-hard problem of INDS. Boland et al. [8] studied the problem of scheduling maintenance on arcs of a capacitated network. They maximized the total flow from a source node to a sink node over a set of time periods and proposed an additional constraint which limits the number of maintenance jobs per time period.

In addition, for D-A-E matching, Altay [9] pointed out the mechanism of distributing rescue resource according to demand in each disaster point, and they built a simple inter programming model to allocate resources. However, no solution was proposed to this model and the notion of nationwide resource inventory listing is too ambiguous.

3   Model Formulation for Rescue Resource Assignment Problem

Assumptions:

The proposed mathematical model is based on the following assumptions

(1)   All rescue units can cover all demand in disaster points.
(2)   The known capacity of the road is no longer changing in post-disaster.
(3)   The transportation of rescue team on road and helicopter, equipments on road transport.

Parameters:

V:	set of disaster points, $V = \{1, \ldots, n\}$
S:	set of rescue teams, $S = \{1, \ldots, m\}$
W:	set of equipment warehouse, $W = \{1, \ldots, w\}$
E:	set of equipment sorts, $E = \{1, \ldots, e\}$
T:	set of time interval. $T = \{1, \ldots, t\}$
R:	set of transport type to send rescue teams
R':	set of transport type to send equipment without helicopter
$P_k(i,j)$:	transporting from support point i to disaster point j on k transport type
$P'_k(i,j)$:	transporting from support point i to disaster point j on k transport type without helicopter
a_{is}:	quantity of equipment s owned by warehouse i
b_{js}:	quantity of equipment s demanded by disaster point j
c_k:	cost of transportation on k transport type
c'_k:	cost of transportation on k transport type without helicopter
t_{ikj}:	time of needed traveling from support point i to disaster point j on k transport type

t'_{ikj}:	time of needed traveling from support point i to disaster point j on k transport type without helicopter
c_d^λ:	cost of the skill of type d in level λ
τ:	maximum tolerable waiting time

$$cs_i^{d\lambda} = \begin{cases} 1 & \text{if rescue team } i \text{ can support skill of type } d \text{ of level } \lambda \\ 0 & \text{otherwise} \end{cases}$$

$$cv_j^{d\lambda} = \begin{cases} 1 & \text{if attacked point } j \text{ needs skill of } d \text{ type of level } \lambda \\ 0 & \text{otherwise} \end{cases}$$

Decision variables:

As mentioned before, we have three decisions in the proposed model. One of these decisions is determining the time of start transporting from each warehouse and is determined by $\theta_{ij}(i \in w, j \in n)$. This decision variable can decide the departure time to conform to the constraint which limits the number of servicing disaster points per unit time. And the others are determining the distributing status of rescue units.

θ_{ij}^{team}:	the moment of start traveling for rescue team i to disaster points j
θ_{ij}^{equ}:	the moment of start traveling for equipment from warehouse i to disaster points j
Q_{ij}^s:	quantity of equipment s delivered to disaster point j from equipment warehouse i

$$x_{ikj} = \begin{cases} 1 & \text{if rescue team } i \text{ visit to disaster point } j \text{ on its } k\text{th type} \\ 0 & \text{otherwise} \end{cases}$$

Auxiliary variables:

The auxiliary variable is decide by Q_{ikj}^s.

$$y_{ikj} = \begin{cases} 1 & \text{if equipments delivered to disaster point } j \text{ from warehouse } i \text{ on its } k\text{th type} \\ 0 & \text{otherwise} \end{cases}$$

3.1   Multi-objective Integer Programming Model

A new mathematical model to solve DAEM problem is proposed based on the assignment problems and transportation problems. Assignment problem is applied to determine which rescue teams and equipment warehouses to complete the mission in each disaster point, and the transportation problem is applied to make decision for type of transport.

$$f_1 = \min \left[\max_{i,k,j \in S} (t_{ikj} + \theta_{ij}^{team}) x_{ikj} \right] \tag{1}$$

$$f_2 = \min \left[\max_{i,k,j \in W} (t'_{ikj} + \theta_{ij}^{equ}) y_{ikj} \right] \tag{2}$$

$$f_3 = \min \sum_{j=1}^{n} \left[\sum_{i=1}^{m} \left(\sum_{\lambda=1}^{a} \sum_{d=1}^{l} c_d^{\lambda} cs_i^{d\lambda} x_{ikj} + \sum_{k=1}^{r} c_k x_{ikj} \right) + \sum_{i=1}^{w} \left(\sum_{s=1}^{e} c_s Q_{ikj}^s + \sum_{k=1}^{r'} c'_k y_{ikj} \right) \right] \tag{3}$$

$$\sum_j Q_{ij}^s \le a_{is} \ \forall i \in W, \forall s \in E, \forall k \in R' \tag{4}$$

$$b_{js} \le \sum_i Q_{ij}^s \ \forall j \in V, \forall s \in E, \forall k \in R' \tag{5}$$

$$Q_{ij}^s \le \sum_k y_{ikj} a_{is} \ \forall i, j, s, \forall k \in R' \tag{6}$$

$$\sum_i \sum_j Q_{ij}^s \ge \sum_k y_{ikj} \ \forall s \in E \tag{7}$$

$$\sum_k y_{ikj} \le 1 \ \forall i \in W, \forall j \in V \tag{8}$$

$$\sum_i \sum_{\lambda=u}^{v} \sum_k x_{ikj} cs_i^{d\lambda} - cv_j^{du} \ge 0 \ \forall w, u, d \in Z^+, \forall j \in V \tag{9}$$

$$\sum_k \sum_j x_{ikj} \le 1 \ \forall i \in S \tag{10}$$

$$-\tau \le x_{ikj}(t_{ikj} + \theta_{ij}^{team}) - x_{fij}(t_{fij} + \theta_{fj}^{team}) \le \tau \ \forall j \in V \tag{11}$$

$$-\tau \le x_{ikj}(t_{ikj} + \theta_{ij}^{team}) - y_{hvj}(t_{hvj} + \theta_{hj}^{equ}) \le \tau \ \forall j \in V \tag{12}$$

$$x_{ikj} \in \{0,1\} \ \forall i,j, \forall k \in R \tag{13}$$

$$y_{ikj} \in \{0,1\} \ \forall i,j, \forall k \in R' \tag{14}$$

$$l \in \{1, \ldots, r\} \tag{15}$$

$$v \in \{1, \ldots, r-1\} \tag{16}$$

$$h \in \{1, \ldots, w\} \tag{17}$$

$$f \in \{1, \ldots, n\} \tag{18}$$

The three objectives are given by Eqs. (1)–(3). The objective function (1) minimizes the time of latest arrival rescue teams. Objective function (2) minimizes the time of latest arrival equipment. The third objective function minimizes the sum of all cost including in the using cost and transporting cost for both rescue teams and equipments. Constraint (4) ensures that the total quantity of equipment s delivered for each disaster point does not exceed the quantity of equipment s available in this warehouse. Constraint (5) ensure that the total quantity of given equipment s in disaster point j delivered from each warehouse must exceed the quantity of demand. Constraints (6 and 7) shows that relationship between decision variable Q_{ij}^s and auxiliary variable y_{ij}. Constraint (8) ensures that there is only one type of transporting equipments which sending from a warehouse to a disaster point. Constraint (9) ensures that the teams assigned from its location can match the demand in disaster point j. Constraint (10) ensures that any rescue team can only be assigned once. Constraints (11 and 12) setting the maximum tolerance value to control the waiting time.

3.2   Model Linearization

The proposed mathematical model is nonlinear because of Constraints (11) and (12). New variables are proposed as Ω_{ij}^{team} and Ω_{ij}^{equ} instead of multiplication of variables θ_{ij}^{team} and x_{ikj}, θ_{hj}^{equ} and y_{ij} respectively, in order to $x_{ikj}\theta_{ij}^{team} = \Omega_{ij}^{team}$, $\forall i,j$ and $y_{ij}\theta_{ij}^{equ} = \Omega_{ij}^{equ}$, $\forall i,j$. Thus, in the mathematical model, Constraints (11) and (12) will be replaced by Constraints (19) and (20), and the additional Constraints (21–24) should be added to the proposed model, and the parameter M is a large enough value to constraint the relationships with two binary variables x_{ikj} and y_{ikj}.

$$-\tau \leq x_{ikj}t_{ikj} + \Omega_{ij}^{team} - x_{fij}t_{fij} + \Omega_{fj}^{team} \leq \tau, \ \forall j \in V \tag{19}$$

$$-\tau \leq x_{ikj}t_{ikj} + \Omega_{ij}^{team} - y_{hvj}t_{hvj} + \Omega_{hj}^{equ} \leq \tau, \ \forall j \in V \tag{20}$$

$$\Omega_{ij}^{team} \leq Mx_{ikj} \ \forall k \in R \tag{21}$$

$$\Omega_{ij}^{team} \geq 0 \tag{22}$$

$$\Omega_{ij}^{equ} \leq My_{ikj} \ \forall k \in R' \tag{23}$$

$$\Omega_{ij}^{equ} \geq 0 \tag{24}$$

4   Exact Solution Method for Demand-Ability-Equipment Matching

4.1   Achieving a Lower Bound of Objective Function

The ε-constraint method finds all the Pareto points of multi-objective integer linear programming problems starting from the one corner Pareto point and finding an adjacent one each iteration [10]. Functions f_1 and f_2 exist their certain range of values, lower bounds of these two object functions must to be find in order to start with a corner Pareto slice in a highly discriminating way.

4.2 The ε-Constraint Method

The ε-constraint method has a good performance for solving non-convex optimization problems, not only restricted to biobjective problem but can also efficiently adapted to multi-objective integer linear programming problems [11]. In this study, the solution method is developed as an exact solution approach based ε-constraint, and the proof of using this method can generate the exact Pareto front is shown in [12].

The Demand-Ability-Equipment matching is a three-objective combinatorial optimization problem for which we proposed the mathematical model. The three objectives f_1, f_2, f_3 which defined by Eqs. (1)–(3), consist in minimizing the traveling time and the totally cost respectively. We choose f_3 as objective function to optimize, and the constrained problem $P(\varepsilon_1, \varepsilon_2)$ is defined by

$$P(\varepsilon_1, \varepsilon_2) \text{ Min } f_3(X) \qquad \text{s.t.} \begin{cases} f_1(X) \leq \varepsilon_1 \\ f_2(X) \leq \varepsilon_2 \\ X = (x, y, \theta, Q) \in D \end{cases}$$

where $X = (x, y, \theta, Q)$ denotes the set of variables defined in the proposed mathematical model; and D is the feasible region defined by Eqs. (4)–(24); ε_1 and ε_2 are the two parameters in iteration to yield the Pareto front.

5 Computational Result

In this section, a simple case is presented to test the application of this distribution model. The Ludian earthquake occurred on August 3, 2014, with its epicenter located in Ludian Country. This disaster caused a large number of infrastracture damaged in Zhaotong, Yunnan Province. According to the HD image of Ludian earthquake area published by National Administrtion of Surveying and Geoinformation (NASG) [13], 36 disaster-affected sites, which located in or near epicenter were identified. Only except Shuifu Country and Weixin Country, disaster-affected sites throughout the city in Zhaotong. Figure 1 displays the location and number of these 36 disaster-affected sites.

To restore lifeline engineering in each disaster-affected site, the lighting equipment is an essential equipment to support constructing. Analyzing the record of using lighting equipment from Yunnan Power Grid Corporation, a subsidiary of China Southern Power Grid, can get parts of information about disaster sites. Combining with the report on direct economic losses of Ludian 6.5 earthquake which described the situation of damage by 8·3 earthquake published by Yunnan Seismological Bureau (YSB), we can easily come to know the classification of damage and their corresponding level for each affected site. Details is shown by Table 12 in Appendix A.

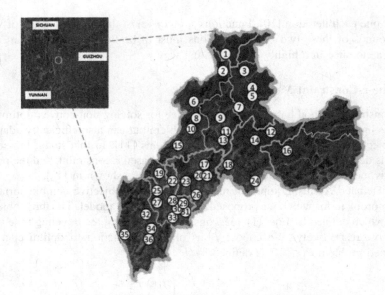

Fig. 1. Disaster-affected sites in Zhaotong, Yunnan.

Table 3. Demand of restoring in each disaster point

Disaster point	Traffic system	Water supply	Oil system	Gas system	Electirc system	Communication system	Water conservancy
Urban district	II	–	I	–	III	–	II
Ludian	IV	I	–	I	IV	III	V
Qiaojia	III	–	–	–	I	II	–
Yanjin	II	I	–	–	–	–	I
Daguan	I	II	–	–	–	–	–
Yongshan	II	–	–	–	–	–	III
Suijiang	–	–	–	–	I	–	–
Yiliang	II	–	–	–	–	–	–
Weixin	I	–	–	–	–	–	–

Taking each administrative district as an disaster point, the demand of restoring in each disaster point an be listed by Table 3.

In addition, the quantity of equipment is generated by estimating the situation of each disaster point. Table 4 shows the demand in each disaster point.

After the earthquake, Zhaotong municipal government and Yunnan provincial government should be in charge of emergency rescue. Surveying the data from available rescue resources in Zhaotong and Kunming, the supporting ability of each rescue team and the quantity of equipment in each warehouse can be obtained and shown as Tables 5 and 6.

Table 4. Large engineering equipment demand in each disaster point

Disaster point	Emergency lighting vehicle	Excavator	Diesel generator	Crane	Water pump
Urban district	3	5	2	0	2
Ludian	13	17	7	7	4
Qiaojia	4	4	3	4	2
Yanjin	2	2	2	0	0
Daguan	3	3	2	0	0
Yongshan	1	2	4	3	0
Suijiang	3	3	1	4	0
Yiliang	0	2	0	0	0
Weixin	3	5	2	0	0

Table 5. Ability of each rescue team can support.

No.	Team affiliation	Location district title	TS	WS	OS	GS	ES	CS	WC
1	Yunnan power grid Zhenxiong power supply company	Zhenxiong	–	–	–	–	III	II	–
2	Yunnan power grid substation in Zhaotong	Zhaotong urban district	–	–	–	–	IV	III	–
3	Zhaotong fire brigade	Zhaotong urban district	–	–	II	IV	–	–	II
4	Zhaotong Highway Bureau	Zhaotong urban district	III	–	–	–	–	–	–
5	Zhaotong municipal engineering company	Zhaotong urban district	II	II	–	–	–	II	–
6	Zhaotong branch of China petroleum & amp; chemical corporation	Zhaotong urban district	–	–	IV	–	–	–	–
7	Zhaotong water conservancy machinery construction team	Zhaotong urban district	–	V	–	–	–	–	III
8	Yunnan armed police detachment of Zhaotong	Zhaotong urban district	IV	–	–	–	–	IV	II
9	Yunnan power grid Yongshan power supply company	Yongshan	–	–	–	–	I	I	–

(*continued*)

Table 5. (*continued*)

No.	Team affiliation	Location district title	TS	WS	OS	GS	ES	CS	WC
10	Yunnan power grid Yiliang power supply company	Yiliang	–	–	–	–	II	I	–
11	Yiliang fire brigade	Yiliang	–	–	II	II	–	–	I
12	Yiliang road administration team	Yiliang	I	–	–	–	–	–	–
13	Kunming fire brigade	Xishan District		–	III	V	–	–	III
14	Kunming Highway Bureau maintenance team	Xishan District	IV	–	–	–	–	–	–
15	Kunming municipal engineering group construction team	Xishan District	IV	IV	–	–	–	III	–
16	Sinopec engineering construction team	Wuhua District	–	–	V	–	–	–	–
17	Yunnan armed police corps 1	Wuhua District	V	III	–	–	–	V	III
18	Yunnan armed police corps 2	Wuhua District	V	III	–	–	–	IV	III
19	Yunnan armed police corps 3	Wuhua District	V	III	–	–	–	V	III
20	Yunnan power grid Weixin power supply company	Weixin	–	–	–	–	III	II	–
21	Weixin road administration team	Weixin	II	–	–	–	–	–	–
22	Yunnan power grid Qiaojia power supply company	Qiaojia	–	–	–	–	I	I	–
23	Qiaojia road administration team	Qiaojia	I	–	–	–	–	–	–
24	Yunnan power grid Ludian power supply company	Ludian	–	–	–	–	III	III	–
25	Ludian highway administration	Ludian	III		III	–	III-	–	III
26	Ludian Zhongcheng gas company	Ludian	–	–	–	II	–	–	–
27	Yunnan Provincial Highway Bureau maintenance team	Guandu District	V	–	–	–	–	–	–
28	The second water conservancy and hydropower construction co., ltd.	Guandu District	–	III	–	–	–	–	V
29	Kunming power supply Bureau of Yunnan power grid of China Southern power grid	Guandu District	–	–	–	–	V	V	–
30	The first water conservancy and hydropower construction co., ltd.	Chenggong	–	V	–	–	–	–	V

TS: Traffic system, **WS:** Water supply, **OS:** Oil system, **GS:** Gas system, **ES:** Electirc system, **CS:** Communication system, **WC:** Water conservancy

Table 6. The quantity of each equipment in each warehouse

No.	Location district title	Emergency lighting vehicle	Excavator	Diesel generator	Crane	Water pump
1	Zhaotong Urban district	20	–	20	–	–
2	Songming	20	–	20	–	–
3	Chenggong	4	30	–	10	2
4	Zhaotong Urban district	0	10	10	0	0
5	Wuhua Distirct	10	5	3	10	1
6	Zhaotong Urban district	5	10	1	10	2
7	Guandu District	–	–	–	–	10

The parameters related to the cost are presented in Tables 7, 8 and 9. The location of rescue teams and equipment warehouses comprises Zhaotong urban district, Ludian, Qiaojia, Yongshan, Suijiang, Zhenxiong, Yiliang, Weixin in Zhaotong and Xishan District, Wuhua District, Guandu District, Chenggong, Songming in Kunming. Table 10 shows the time matrix between any pair of points is determined by a geography information system (GIS). The data of time spent on road transportation and highway transportation are export from GIS, and the data of time spent on helicopter transportation is estimated by the distance between any pair of points with the 250 km/h ordinary speed for helicopter.

Table 7. Cost spent on using equipment (T$).

Emergency lighting vehicle	Excavator	Diesel generator	Crane	Water pump
10	30	100	40	30

Table 8. Cost of restoring lifeline engineering in each level (T$).

	I	II	III	IV	V
Traffic system	16	32	48	144	216
Water supply	8	16	24	72	108
Oil system	24	48	72	216	324
Gas system	10	20	30	90	136
Electric system	16	32	48	144	216
Communication system	24	50	74	224	336
Water conservancy	40	80	120	360	540

Table 9. Cost of each transportation type (T$).

Road	Highway	Helicopter
10	50	500

Table 10. Time spent on road/highway/helicopter transportation (h).

	Urban district	Ludian	Qiaojia	Yanjin	Daguan	Yongshan	Suijiang	Yiliang	Weixin
Urban district	1/1/1	2/2/2	7/4/2	4/2/2	3/2/2	7/4/2	7/4/2	3/2/2	6/3/2
Ludian	2/2/2	1/1/1	7/4/2	5/3/2	3/2/2	7/4/2	7/4/2	4/2/2	6/3/2
Qiaojia	7/3/2	7/4/2	1/1/1	10/5/2	9/5/2	12/6/2	13/7/3	8/4/2	11/6/3
Yongshan	7/4/2	7/2/2	12/6/2	7/4/2	6/3/2	1/1/1	9/5/2	6/3/2	9/5/2
Zhenxiong	9/5/2	7/4/2	12/6/3	6/3/2	5/3/2	9/5/2	8/4/2	7/4/2	3/2/2
Yiliang	3/2/2	4/2/2	8/4/2	4/2/2	3/2/2	6/3/2	8/4/2	1/1/1	6/3/2
Weixin	6/3/2	6/3/2	11/6/3	5/3/2	5/3/2	9/5/2	8/4/2	6/3/2	1/1/1
Xishan District	6/3/3	6/2/3	9/5/3	8/4/3	7/4/3	10/5/3	11/6/4	9/5/3	11/6/3
Wuhua District	6/3/3	6/3/3	9/5/3	8/4/3	7/4/3	10/5/3	11/6/4	9/5/3	11/6/3
Guandu District	6/3/3	6/3/3	9/5/3	8/4/3	7/4/3	10/5/3	11/6/4	9/5/3	11/6/3
Chenggong	6/3/3	6/3/3	9/5/3	8/4/3	7/4/3	10/5/3	11/6/4	8/4/3	11/6/4
Songming	3/2/2	3/2/2	4/2/2	5/3/2	4/2/2	8/4/4	8/4/4	5/3/2	8/4/4

Table 11. Detailed Pareto fronts statistics of the ε-constraint method

ε_1	ε_2	Cost of rescue teams (T$)	Cost of equipment (T$)	min f_3 (T$)
3	3	6662	5090	11752
3	4	6662	4850	11512
4	3	5200	5090	10290
4	4	5200	4850	10050
4	5	5200	4810	10010
5	4	5096	4850	9946
5	5	5096	4810	9906
5	6	5096	4770	9866
6	5	4902	4810	9712
6	6	4902	4770	9672
7	6	4782	4770	9552
7	8	4782	4760	9542
8	8	4774	4760	9534
9	8	4702	4760	9462
11	10	4662	4760	9422

At the beginning of the computation, a tight lower bound have be find, $f_1^{\min} = 2$ and $f_2^{\min} = 3$. Therefore, this example is implemented in MATLAB R2015a and run on a 2.93 GHz workstation with 2 GB of RAM. Gurobi 7.0.2 is the solver used. The result for exact Pareto fronts are given in Table 11, and the parameter τ have been set to 1.

When the maximum time of objective f_1 amplify 3 to 4, the total cost of objective f_3 drops a lot. From the results we can see that, only **rescue team 25** (*Kunming municipal engineering group construction team*) and **rescue team 9** *(Yunnan power grid Yong-shan power supply company)* go to disaster points by helicopter since the time of latest arrival rescue teams is limited. It fits the actual character which helicopter only can be used in costly short and long distance transportation. The **rescue team 25** is assigned in any determined time, because it has various skills with outstanding level and located within an appropriate distance between these disaster points.

6 Conclusion

An efficient auxiliary decision-making method, DAEM method to solve the transportation and assignment problem in the post-disaster in introduced in this paper. With this method, an improved strategy with novel decision-making basis is developed. The method includes two components, a 0–1 metric for assigning rescue teams and an integrated metric for assigning equipment, building a reasonable optimization of rescue process based on the abundant available rescue resources in a given disaster area.

In post-disaster allocation process, the dominating characteristics about suddenly-occurring demand in very large amounts and the short lead times for a wide variety of supplies [14]. It means that the information in the first time after a disaster happened may be ambiguous, being different from accurate simple case. Thus, how to identify the demand in disaster points accurately is of vital importance in the next research, a quantitative estimation method which is the premise to determine the condition of disaster point must be explored. It can provide great practical significance for this research.

Acknowledgment. This work was supported in part by the National Natural Science Foundation of China under Grant Nos. 91324012, 91024031.

A Appendix

Table 12. The classification and hierarchy in each disaster-affected site.

No.	District title	Damaged infrastructure	Classification	Level
1	Suijiang	Center control room	Electirc system	I
2	Yongshan	Road	Traffic system	I
3	Yanjin	Road	Traffic system	I
4	Yanjin	Road	Traffic system	II
5	Yanjin	Distribution network	Water supply	I
6	Yongshan	Earth and rockfill dam.	Water conservancy	I
7	Yanjin	Earth and rockfill dam.	Water conservancy	I
8	Yongshan	Bridge	Traffic system	II
9	Daguan	Distribution network	Water supply	II
10	Yongshan	Earth and rockfill dam.	Water conservancy	I
11	Daguan	Tunnel	Traffic system	I
12	Yiliang	Road	Traffic system	I
13	Daguan	Railway line	Traffic system	II
14	Yiliang	Road	Traffic system	I
15	Yongshan	Earth and rockfill dam.	Water conservancy	III
16	Weixin	Road	Traffic system	I
17	Urban district	Transformer substation	Electirc system	III
18	Yiliang	Railway line	Traffic system	II
19	Urban district	Earth and rockfill dam.	Water conservancy	II
20	Urban district	Oil depot	Oil system	I
21	Urban district	Transmission line.	Electirc system	II
22	Ludian	Transformer substation	Communication system	III
23	Urban district	Bridge	Traffic system	II
24	Yiliang	Bridge	Traffic system	I
25	Ludian	Road	Traffic system	II
26	Urban district	Highway	Traffic system	II
27	Ludian	Highway	Traffic system	IV
28	Ludian	Gas pipeline.	Gas system	I
29	Ludian	Distribution network	Water supply	I
30	Ludian	Transmission line.	Electirc system	IV
31	Ludian	Transmission line	Electirc system	II
32	Qiaojia	Transmission line	Electirc system	I
33	Ludian	Earth and rockfill dam.	Water conservancy	V
34	Qiaojia	Tunnel	Traffic system	III
35	Qiaojia	Center control room	Communication system	II
36	Qiaojia	Road	Traffic system	II

References

1. Gunn, S.: The language of disasters: a brief terminology of disaster management and humanitarian action. In: Basics of International Humanitarian Missions, pp. 37–40 (2003)
2. Yi, W., Özdamar, L.: A dynamic logistics coordination model for evacuation and support in disaster response activities. Eur. J. Oper. Res. **179**, 1177–1193 (2007)
3. Tofighi, S., Torabi, S.A., Mansouri, S.A.: Humanitarian logistics network design under mixed uncertainty. Eur. J. Oper. Res. **250**, 239–250 (2016)
4. Torabi, S.A., Baghersad, M., Mansouri, S.A.: Resilient supplier selection and order allocation under operational and disruption risks. Transp. Res. Part E: Logist. Transp. Rev. **79**, 22–48 (2015)
5. Wallace, W.A.: Restoration of Services in Interdependent Infrastructure Systems: A Network Flows Approach
6. Campbell, A.M., Vandenbussche, D., Hermann, W.: Routing for relief efforts. Transp. Sci. **42**, 127–145 (2008)
7. Nurre, S.G., Sharkey, T.C.: Integrated network design and scheduling problems with parallel identical machines: complexity results and dispatching rules. Networks **63**, 306–326 (2014)
8. Boland, N., Kalinowski, T., Kaur, S.: Scheduling arc shut downs in a network to maximize flow over time with a bounded number of jobs per time period. J. Comb. Optim. **32**, 885–905 (2016)
9. Altay, N.: Capability-based resource allocation for effective disaster response. IMA J. Manag. Math. **24**, 253–266 (2012)
10. Soylu, B., Yildiz, G.B.: An exact algorithm for biobjective mixed integer linear programming problems. Comput. Oper. Res. **72**, 204–213 (2016)
11. Özlen, M., Azizoğlu, M.: Multi-objective integer programming: a general approach for generating all non-dominated solutions. Eur. J. Oper. Res. **199**, 25–35 (2009)
12. Abounacer, R., Rekik, M., Renaud, J.: An exact solution approach for multi-objective location-transportation problem for disaster response. Comput. Oper. Res. **41**, 83–93 (2014)
13. http://www.sbsm.gov.cn/article/chyw/
14. Balcik, B., Beamon, B.M.: Facility location in humanitarian relief. Int. J. Logist. **11**, 101–121 (2008)

Queueing Applications II

Tail Probabilities of the Delay Under a Reservation-Based Scheduling Mechanism

Michiel De Muynck[✉], Herwig Bruneel, and Sabine Wittevrongel

Department TELIN, Ghent University,
Sint-Pietersnieuwstraat 41, 9000 Ghent, Belgium
{MichielR.DeMuynck,Herwig.Bruneel,Sabine.Wittevrongel}@UGent.be

Abstract. We study the tail probabilities of the customer delay for a buffer operating under a reservation-based scheduling discipline known as R-scheduling. Previous numerical work on this model has led to a hypothesized meta-model, which was tested using simulations but not proven analytically. In this paper, we prove the correctness of this meta-model analytically, and extend it to more general arrival processes. The results are also compared to simulations for several example scenarios.

Keywords: Delay differentiation · Traffic classes · Reservation places Delay quantile spacing

1 Introduction

There are many different kinds of queueing phenomena where multiple classes of customers with different quality of service (QoS) requirements must be handled by one server: packets with widely varying priorities may be transmitted over the same channel, one web server may be used to process various kinds of requests, one clerk may have to perform multiple administrative jobs for various people, etc. In all of these cases, the order in which the customers receive service plays an important role in the delays experienced by these customers. As such, the effects of this ordering have received much attention in the literature, and many scheduling mechanisms have been proposed (see e.g. [1] for a survey). Examples include scheduling mechanisms where the next customer to be served is chosen probabilistically [2] or using a (weighted) round-robin process [3], scheduling mechanisms where high-priority customers may only overtake low-priority customers that arrive in the same batch [4] or within a certain window [5], scheduling mechanisms where low-priority customers can become high-priority customers under certain conditions [6,7], etc.

However, in general, no scheduling mechanism can be said to be strictly better than any other scheduling mechanism, as choosing to serve customer A before customer B typically increases the delay experienced by customer B, and vice versa. As an example, choosing to service all customers in first-in first-out (FIFO) order is fair for all customers, but may lead to unacceptable delays for

© Springer International Publishing AG 2017
W. Yue et al. (Eds.): QTNA 2017, LNCS 10591, pp. 265–281, 2017.
https://doi.org/10.1007/978-3-319-68520-5_16

delay-sensitive customers, whereas giving absolute priority (AP) to these delay-sensitive customers may lead to extremely high delays for the low-priority packets (see e.g. [8,9]), a phenomenon sometimes referred to as "packet starvation". In between these two extremes of FIFO and AP lies an entire spectrum of scheduling mechanisms, each with different trade-offs.

Somewhere on that spectrum lies the reservation-based scheduling mechanism proposed by Wittevrongel et al. in [10] referred to as "R-scheduling", which is the object of study in this paper. In that scheduling mechanism, each traffic class is given a fixed number of "reservation places" in the queue, that can be thought of as "holes" in the queue that can later be filled in by arriving customers of that traffic class. The reservation places need not necessarily be actual holes in the queue. For example, in a queueing system where customers take tickets with a ticket number on it and wait until the number on their ticket is shown, a reservation place may also be a ticket number that is "skipped" and not given out until a later time when a delay-sensitive customer arrives.

The behavior of the R-scheduling mechanism depends on the number of reservation places of each traffic class as well as the traffic intensity of each traffic class. If a high-priority traffic class has many reservation places in the queue and the queue is short due to a period of low traffic intensity, then R-scheduling behaves very similarly to AP, since new arrivals of the high-priority traffic class will almost always find a reservation place waiting at the front of the queue that they can fill in. However, if there is a period of high intensity traffic that causes the queue to become very long, then this will no longer be the case, since the reservation places will not be able to make it all the way to the front of the queue before being filled in. This may however be beneficial, as AP would in this case lead to very long delays of the low-priority packets due to packet starvation, while under R-scheduling, a low-priority packet can not be overtaken by more high-priority packets than there are reservation places in front of it in the queue, effectively preventing packet starvation.

As mentioned earlier, which scheduling mechanism is best suited for a given situation depends on the QoS requirements of the specific classes of customers. One commonly used metric in specifications of QoS requirements is the probability that the delay of a customer of a given class exceeds a certain large value. These so-called "tail probabilities" are especially useful in real-time applications where, for instance, a packet arriving unacceptably late is as bad as if it were lost entirely [11].

In [10], the tail probabilities of the delay of the R-scheduling mechanism were studied using extensive simulations. Under the assumptions that the arrival process is a discrete-time independent Poisson arrival process and the classes of consecutive customers form an independent and identically distributed (i.i.d.) sequence, a heuristic formula for these tail probabilities was proposed that closely matched the simulation results. In this paper, we prove analytically that this formula is indeed (almost) correct, and we derive the exact values for the constants used in that formula, by first studying a generalization of the model that relaxes the requirements of Poisson arrivals and independent customer classes. The proof of the heuristic formula from [10] then follows as a special case of the general model.

In other related work on R-scheduling, analytical techniques using probability-generating functions (pgfs) have been used to obtain expressions for the pgf of the customer delay for each traffic class (see e.g. [12,13] for the most-recently published results). These are very thorough and complete results, as the pgf of a distribution fully describes that distribution (all probabilities, moments, and of course also the tail probabilities). However, due to the difficulty of the analysis, several simplifying assumptions typically need to be made on the model to keep the analysis feasible. In both [12,13], the model was restricted to 2 traffic classes, with only a single reservation place in the queue for the high-priority traffic class. This reduces the applicability of these results in practice, as the possible delay differentiation with only one reservation place is limited.

In Sect. 2, we describe the queueing model in detail, as well as the mathematical notation used in this paper. We study the tail probabilities of the delay of this model analytically in Sect. 3. The special case of uncorrelated traffic classes is studied in Sect. 4. In Sect. 5, we briefly describe possible computationally efficient implementations of R-scheduling. Finally, in Sect. 6, we explore several numerical examples to demonstrate the behavior of the queueing model.

2 Queueing Model

In this section, we will describe the operation of the queueing model studied in this paper, and the mathematical notation used to study it. While this queueing model is applicable to many kinds of queueing phenomena, we will consider the specific queueing phenomenon where packets of various classes are transmitted across a single channel.

We consider a queueing model with m classes of packets. These packets are all stored in a single queue, but in addition to these packets, the queue also holds several "reservation places". Each traffic class i ($i = 1, ..., m$) has n_i such reservation places in the queue. The total number of reservation places is denoted as $n = n_1 + ... + n_m$. These reservation places behave exactly like packets in the sense that they take up a space in the queue and move forwards in lockstep with the packets in the queue. Reservation places, however, are never transmitted from the queue.

Packets arriving at the queue are placed in the queue one by one in the order in which they arrived according to the following mechanism: When a packet of class i enters the queue, if $n_i = 0$, then the packet enters the queue at the back, behind all other packets and all reservation places. If $n_i > 0$, then the packet takes the place of the frontmost reservation place of class i, i.e., the one closest to the head of the queue. That reservation place is destroyed and a new reservation place for class i is immediately created and placed at the back of the queue, behind all packets and all other reservation places. In this way, the number of reservation places for class i always remains constant, equal to n_i.

We assume that the output channel allows only one packet to be transmitted at the same time, and that the transmission of each packet takes the same amount of time, exactly one unit of time. If, at any time, there is at least one packet in the queue but no packet in transmission, the frontmost packet in the queue immediately begins its transmission. Any reservation places that were in front of that packet remain in place.

The arrival process in this queueing model is relatively general. In fact, we make no assumptions about the arrival process other than that if two packets arrive at the same instant, they arrive in a certain order so that one can still be said to arrive "before" the other. The arrivals may occur in batches at regular intervals ("slot boundaries"), as in discrete-time queueing models, or they may occur irregularly, as in continuous-time models.

To allow for correlations between the classes of subsequent arriving packets, we assume that there is a background Markov chain with a finite number n_{MC} of states and transition matrix $\mathbf{R}_{\mathrm{MC}}$. This Markov chain is assumed to be irreducible and aperiodic, and we denote its equilibrium distribution by the probability vector $\boldsymbol{\pi}$, so that $\boldsymbol{\pi} = \boldsymbol{\pi}\mathbf{R}_{\mathrm{MC}}$. Whenever a packet P arrives, this Markov chain performs one transition, and then the traffic class T_P of the arriving packet is determined by the following probabilities:

$$P(T_P = i | S_{P-1} = j, S_P = k) \triangleq p_{i,j,k}, \quad 1 \leq i \leq m, \ 0 \leq j, k < n_{\mathrm{MC}}, \quad (1)$$

where S_{P-1} and S_P denote the state of the background Markov chain right before and right after the packet arrived respectively. Given S_{P-1} and S_P, the traffic class of P is assumed to be independent from the time of P's arrival, the arrival times or the classes of all other packets, the state of the system at the moment of P's arrival, and all other past events. Note that in particular, when multiple packets arrive at the same time, there may be no reordering based on the classes of the packets, i.e., there may be no "slot-bound priority" [4].

We denote the delay experienced by an arbitrary packet of class i in steady state as D_i. This delay is defined as the time between the arrival instant and the end of the transmission of that packet. We denote the delay that the packet would have experienced if the scheduling mechanism were FIFO, i.e., if $n_1 = \ldots = n_m = 0$, as D_{FIFO}, and assume that the probability that $D_{\mathrm{FIFO}} \geq x$ decays exponentially as $x \to \infty$. That is, we assume that there are two constants c_{FIFO} and s_{FIFO} such that $P(D_{\mathrm{FIFO}} \geq x)s_{\mathrm{FIFO}}^{-x}$ approaches c_{FIFO} as $x \to \infty$ (considering only integer x in case of a discrete-time arrival process). Methods to obtain these constants for a wide variety of models are available in the queueing literature.

Another quantity of interest in this paper is the α-quantile d_i^α of the delay D_i perceived by packets of class i in steady state. This α-quantile d_i^α is the value of x for which $P(D_i \geq x) = \alpha$, with appropriate interpolation between values of $P(D_i \geq x)$ in case of a discrete-time arrival process.

3   Analysis

In this section, we analyze the queueing model described in the previous section in order to obtain an expression for the tail probabilities of the delay of an arbitrary packet of class i in steady state. We begin the analysis by studying the positions of the reservation places in the queue. Defining N_R as the number of packets and/or reservation places in the queue *behind* a given reservation place R of class i, we look at all the events that can cause N_R to change. Whenever a reservation place R is created, it is placed at the back of the queue, so at that moment N_R is 0. Afterwards, N_R changes only when these events occur:

1. When a new packet belonging to a class $j \neq i$ arrives,
   (a) if $n_j = 0$, then the new packet is placed at the back of the queue,
   (b) otherwise, the new packet takes the place of a different reservation place than R and a new reservation place of class j is created at the back of the queue.
   In both cases, N_R increases by 1.
2. When a new packet belonging to class i arrives,
   (a) if R is the frontmost reservation place of class i, it is destroyed,
   (b) otherwise, the arriving packet replaces the frontmost reservation place of class i, and a new reservation place is created at the back of the queue, increasing N_R by 1.
3. When a packet begins its transmission,
   (a) if that packet was in front of R in the queue, N_R remains unchanged,
   (b) otherwise, N_R decreases by 1.

Determining the distribution of N_R at the moment that R is destroyed is difficult in general and essentially requires solving the entire queueing system. However, note that if R is filled in by a packet P that has a delay $D_i > n + 1$, event 3 (b) above cannot occur for R, since event 3 (b) can only occur if there are only other reservation places in front of R in the queue. Since the goal of this analysis is to find an expression for the tail probabilities $P(D_i \geq x)$ for large x, we will from now restrict our analysis to reservation places R for which the event 3 (b) never occurs.

If we denote the value of N_R at the moment R is destroyed as V_R, we find that V_R is simply equal to the number of packets that arrived between the creation and destruction of R, not including the packets whose arrival caused the creation or destruction of R. Alternatively, V_R is the number of transitions experienced by the background Markov chain during that same time. But the destruction of R happens precisely at the n_ith transition of that Markov chain since the creation of R that produced a packet of class i. If we can rewrite this description of V_R as the number of transitions of a certain terminating Markov chain with 1 terminating state until that terminating state is reached, then we know that V_R is a discrete phase-type distribution (by definition, see e.g. [14]). We can do this by choosing the states of this terminating Markov chain to correspond to the combination of the state of the background Markov chain and the number of class-i arrivals since the creation of R. The terminating state corresponds to all situations where n_i or more class-i packets have arrived since the creation of R.

However, we have to be slightly careful. In the definition of a discrete phase-type distribution, the first transition to the absorbing state is included, whereas the n_ith arrival of class i does not contribute to V_R. We therefore find that $V_R + 1$

follows the discrete phase-type distribution $\text{PH}_d(\boldsymbol{\alpha}^{(i)}, \mathbf{T}^{(i)})$, where $\boldsymbol{\alpha}^{(i)}$ will be determined later and $\mathbf{T}^{(i)}$ is the $n_i \times n_i$ block matrix

$$
\mathbf{T}^{(i)} = \begin{bmatrix}
\mathbf{T}_0^{(i)} & \mathbf{T}_1^{(i)} & 0 & \dots & 0 & 0 \\
0 & \mathbf{T}_0^{(i)} & \mathbf{T}_1^{(i)} & \dots & 0 & 0 \\
0 & 0 & \mathbf{T}_0^{(i)} & \dots & 0 & 0 \\
\vdots & \vdots & \vdots & \ddots & \vdots & \vdots \\
0 & 0 & 0 & \dots & \mathbf{T}_0^{(i)} & \mathbf{T}_1^{(i)} \\
0 & 0 & 0 & \dots & 0 & \mathbf{T}_0^{(i)}
\end{bmatrix}. \tag{2}
$$

Here, $\mathbf{T}_0^{(i)}$ and $\mathbf{T}_1^{(i)}$ are the $n_{\text{MC}} \times n_{\text{MC}}$ matrices defined as

$$
[\mathbf{T}_0^{(i)}]_{j,k} = [\mathbf{R}]_{j,k}(1 - p_{i,j,k}), \qquad 0 \le j, k < n_{\text{MC}},
$$
$$
[\mathbf{T}_1^{(i)}]_{j,k} = [\mathbf{R}]_{j,k} p_{i,j,k}, \qquad 0 \le j, k < n_{\text{MC}}.
$$

In the above phase-type distribution, phase j $(0 \le j < n_{\text{MC}})$ in block-phase k $(0 \le k < n_i)$ corresponds to the background Markov chain being in state j while there have already been k arrivals of traffic class i since reservation place R was created. The number of packet arrivals until the n_ith packet of class i arrives (including that last packet, which should not be counted when calculating V_R) is precisely the number of transitions until the absorbing state in the above phase-type distribution is reached, i.e.,

$$
V_R + 1 \sim \text{PH}_d(\boldsymbol{\alpha}^{(i)}, \mathbf{T}^{(i)}). \tag{3}
$$

The probability vector $\boldsymbol{\alpha}^{(i)}$ describes the state of the system immediately after the reservation place R is created. Trivially, there are no class-i arrivals at that time since the creation of R yet, so $\boldsymbol{\alpha}^{(i)}$ is a $1 \times n_i$ block row-vector

$$
\boldsymbol{\alpha}^{(i)} = \begin{bmatrix} \boldsymbol{\alpha}_0^{(i)} & 0, \dots, 0 \end{bmatrix}, \tag{4}
$$

where $\boldsymbol{\alpha}_0^{(i)}$ is the $1 \times n_{\text{MC}}$ probability vector that describes the distribution of the state S_R of the background Markov chain at the moment R is created. Since we know that R was created as a result of a packet of class i arriving at the queue, and since R is an arbitrarily chosen reservation place in steady state, we find after some applications of Bayes' theorem that

$$
[\boldsymbol{\alpha}_0^{(i)}]_j \triangleq P(S_R = j) = \frac{\displaystyle\sum_{l=0}^{n_{\text{MC}}} [\boldsymbol{\pi}]_l [\mathbf{R}]_{l,j} p_{i,l,j}}{\displaystyle\sum_{k=0}^{n_{\text{MC}}} \sum_{l=0}^{n_{\text{MC}}} [\boldsymbol{\pi}]_l [\mathbf{R}]_{l,k} p_{i,l,k}}. \tag{5}
$$

We can now use the distribution of V_R to calculate the tail probabilities $P(D_i \ge x)$ for large x. Consider an arbitrary packet P of class i with delay

$D_i \geq x$. As mentioned previously, if $x > n + 1$ then event 3 (b) can cannot have occurred for the reservation place R that packet P filled. Secondly, since we are only interested in large x, we may also make the simplifying assumption that all reservation places that existed at the moment P arrived are destroyed before P begins its transmission, i.e., in at most $x - 1$ units of time. Since the total number of reservation places n is finite and fixed, the probability of this assumption being true approaches 1 as x goes to infinity.

With this assumption, finding the delay D_i given D_{FIFO} is easy. We will consider the cases where $n_i = 0$ and $n_i > 0$ separately. If $n_i = 0$, then the delay D_i is simply equal to $D_{\text{FIFO}} + n$, since packet P enters the queue at the back, behind all n reservation places, which will be filled before P begins its transmission. We therefore have for large x that

$$P(D_i \geq x) = P(D_{\text{FIFO}} + n \geq x) \approx c_{\text{FIFO}} s_{\text{FIFO}}^{x-n}, \tag{6}$$

in view of the assumed exponential delay of D_{FIFO}.

If $n_i > 0$, then at the moment when packet P fills in reservation place R, but before R is recreated, the total (remaining) transmission time of all packets and reservation places (counting them as 1 unit of time each) in the queue is $D_{\text{FIFO}} + n - 1$, of which V_R units of time are behind P in the queue, so we have that $D_i = D_{\text{FIFO}} + n - 1 - V_R$. From this, $P(D_i \geq x)$ for large x easily follows as

$$P(D_i \geq x) = P(D_{\text{FIFO}} + n - 1 - V_R \geq x) \tag{7}$$

$$= \sum_{k=1}^{\infty} P(D_{\text{FIFO}} \geq x + k - n) P(V_R + 1 = k)$$

$$\approx \sum_{k=1}^{\infty} c_{\text{FIFO}} s_{\text{FIFO}}^{x+k-n} \boldsymbol{\alpha}^{(i)} (\mathbf{T}^{(i)})^{k-1} (\mathbf{I} - \mathbf{T}^{(i)}) \mathbf{1}$$

$$= c_{\text{FIFO}} s_{\text{FIFO}}^{x+1-n} \boldsymbol{\alpha}^{(i)} (\mathbf{I} - s_{\text{FIFO}} \mathbf{T}^{(i)})^{-1} (\mathbf{I} - \mathbf{T}^{(i)}) \mathbf{1}, \tag{8}$$

where $\mathbf{1}$ denotes a column vector of the appropriate dimension whose elements are all equal to 1 and $\mathbf{I}$ denotes the identity matrix.

It can be verified that the inverse of $\mathbf{I} - s_{\text{FIFO}} \mathbf{T}^{(i)}$ is given by

$$(\mathbf{I} - s_{\text{FIFO}} \mathbf{T}^{(i)})^{-1} = \begin{bmatrix} \boldsymbol{\Gamma} & \boldsymbol{\Lambda}\boldsymbol{\Gamma} & \boldsymbol{\Lambda}^2\boldsymbol{\Gamma} & \boldsymbol{\Lambda}^3\boldsymbol{\Gamma} & \dots & \boldsymbol{\Lambda}^{n_i-1}\boldsymbol{\Gamma} \\ & \boldsymbol{\Gamma} & \boldsymbol{\Lambda}\boldsymbol{\Gamma} & \boldsymbol{\Lambda}^2\boldsymbol{\Gamma} & \dots & \boldsymbol{\Lambda}^{n_i-2}\boldsymbol{\Gamma} \\ & & \boldsymbol{\Gamma} & \boldsymbol{\Lambda}\boldsymbol{\Gamma} & \dots & \boldsymbol{\Lambda}^{n_i-3}\boldsymbol{\Gamma} \\ & & & \ddots & \ddots & \vdots \\ & & & & \boldsymbol{\Gamma} & \boldsymbol{\Lambda}\boldsymbol{\Gamma} \\ & & & & & \boldsymbol{\Gamma} \end{bmatrix}, \tag{9}$$

where

$$\boldsymbol{\Gamma} \triangleq (\mathbf{I} - s_{\text{FIFO}} \mathbf{T}_0^{(i)})^{-1},$$

$$\boldsymbol{\Lambda} \triangleq (\mathbf{I} - s_{\text{FIFO}} \mathbf{T}_0^{(i)})^{-1} s_{\text{FIFO}} \mathbf{T}_1^{(i)}.$$

Indeed, it is easily seen that multiplying the above result for $(\mathbf{I} - s_{\text{FIFO}}\mathbf{T}^{(i)})^{-1}$ by $\mathbf{I} - s_{\text{FIFO}}\mathbf{T}^{(i)}$ results in the identity matrix.

Since $(\mathbf{I} - \mathbf{T}^{(i)})\mathbf{1}$ is given by

$$(\mathbf{I} - \mathbf{T}^{(i)})\mathbf{1} = \begin{bmatrix} 0 \\ 0 \\ \vdots \\ \mathbf{T}_1^{(i)}\mathbf{1} \end{bmatrix}, \tag{10}$$

we find that

$$(\mathbf{I} - s_{\text{FIFO}}\mathbf{T}^{(i)})^{-1}(\mathbf{I} - \mathbf{T}^{(i)})\mathbf{1} = \frac{1}{s_{\text{FIFO}}} \begin{bmatrix} \boldsymbol{\Lambda}^{n_i}\mathbf{1} \\ \boldsymbol{\Lambda}^{n_i-1}\mathbf{1} \\ \vdots \\ \boldsymbol{\Lambda}\mathbf{1} \end{bmatrix}, \tag{11}$$

so that (8) reduces to the very simple expression

$$P(D_i \geq x) \approx c_{\text{FIFO}} s_{\text{FIFO}}^{x-n} \boldsymbol{\alpha}_0^{(i)} \boldsymbol{\Lambda}^{n_i}\mathbf{1}. \tag{12}$$

It can be verified that (12) simplifies to (6) when $n_i = 0$, so that (12) is valid for all $n_i \geq 0$.

4   Special Case: Uncorrelated Traffic Classes

It is interesting to study the special case where there is no correlation between the classes of the arriving packets, since this occurs in many queueing phenomena and since this special case is also studied in [10]. In this special case, $n_{\text{MC}} = 1$, and we define

$$p_i \triangleq p_{i,0,0}. \tag{13}$$

Then (12) becomes

$$P(D_i \geq x) \approx c_{\text{FIFO}} s_{\text{FIFO}}^{x+n_i-n} \left(\frac{p_i}{1 - s_{\text{FIFO}}(1 - p_i)} \right)^{n_i}. \tag{14}$$

As a verification of the correctness of (14), we note that it may also be derived directly from (7). Indeed, when the classes of arriving packets are independent from each other, we find that between the creation and destruction of the arbitrarily chosen reservation place R, the number of arrivals of classes other than i is negative binomially distributed with parameters n_i and $1 - p_i$. The number of class-i arrivals in that same period is equal to $n_i - 1$, so in total we find that $V_k - (n_i - 1)$ is negative binomially distributed as

$$P(V_R - (n_i - 1) = k) = \binom{k + n_i - 1}{k} p_i^{n_i}(1 - p_i)^k, \quad k \geq 0. \tag{15}$$

Substituting (15) into (7) results in (14), as expected.

We can use (14) to find the delay quantile d_i^α. To do this, we solve the equation $\log P(D_i \geq x) = \log \alpha$ for x and obtain

$$d_i^\alpha = \frac{\log \alpha}{\log s_{\text{FIFO}}} - \frac{\log c_{\text{FIFO}}}{\log s_{\text{FIFO}}} + (n - n_i) + \frac{\log \left(\frac{1 - s_{\text{FIFO}}(1 - p_i)}{p_i} \right)}{\log s_{\text{FIFO}}} n_i. \qquad (16)$$

The above expression for d_i^α is linear in the n_i's, with coefficients that are very simple to calculate. This is very useful in practice if the arrival process (and, as a consequence, c_{FIFO} and s_{FIFO}) and the various relative loads p_i are fixed, and the goal is to achieve certain given asymptotic delay probabilities for each class. In that case, the R-scheduling mechanism may be used, and finding the optimal n_i's to achieve the required tail probabilities is easy, as the effects of adding or removing a reservation place from a traffic class on the tail probabilities of the delay are evident.

We can compare (16) with the meta-model found experimentally in [10]. Specifically, in [10], the following form for d_i^α was hypothesized (note that $\log s_{\text{FIFO}}$ was denoted as r_{FIFO} in [10]):

$$d_i^\alpha = \frac{\log \alpha}{\log s_{\text{FIFO}}} + C_1 + C_2 \cdot (n - n_i) + C_3(s_{\text{FIFO}}, p_i) \cdot n_i. \qquad (17)$$

We see that (16) is almost of the form (17). The only difference is that C_1 is not a constant but instead depends on c_{FIFO} and s_{FIFO}, which in turn depend on the total load and the arrival process. In [10], the arrival process was restricted to discrete-time independent arrival processes where the number of arrivals per time slot follows a Poisson distribution with mean λ. For this case, c_{FIFO} and s_{FIFO} can be calculated as

$$c_{\text{FIFO}} = -\frac{(1 - \lambda)z_d}{\lambda(1 - \lambda z_d)}, \quad s_{\text{FIFO}} = \frac{1}{z_d}, \qquad (18)$$

where z_d is the smallest real-valued zero for $z > 1$ of $z - e^{\lambda(z-1)}$. In Fig. 1 we show the exact value of C_1 for this arrival process as a function of λ. Note that even though C_1 grows unbounded as $\lambda \to 0$, C_1 varies relatively little for realistic

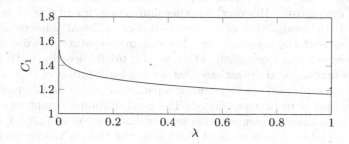

Fig. 1. C_1 versus λ for an independent discrete-time Poisson arrival process

values of λ and can almost be considered a constant. In [10], the value of this "constant" was reported as approximately 1.1, which roughly agrees with Fig. 1.

The value of the constant C_2 was also estimated from simulations in [10] and found to be approximately 1.0. Equation (16) confirms that C_2 is in fact exactly equal to 1. The value of $C_3(s_{\mathrm{FIFO}}, p_i)$ for a discrete-time Poisson arrival process was reported in [10, table I] for several values of the total load and the relative load p_i (see also Table 1). We find that these values agree very closely with those given by (16) (see Table 2).

Table 1. Values of $C_3(s_{\mathrm{FIFO}}, p_i)$ for a discrete-time Poisson arrival process, as reported in [10, table I]

Total load	s_{FIFO}	p_i					
		10%	20%	30%	40%	50%	60%
95%	0.903277	−6.1	−3.2	−2.0	−1.3	−0.9	−0.7
90%	0.812902	−4.7	−2.7	−1.7	−1.2	−0.8	−0.6
85%	0.728560	−3.8	−2.3	−1.5	−1.1	−0.8	−0.6

Table 2. Values of $C_3(s_{\mathrm{FIFO}}, p_i)$ for a discrete-time Poisson arrival process, as given by (16)

Total load	s_{FIFO}	p_i					
		10%	20%	30%	40%	50%	60%
95%	0.903277	−6.16	−3.22	−2.00	−1.33	−0.91	−0.61
90%	0.812902	−4.77	−2.70	−-1.75	−1.19	−0.83	−0.56
85%	0.728560	−3.90	−2.32	−1.55	−1.08	−0.76	−0.53

5 A Note About Implementation

In the previous sections, we analyzed queueing characteristics of the studied scheduling mechanism. However, another important quality of a scheduling mechanism is the availability of a computationally efficient implementation. It therefore deserves to be noted that two efficient implementations are available for the studied scheduling mechanism, which we will briefly describe in this section.

The two fundamental operations that an implementation of this scheduling mechanism must provide are enqueueing a packet of class i and dequeueing the next packet that is to be transmitted. The computational complexity of these operations for a naive implementation may be linear in N, i.e., $O(N)$, where N denotes the number of packets in the queue at the time of the operation. This may be prohibitively expensive if the load is high and the average queue length is large.

Fortunately, a more efficient implementation is possible, as follows. When a reservation place R is created, we denote the number of reservation places that have already been created up to that point as the *index* of R. When a packet takes the place of R, its index is the same as the index of R. When a packet of class i for which $n_i = 0$ enters, we assume that it creates a temporary reservation place and then instantly takes its place, effectively increasing the index of the next reservation places that will be created by 1. This way, no two reservation places or packets ever have the same index. Note that the "index" of a packet is actually the same as the "ticket number" of a customer in the example given in Sect. 1.

By storing the indices of the reservation places of traffic class i in a separate queue (for instance a linked list) for each i, and storing the indices of all the packets in the queue in one balanced search tree, the enqueueing and dequeueing operations can both be implemented with computational complexity $O(\log N)$. By storing the reservation places the same way but instead storing the indices of the packets of each class in a separate queue for each traffic class, the enqueueing and dequeueing operations can be implemented with respective complexities $O(1)$ and $O(m)$. Which of these implementations is more efficient depends on the magnitude of m and the average length N.

6   Numerical Examples

In this section, we study several numerical examples to demonstrate the behavior of the queueing model and the correctness of the analysis.

In the first numerical example, we examine the impact of correlation in the arrival process on the packet delay under the R-scheduling mechanism. In Figs. 2 and 3, we show the cumulative distribution function of the packet delay for each traffic class, as estimated by our model and by a Monte-Carlo simulation of 10^9 time slots. There are $m = 3$ traffic classes, (independent) probabilities $p_1 = 0.3$, $p_2 = 0.3$ and $p_3 = 0.4$ of a customer belonging to each class. The three classes respectively have $n_1 = 0$, $n_2 = 5$, and $n_3 = 15$ reservation places. In Fig. 2, the arrival process is a discrete-time Poisson process (as considered in [10]) with $\lambda = 8\ln(9/8) \approx 0.94$. Why this specific value of λ was chosen will be explained later. From (18), we find that $s_{\text{FIFO}} = 8/9$ and $c_{\text{FIFO}} \approx 1.148$.

In Fig. 3, the arrival process is a discrete-time batch Markovian arrival process (D-BMAP), which is suitable to model a large class of correlated processes and variable bit rate sources (see e.g. [15,16]). In a D-BMAP, the arrival process is governed by a Markov chain with n_{DBMAP} states. Each time slot, this Markov chain performs one transition, and generates a number of arrivals that depends on the states that the Markov chain transitioned from and to. The probability that this Markov chain transitions to state j and generates k arrivals, given that it was in state i during the previous slot, is denoted as $[\mathbf{D}_k]_{i,j}$. A D-BMAP arrival process is therefore determined by the sequence of $n_{\text{DBMAP}} \times n_{\text{DBMAP}}$ matrices $\mathbf{D}_0$, $\mathbf{D}_1$, $\mathbf{D}_2$, etc.

We choose the parameter matrices of the D-BMAP arrival process in this example to be

$$\mathbf{D}_0 = \begin{bmatrix} 0.9 & 0 \\ 0 & 0 \end{bmatrix}, \quad \mathbf{D}_1 = \begin{bmatrix} 0 & 0.1 \\ 0.2 & 0 \end{bmatrix}, \quad \mathbf{D}_2 = \begin{bmatrix} 0 & 0 \\ 0 & 0.8 \end{bmatrix}, \tag{19}$$

and $\mathbf{D}_k = \mathbf{0}$ for $k > 2$. This arrival process generates either 0, 1 or 2 arrivals each slot, while never increasing or decreasing the number of arrivals by more than 1 per time slot. Using the methods described in [16] we find $s_{\text{FIFO}} = 8/9$ and $c_{\text{FIFO}} = 9/8$. Note that the value of s_{FIFO} is the same in Figs. 2 and 3. The value of λ in Fig. 2 was chosen specifically so that this would be the case, as it allows for a cleaner comparison between the two different arrival processes.

From Figs. 2 and 3, we notice that the time correlation in the arrival process has no impact on the delay quantile spacing, as predicted since the tail probabilities of the delay only depend on the arrival process through c_{FIFO} and s_{FIFO} (see (18)). However, the correlation in the arrival process in Fig. 3 does cause the cumulative distribution function of the delay to approach its asymptote more slowly. In Fig. 2, the model accurately predicts d_i^α for $\alpha < 10^{-2}$, while in Fig. 3, the model only accurately predicts d_i^α for $\alpha < 10^{-3}$.

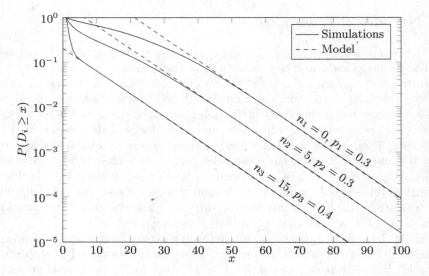

Fig. 2. Tail probabilities of the packet delay for independent customer classes and a discrete-time Poisson arrival process

In our second numerical example, shown in Fig. 4, we study the impact of correlation between the classes of arriving customers. As the arrival process we choose a discrete-time Poisson process with $\lambda = 0.95$. There are 3 customer

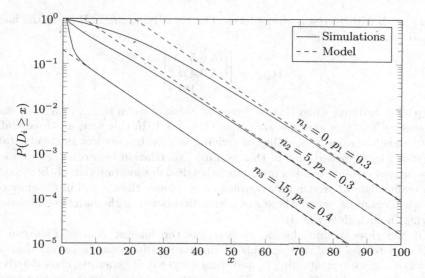

Fig. 3. Tail probabilities of the packet delay for independent customer classes and a D-BMAP arrival process

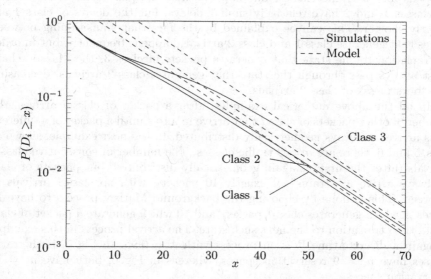

Fig. 4. Tail probabilities of the packet delay for correlated customer classes and a discrete-time Poisson arrival process

classes, each with 10 reservation places. The transition matrix $\mathbf{R}_{MC}$ of the background Markov chain is

$$\mathbf{R}_{MC} = \begin{bmatrix} 0.5 & 0.5 & 0 \\ 0.5 & 0.4 & 0.1 \\ 0 & 0.1 & 0.9 \end{bmatrix} . \tag{20}$$

The packet arriving when this background Markov chain transitions from state i to state j is in this example simply of class $j + 1$. In this way, all three traffic classes have the same probability of arriving in an arbitrary slot in steady state.

From Fig. 4, we see that in this example, correlation between the classes of consecutive customers makes the cumulative distribution function of the delay of each traffic class approach its asymptote even slower than it did in the previous numerical example, where there was correlation between the numbers of packets arriving in each slot (Fig. 3).

Of the three traffic classes, class 3 has the longest delays. This can be attributed to the fact that arrivals of class 3 mostly occur in bursts that are spread out, since transitioning to and from a state that generates class-3 arrivals occurs with very low probability. These bursts quickly consume the 10 reservation places, while during a long period with no class-3 arrivals, the reservation places may reach the front of the queue and be unable to advance further. This happens for all traffic classes, but happens more frequently for class 3. Traffic classes 1 and 2 have relatively similar delays, but the delays of class 1 are slightly shorter. This can be explained by the fact that transitioning between states that generate class-1 and class-2 arrivals happens frequently, but in order to transition to the state that generates packets of class 3, the Markov chain must always pass through the state that generates class-2 arrivals, decreasing the "burstiness" of class-2 arrivals.

In all the above numerical examples, after a packet of class i arrives, the number of other packets of class i that arrive in a row until a packet of a different class arrives is always geometrically distributed. In the next example, shown in Figs. 5 and 6, there are $m = 2$ traffic classes. The number of consecutive class-1 arrivals (after the first) is again geometrically distributed, but packets of class 2 always arrive in "trains" of exactly 10 packets, with no class-1 arrivals in between. This is done by choosing the background Markov process to have 11 states, 1 which generates class-1 packets and 10 which generate a packet of class 2 and then transition to the next such state. The arrival process in this example is again a discrete-time Poisson process with $\lambda = 0.95$. In Fig. 5, both traffic classes have $n_i = 9$ reservation spaces, whereas in Fig. 6, both have $n_i = 10$ reservation spaces.

It may seem beneficial for class-2 packets if the length of each train matches the number of available reservation places for that class (i.e., 10), as then each packet in a class-2 train will use exactly one reservation place, and the reservation places of class 2 stay clumped together. On the contrary, for 9 reservation places, the 10th arriving packet of a class-2 train will always have to use a newly created reservation place near the back of the queue. However, from Figs. 5 and 6 we see that the number of reservation places has almost no impact on the delays of both

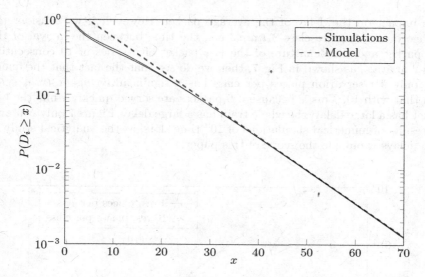

Fig. 5. Tail probabilities of the packet delay for a system with 2 classes, where the arrivals of both classes are trains of geometric length and fixed length 10 respectively. Each traffic class has 9 reservation slots.

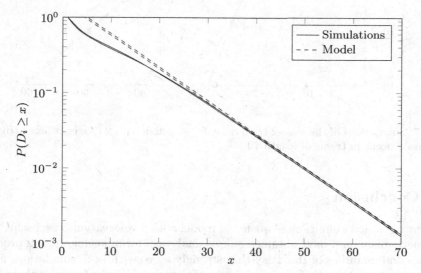

Fig. 6. Tail probabilities of the packet delay for a system with 2 classes, where the arrivals of both classes are trains of geometric length and fixed length 10 respectively. Each traffic class has 10 reservation slots.

classes, which have an almost identical distribution. This is caused by the fact that this grouping has little effect on the delay of an average *packet*, because if 9 packets are served quickly while the last packet has a longer delay, that has little impact on the delay of the average packet. However, if we consider the average delay of entire class-2 *trains*, i.e., the time between the arrival of the first packet and the departure of the last packet of a series of 10 consecutive class-2 arrivals, as shown in Fig. 7, then we do see that the fact that the queue with only 9 reservation places per class has a significantly higher *train delay* than that with 10. This is because if 9 packets are served quickly while the last packet has a large delay, the whole train has a large delay. Figure 7 only contains the results of numerical simulations of 10^9 time slots, as the analytical study of train delays is outside the scope of this paper.

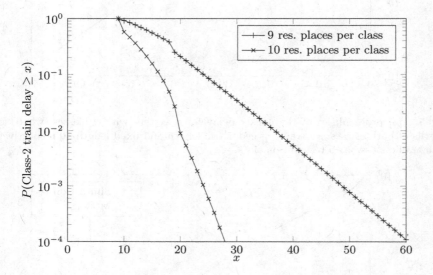

Fig. 7. Simulations of the class-2 train delay for a system with 2 classes, where arrivals of class 2 occur in trains of length 10

7   Conclusion

We have studied a multi-class queueing model with a reservation-based scheduling mechanism, and have obtained simple analytical formulas for the tail probabilities and quantiles of the delay that strongly agree with both simulations and earlier numerical results for a special case of the model. These formulas show the clear relationships between the system's parameters and its performance, and as such, are very useful in practice for determining the optimal system parameters to achieve given asymptotic delay probabilities for each desired traffic class.

Acknowledgment. This research has been partly funded by the Interuniversity Attraction Poles Programme initiated by the Belgian Science Policy Office.

References

1. Wierman, A.: Fairness and scheduling in single server queues. Surv. Oper. Res. Manag. Sci. **16**, 39–48 (2011)
2. Li, C.-C., Tsao, S.-L., Chen, M.C., Sun, Y., Huang, Y.-M.: Proportional delay differentiation service based on weighted fair queuing. In: 9th Computer Communications and Networks (2000)
3. Liu, K.Y., Petr, D.W., Frost, V.S., Zhu, H.B., Braun, C., Edwards, W.L.: Design and analysis of a bandwidth management framework for ATM-based broadband ISDN. IEEE Commun. Mag. **35**(5), 138–145 (1997). In: ICC 1996
4. De Clercq, S., Steyaert, B., Bruneel, H.: Delay analysis of a discrete-time multiclass slot-bound priority system. 4OR **10**, 67–79 (2012)
5. De Clercq, S., Steyaert, B., Wittevrongel, S., Bruneel, H.: Analysis of a discrete-time queue with time-limited overtake priority. Ann. Oper. Res. **238**, 69–97 (2016)
6. Maertens, T., Walraevens, J., Bruneel, H.: Performance comparison of several priority schemes with priority jumps. Ann. Oper. Res. **162**, 109–125 (2008)
7. Melikov, A., Ponomarenko, L., Kim, C.: Approximate method for analysis of queuing models with jump priorities. Autom. Remote Control **74**, 62–75 (2013)
8. Ndreca, S., Scoppola, B.: Discrete time GI/Geom/1 queueing system with priority. Eur. J. Oper. Res. **189**, 1403–1408 (2008)
9. Walraevens, J., Steyaert, B., Bruneel, H.: Delay characteristics in discrete-time GI-GI-1 queues with non-preemptive priority queueing discipline. Perform. Eval. **50**, 53–75 (2002)
10. Wittevrongel, S., De Vuyst, S., Sys, C., Bruneel, H.: A reservation-based scheduling mechanism for fair QoS provisioning in packet-based networks. In: 26th International Teletraffic Congress (ITC), pp. 1–8 (2014)
11. Vogel, A., Kerherve, B., von Bochmann, G., Gecsei, J.: Distributed multimedia and QOS: a survey. IEEE Multimedia **2**, 10–19 (1995)
12. Wittevrongel, S., Feyaerts, B., Bruneel, H., De Vuyst, S.: Delay analysis of a queue with reservation-based scheduling and class-dependent service times. In: 11th International Conference on Queueing Theory and Network Applications (QTNA). ACM (2016)
13. Feyaerts, B., De Vuyst, S., Bruneel, H., Wittevrongel, S.: Delay analysis of a discrete-time GI-GI-1 queue with reservation-based priority scheduling. Stochast. Models **32**, 179–205 (2016)
14. Latouche, G., Ramaswami, V.: Introduction to Matrix Analytic Methods in Stochastic Modeling. Society for Industrial and Applied Mathematics, ASA-SIAM Series on Statistics and Applied Probability (1987)
15. Blondia, C., Casals, O.: Statistical multiplexing of VBR sources - a matrix-analytical approach. Perform. Eval. **16**(1–3), 5–20 (1992)
16. Steyaert, B., Walraevens, J., Fiems, D., Bruneel, H.: The $N \times D$-BMAP/G/1 Queueing Model: Queue Contents and Delay Analysis. Mathematical Problems in Engineering 401365 (2011)

The Impact of Priority Policy in a Two-Queue Markovian Polling System with Multi-Class Priorities

Yuqing Chu[1]([⊠]) and Zaiming Liu[2]

[1] School of Science, Wuhan University of Technology, Wuhan 430070, China
chuyuqingcumt@163.com
[2] School of Mathematics and Statistics, Central South University,
Changsha 410083, China
math_lzm@csu.edu.cn

Abstract. In this paper, we consider a two-queue Markovian polling system attended by a single server. For this model, multiple-class customers with priorities are concerned in each queue. We first present the exact LST (Laplace-Stieltjes transformation) expressions and means of the waiting times of each type customers by utilizing the concept of delay-cycle. Furthermore, we prove that prioritizing customers with smaller mean service times could shorten the mean response time, especially in the heavy traffic regime. With the LSTs of waiting times, we also concentrate on the derivation of the exact asymptotics of the scaled delay in the heavy-traffic scenario. It is illustrated that the priority policy generates a mixture distribution of the limiting scaled delay in comparison with the non-priority policy. Lastly, simulations are used for validation of the limiting results and the impact of priority policy.

Keywords: Polling system · Markovian routing · Priority Waiting time · Simevents

1 Introduction

Polling models occur naturally in the modelling of applications where different types of customers compete for access to common resource, such as computer-communication systems and manufacturing systems. Traditionally, simple scheduling policies such as FCFS (within a queue) and cyclic order policy (between queues), which shares the service capacity equally among all jobs in the system, have been applied most frequently and thus dominate the literature on polling systems. However, recently, policies that give priority to jobs with small service demands have been used in a variety of application domains, e.g. web servers and ATM systems in [11]. In [15], it has illustrated that a large class of priority policies (such as LCFS, m-class priority) in exhaustive polling systems could reduce the mean response time.

© Springer International Publishing AG 2017
W. Yue et al. (Eds.): QTNA 2017, LNCS 10591, pp. 282–296, 2017.
https://doi.org/10.1007/978-3-319-68520-5_17

Within a polling system, the system operator mainly has three ways to introduce priority policies: (i) routing mechanism (such as random routing with more frequencies to serve certain queues), (ii) service discipline (such as exhaustive policy), and (iii) service order within each queue (such as multiple-class priorities). The present paper concerns a two-queue polling system with Markovian routing policy and introduces multiple-class priorities in each queue, that combines the priority policies in (i) and (iii).

Polling systems with random server routing were introduced in [10] and there have been a growing number of applications that can be modelled by polling systems with random routing policy, particularly in computer communications, geostationary satellite communication and manufacturing. In these models the queue to be visited next always depend on the random environments or access demand, such as CSMA-CA (Carrier-Sense Multiple-Access Collision-Avoidance) algorithms and slotted ALOHA. More applications can refer to [4,6] and references therein.

Boxma et al. presented a pseudo-conservation law for mean waiting times in [1] and the performance equations concerning more general Markovian polling systems were given in [6,14], while no exact expressions were obtained. Recently, Dorsman et al. studied a two-queue polling system specially in [5]. The exact expressions of the PGFs (Probability Generating Functions) of the joint queue length distributions at polling epochs were derived and a DSA (Descendant Set Approach) interpretation was presented. The heavy-traffic behavior were also extended to the Markovian polling systems.

In the polling model considered in the present paper, a single server visits 2 queues according to a discrete time Markov chain and each queue contains multiple-class customers with priorities. We are primarily motivated by the desire to exploit the impact of the multi-class priority policy on the waiting times within each queue, including the mean waiting times and the limiting scaled delay in the heavy traffic.

The remainder paper is organized as follows. We first give a detailed description of the model with multiple-class priority policies in Sect. 2. In Sect. 3, we are dedicated to the derivation of the LST of the waiting times and the impact of priority policy on the mean waiting times. To proceed, we discuss the limiting scaled delay in the heavy traffic in Sect. 4. In Sect. 5, simulations with Simevents toolbox of Matlab are undertaken to test the validity of limiting behaviors of the scaled delays and the impact of the priority policy. We finally propose some topics for further research in Sect. 6.

2   Model Description

We consider a single server Markovian polling system consisting of two queues (Q_1 and Q_2) (cf. Fig. 1). Each queue contains multiple priority classes customers (K_i classes in Q_i). It is assumed that customers of class k in Q_i, denoted by type-ik customers, receive a higher non-preemptive priority over class $k+1$ ($k = 1, 2, \ldots, K_i - 1$). Type-$ik$ customers arrive independently according to a Poisson process with rate λ_{ik} and have a generally distributed service requirement B_{ik}. Set $\rho_{ik} = \lambda_{ik}\mathbb{E}B_{ik}$. The buffer capacity of each queue is infinite and customers within each class are served in FCFS discipline. It is assumed that each queue is served exhaustively. Let $\rho_i = \sum_{k=1}^{K_i} \rho_{ik}$. We assume the stability condition $\rho = \rho_1 + \rho_2 < 1$ is satisfied (see [7]).

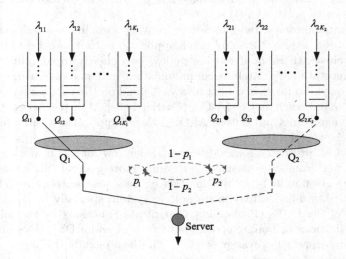

Fig. 1. The schematic diagram of the model

Once the server completes service at Q_i, it begins a switch-over time S_{ii} for another poll at Q_i with probability $p_i < 1$ and begins a switch-over time S_{ij} ($j \neq i$) for a poll at Q_j with probability $1 - p_i$ ($i = 1, 2$). Apparently, the routing mechanism is determined by a discrete time Markov chain $M = \{d_n, n \geq 0\}$ with state space $I = \{1, 2\}$, where $d_n = i$ means the nth polled queue after $t = 0$ is Q_i. Define

$$\pi_i = \lim_{n \to \infty} Pr\{d_n = i\}, \quad i \in I,$$
$$r_{ij} = Pr\{d_n = i | d_{n+1} = j\}, \quad i, j \in I, \ n = 0, 1, \ldots$$

Then

$$\pi_1 = \frac{1 - p_2}{2 - p_1 - p_2}, \qquad\qquad \pi_2 = \frac{1 - p_1}{2 - p_1 - p_2},$$
$$r_{11} = p_1, \qquad\qquad r_{12} = 1 - p_2,$$
$$r_{21} = 1 - p_1, \qquad\qquad r_{22} = p_2.$$

With the notations, the average duration of an arbitrary switch-over time is given by $\sigma = \sum_{i=1}^{2} \sum_{j=1}^{2} r_{ij}\pi_j \mathbb{E}S_{ij}$.

Throughout the paper, we shall employ the notations:

- For a random variable X, $\widetilde{X}(\cdot)$ denotes its PGF or LST.
- θ_i and $W_{i,M/G/1}$ denote a busy period and the waiting time of an arbitrary customer in an $M/G/1$ queue with arrival rate λ_i and general service time B_i.
- For a random variable X (denoting a period), X^{res} denotes its residual life time.
- $\Gamma_{(\alpha,\mu)}$ denotes a random variable of Gamma distributed with shape parameter α and rate parameter μ.
- U denotes a random variable of standard uniformly distributed in $[0,1]$.

3 LSTs and Means of Waiting Time

Since the priority policy within Q_i only influences the system performance locally, leaving the amount of time spent outside Q_i unaffected, such as the intervisit times. Hence, we mainly concern on the waiting time of an arbitrary type-ik customer, denoted by W_{ik} ($k = 1, 2, \ldots, K_i$).

3.1 LSTs of Waiting Time

As for a type-ik customer, the polling system is an $M/G/1$ queue with a multiple vacation I_i under the non-preemptive priority regime. Hence, the framework of the delay-cycle (see [9]) in conjunction with the Fuhrmann-Cooper decomposition Theorem [8] are applied here.

For type-ik customers, we denote all the customers with higher priorities by "type-iH customer" collectively. Then type-iH customers arrive according to a Poisson process with intensity $\lambda_{iH} = \sum_{j=1}^{k-1} \lambda_{ij}$ and have service requirement B_{iH} with LST $\widetilde{B}_{iH}(s) = \sum_{j=1}^{k-1} \frac{\lambda_{ij}}{\lambda_{iH}} \widetilde{B}_{ij}(s)$. Similarly, denote customers with lower priorities by "type-iL customers" collectively, with arriving intensity $\lambda_{iL} = \sum_{j=k+1}^{K_i} \lambda_{ij}$ and service requirement B_{iL} of LST $\widetilde{B}_{iL}(s) = \sum_{j=k+1}^{K_i} \frac{\lambda_{ij}}{\lambda_{iL}} \widetilde{B}_{ij}(s)$. Meanwhile, let $\rho_{iH} = \lambda_{iH} \mathbb{E} B_{iH} = \sum_{j=1}^{k-1} \rho_{ij}$, $\rho_{iL} = \lambda_{iL} \mathbb{E} B_{iL} = \sum_{j=k+1}^{K_i} \rho_{ij}$.

The time intervals between two consecutive services of type-ik customers (denoted by $B_{ik,H}$) consists of the service time of a type-ik customer and the service times of all his type-iH descendants. Then

$$\widetilde{B}_{ik,H}(s) = \widetilde{B}_{ik}(s + \lambda_{iH}(1 - \widetilde{\theta}_{iH}(s))) \quad \text{and} \quad \mathbb{E} B_{ik,H} = \frac{\mathbb{E} B_{ik}}{1 - \rho_{iH}}. \quad (1)$$

For an arbitrary type-ik customer, define $T_{X,iHk}$-cycle and $T_{X,iH}$-cycle as follows.

1. A $T_{X,iHk}$-cycle is a cycle that starts with a certain initial delay X that no type-iH or type-ik customer is waiting in line, continued with type-iH or type-ik customers served only and terminates at the moment that no type-iH or type-ik customer is present in the system. Hence, the mean length of a $T_{X,iHk}$-cycle equals $\frac{\mathbb{E} X}{1 - \rho_{iH} - \rho_{ik}}$.

2. A $T_{X,iH}$-cycle is a cycle that starts with a delay X that no type-iH customer is waiting in line, continued with type-iH customers served only and terminates at the moment that no type-iH customer is present in the system. Hence, the mean length of a $T_{X,iH}$-cycle equals $\frac{\mathbb{E}X}{1-\rho_{iH}}$.

From the above definitions, for a tagged type-ik customer arriving within a $T_{X,iHk}$-cycle, he will view the cycle as a standard $M/G/1$ queue with multiple server vacation $T_{X,iH}$-cycle, in which the arrival rate equals λ_{ik} and service time equals $B_{ik,H}$.

Denote the intervisit time of Q_i by I_i. It is noted that the arrival of a type-ik customer always takes place within a $T_{I_i,ik}$-cycle or a $T_{B_{iL},ik}$-cycle. The mean length of a $T_{B_{iL},ik}$-cycle equals $\frac{\mathbb{E}B_{iL}}{1-\rho_{iH}-\rho_{ik}}$. Hence, the fraction of the time that the system is in a $T_{B_{iL},ik}$-cycle equals $\lambda_{iL} \cdot \frac{\mathbb{E}B_{iL}}{1-\rho_{iH}-\rho_{ik}} = \frac{\rho_{iL}}{1-\rho_{iH}-\rho_{ik}}$.

This key observation enables us to give the waiting time of a type-ik customer by utilizing the Fuhrmann-Cooper decomposition Theorem within each $T_{X,iHk}$-cycle:

Theorem 1. *For an arbitrary type-ik customer, the LST of the waiting time equals*

$$\widetilde{W}_{ik}(s) = \frac{(1-\rho_{ik,H})s}{s-\lambda_{ik}(1-\widetilde{B}_{ik,H}(s))}\Big[(1-\rho_{iL,Hk})\frac{1-\widetilde{T}_{I_i,iH}(s)}{s\mathbb{E}T_{I_i,iH}} + \rho_{iL,Hk}\frac{1-\widetilde{T}_{B_{iL},iH}(s)}{s\mathbb{E}T_{B_{iL},iH}}\Big],$$
$$(2)$$

where $B_{ik,H}$ is defined in (1) and $\rho_{ik,H}$, $\rho_{iL,Hk}$ equal

$$\rho_{ik,H} = \frac{\rho_{ik}}{1-\rho_{iH}} \qquad and \qquad \rho_{iL,Hk} = \frac{\rho_{iL}}{1-\rho_{iH}-\rho_{ik}}.$$

Besides,

$$\widetilde{T}_{I_i,iH}(s) = \widetilde{I}_i(s+\lambda_{iH}(1-\theta_{iH}(s))) \qquad and \qquad \mathbb{E}(T_{I_i,iH}) = \frac{\mathbb{E}I_i}{1-\rho_{iH}},$$

$$\widetilde{T}_{B_{iL},iH}(s) = \widetilde{B}_{iL}(s+\lambda_{iH}(1-\theta_{iH}(s))) \qquad and \qquad \mathbb{E}(T_{B_{iL},iH}) = \frac{\mathbb{E}B_{iL}}{1-\rho_{iH}},$$

where $\widetilde{I}_1(s) = \widetilde{F}_1(1-\frac{s}{\lambda_1},1)$ and $\widetilde{I}_2(s) = \widetilde{F}_2(1,1-\frac{s}{\lambda_2})$. The expressions of $\widetilde{F}_1(z_1,z_2)$ and $\widetilde{F}_2(z_1,z_2)$ refer to equations (11) and (12) in [5].

3.2   Mean of Waiting Time

Following the above methodology or taking the derivatives directly leads to the means of the waiting time. Here we give a more simple derivation by utilizing Little's formula.

In the non-priority case, when an arbitrary customer arrives at Q_i, the server may be serving a customer at Q_i or be working within an intervisit time I_i. For brevity, denote the residual service time and the residual intervisit time collectively by residual delay $W_{i,0}$. The waiting time $W_{i,FCFS}$ of the tagged customer equals the residual delay $W_{i,0}$ plus the service time of customers before him in the queue, the mean of which equals $\rho_i \mathbb{E} W_{i,FCFS}$ by Little's formula. Hence,

$$\mathbb{E} W_{i,FCFS} = \mathbb{E} W_{i,0} + \rho_i \mathbb{E} W_{i,FCFS}.$$

Similarly, for an arbitrary type-iH customer in the priority case, we have

$$\mathbb{E} W_{iH} = \mathbb{E} W_{i,0} + \rho_{iH} \mathbb{E} W_{iH}.$$

Combining the above two equations leads to

$$\mathbb{E} W_{iH} = \frac{1 - \rho_i}{1 - \rho_{iH}} \mathbb{E} W_{i,FCFS}.$$

For an arbitrary type-iL customer, the waiting time consists of the residual delay and the workload of customers waiting before as well as all their type-iH and type-ik descendants. Therefore,

$$\mathbb{E} W_{iL} = \frac{\mathbb{E} W_{i,0} + \rho_i \mathbb{E} W_{i,FCFS}}{1 - \rho_{iH} - \rho_{ik}} = \frac{1}{1 - \rho_{iH} - \rho_{ik}} \mathbb{E} W_{i,FCFS}.$$

For an arbitrary type-ik customer, the waiting time equals the residual delay, plus the workload of type-iH and type-ik customers waiting before and all their type-iH descendants. Therefore,

$$\mathbb{E} W_{ik} = \frac{\mathbb{E} W_{i,0} + \rho_i \mathbb{E} W_{i,FCFS} - \rho_{iL} \mathbb{E} W_{iL}}{1 - \rho_{iH}}$$

$$= \frac{(1 - \rho_i)}{(1 - \sum_{l \leq k} \rho_{il})(1 - \sum_{l < k} \rho_{il})} \mathbb{E} W_{i,FCFS}. \tag{3}$$

3.3 Impact of Priority Policy on Mean Waiting Times

In this subsection, we consider the impact of the priority policy on the mean waiting time of an arbitrary customer (to distinguish with the non-priority case, denoted by $W_{i,priority}$, $i = 1, 2$). For ease of presentation, we need more notations:

1. X_i: the type of the tagged arriving customer at Q_i. Then

$$Pr\{X_i = ik\} = \frac{\lambda_{ik}}{\lambda_i}.$$

2. $\Delta_{ij,(\text{preempt } ik)}$, $(j < k \leq K_i)$: the decrement of the waiting time of a type-ij customer induced by preempting type-ik customers, i.e.

$$\Delta_{ij,(\text{preempt } ik)} = -\rho_{ik} \mathbb{E} W_{ik} = -\frac{(1 - \rho_i)\rho_{ik}}{(1 - \sum_{l \leq k} \rho_{il})(1 - \sum_{l < k} \rho_{il})} \mathbb{E} W_{i,FCFS} \leq 0.$$

3. $\Delta_{ik,\text{(preempted by } ij)}, (j < k \le K_i)$: the increment of the waiting time of a type-ik customer induced by being preempted by type-ij customers (including all the type-ij descendants belonging to customers waiting before). Hence,

$$\Delta_{ik,\text{(preempted by } ij)} = (\mathbb{E}W_{i,FCFS} - \sum_{l>k} \rho_{il}\mathbb{E}W_{il})\frac{\rho_{ij}}{1 - \sum_{l<k} \rho_{il}}$$

$$= \frac{(1-\rho_i)\rho_{ij}}{(1 - \sum_{l \le k} \rho_{il})(1 - \sum_{l<k} \rho_{il})}\mathbb{E}W_{i,FCFS} \ge 0.$$

4. $\Delta_{net,(ij,ik)}$: the net preempting time between type-ij customers and type-ik customers defined by

$$\Delta_{net,(ij,ik)} = \Delta_{ij,\text{(preempt } ik)}I_{\{X_i=ij\}} + \Delta_{ik,\text{(preempted by } ij)}I_{\{X_i=ik\}}$$

$$= \frac{\lambda_{ij}\lambda_{ik}}{\lambda_i}\frac{(1-\rho_i)}{(1 - \sum_{l \le k} \rho_{il})(1 - \sum_{l<k} \rho_{il})}\left[\frac{1}{\mu_{ij}} - \frac{1}{\mu_{ik}}\right]\mathbb{E}W_{i,FCFS}.$$

For an arbitrary customer arriving at Q_i in priority case, we have

$$\mathbb{E}W_{i,priority} - \mathbb{E}W_{i,FCFS} = \sum_{j<k \le K_i} \Delta_{net,(ij,ik)},$$

from which it follows that, if $\mu_{ij} \ge \mu_{ik}$ for all $j < k \le K_i$, then $\mathbb{E}W_{i,priority} \le \mathbb{E}W_{i,FCFS}$. Therefore, prioritizing small service times is an effective scheduling policy not only to satisfy different types of QoS (Quality of service) standard but also to reduce the mean delay, which is also in line with the conclusion in [15].

Remark 1. From the derivation of the mean waiting times and the methodology of the net preempting time, it is easy to see that this conclusion also applies for each $M/G/1$ queue with generalized vacations given that the vacation begins once the system gets empty.

4   Heavy Traffic Asymptotics

Observing the exact LSTs of the waiting time given in Theorem 1, It is too intractable to analyze the impact of priority policy on the distributions of waiting times since the existence of some infinite number of products. In this section, we turn to study the limiting distributions of the scaled delay $\mathcal{W}_{ik} = (1-\rho)W_{ik}, i = 1, 2; k = 1, 2, \ldots, K_i$, when $\rho \to 1$. The results are rather intriguing and provide new fundamental insight in the impact of the priority policies. Throughout the remainder paper, we introduce a notation $\hat{x}$ such that $\hat{x} = \frac{x}{\rho}$. For example, $\hat{\rho}_1 = \frac{\rho_1}{\rho}$.

We introduce three Lemmas first. Lemma 1 is used to prove a Gamma distributed limiting random variable. Lemma 2 (see [13] Theorem 1) is devoted to the expressions of the kth derivative $[f(g)]^{(k)}(x)$ of a general composite function $f(g(x))$. Lemma 3 follows by applying Lemmas 1 and 2 to the DSA expression of G_1 in [5].

Lemma 1 (Method of Moments [12]). *Let $\{Y_n\}$ be a sequence of random variables with finite moments, satisfying*

$$\lim_{n\to\infty} \mathbb{E}Y_n^k = \mathbb{E}\Gamma_{(\alpha,\mu)}^k, \quad k = 1, 2, \ldots$$

Then $Y_n \xrightarrow{d} \Gamma_{(\alpha,\mu)}$.

Lemma 2. *For $k = 1, 2, \ldots$, the kth derivative of composite function $f(g(x))$ is given by*

$$[f(g)]^{(k)}(x) = \sum_{\mathbf{m}^{(k)} \in S_k} c_k\left(\mathbf{m}^{(k)}\right) f^{(l_k)}(g(x)) \prod_{i=1}^{k} \left(g^{(i)}(x)\right)^{m_i}, l_k = \sum_{j=1}^{k} m_j,$$

where

$$S_k := \left\{ \mathbf{m}^{(k)} = (m_1, \ldots, m_k) : m_j \text{ are non-negative integers with} \sum_{j=1}^{k} jm_j = k \right\},$$

and $c_k\left(\mathbf{m}^{(k)}\right)$ can be calculated in the following recursive way: $c_1(1) = 1$ and for $k = 2, 3, \ldots$

$$c_k\left(\mathbf{m}^{(k)}\right) = c_{k-1}(m_1 - 1, m_2, \ldots, m_{k-1})I_{\{m_1 > 0\}} + \sum_{j=1}^{k-1}(m_j + 1)$$

$$\times c_{k-1}(m_1, \ldots, m_{j-1}, m_j + 1, m_{j+1} - 1, m_{j+2}, \ldots, m_{k-1})I_{\{m_{j+1} > 0\}}.$$

Lemma 3. *For $i = 1, 2$, as $\rho \to 1$, we have*

$$\lim_{\rho \to 1} \tilde{I}_i((1 - \rho)s) = p_i + (1 - p_i) \left(\frac{\nu_i}{\nu_i + s}\right)^\alpha,$$

where α and ν_i are defined as follows:

$$\alpha = \frac{2\hat{\rho}_1\hat{\rho}_2\mathbb{E}S^{tot}}{\hat{\lambda}_1\mathbb{E}B_1^2 + \hat{\lambda}_2\mathbb{E}B_2^2}, \qquad \nu_i = \frac{2\hat{\rho}_i}{\hat{\lambda}_1\mathbb{E}B_1^2 + \hat{\lambda}_2\mathbb{E}B_2^2}, \quad i = 1, 2.$$

Equivalently,

$$\lim_{\rho \to 1} Pr\{(1 - \rho)I_i \le t\} = p_i + (1 - p_i)Pr\{\Gamma_{(\alpha,\nu_i)} \le t\}.$$

Proof. According to [5], denote by $\tilde{G}_1(z) = \tilde{S}_{21}(\lambda_1(1 - z)) \prod_{j=0}^{\infty} a_2(f_2^{(j)}(z))$. Then

$$\tilde{I}_1(s) = r_{11}\tilde{S}_{11}(s) + r_{21}\tilde{G}_1(1 - \frac{s}{\lambda_1}).$$

Hence, for $i = 1$, the limiting distribution for $(1 - \rho)I_1$ follows since

$$\lim_{\rho \to 1} \tilde{I}_1((1 - \rho)s) = \lim_{\rho \to 1} \left(r_{11}\tilde{S}_{11}((1 - \rho)s) + r_{21}\tilde{G}_1\left(1 - \frac{(1 - \rho)s}{\lambda_1}\right)\right)$$

$$= r_{11} + r_{21}\left(\frac{\nu_1}{\nu_1 + s}\right)^\alpha = p_1 + (1 - p_1)\left(\frac{\nu_1}{\nu_1 + s}\right)^\alpha,$$

where the second equation follows from Lemma 4.4 in [5].

Proposition 1. *For $i = 1, 2$, as $\rho \to 1$, we have*

$$\lim_{\rho \to 1} \widetilde{T}_{I_i, iH}\big((1-\rho)s\big) = p_i + (1-p_i)\left(\frac{(1-\hat{\rho}_{iH})\nu_i}{(1-\hat{\rho}_{iH})\nu_i + s}\right)^{\alpha}.$$

Equivalently,

$$\lim_{\rho \to 1} Pr\{(1-\rho)T_{I_i, iH} \leq t\} = p_i + (1-p_i)Pr\{\Gamma_{(\alpha, (1-\hat{\rho}_{iH})\nu_i)} \leq t\}.$$

Proof. To prove the limiting distribution of $(1-\rho)T_{I_i, iH}$, we first construct a composite function $f(g(s))$. Set

$$f(g(s)) = \widetilde{T}_{I_i, iH}\big((1-\rho)s\big) = \widetilde{I}_i\left((1-\rho)s + \lambda_{iH}\left(1 - \tilde{\theta}_{iH}\big((1-\rho)s\big)\right)\right),$$

where $f(s) = \widetilde{I}_i(s)$ and $g(s) = (1-\rho)s + \lambda_{iH}\left(1 - \tilde{\theta}_{iH}\big((1-\rho)s\big)\right)$. It is easy to obtain

$$f^{(k)}(0) = (-1)^k \mathbb{E}I_i^k, \quad k = 1, 2, \ldots,$$

$$g^{(1)}(0) = 1 - \rho + \lambda_{iH}(1-\rho)\mathbb{E}\theta_{iH} = \frac{1-\rho}{1-\rho_{iH}},$$

$$g^{(k)}(0) = (-1)^{(k+1)}\lambda_{iH}(1-\rho)^k \mathbb{E}\theta_{iH}^k, \quad k = 2, 3, \ldots.$$

Then

$$\mathbb{E}\big((1-\rho)I_{i,H}\big)^k = (-1)^k [f(g)]^{(k)}(0)$$

$$= \sum_{\mathbf{m}^{(k)} \in S_k} (-1)^{k+l_k+\sum_{j=2}^{k}(j+1)m_j} c_k\left(\mathbf{m}^{(k)}\right) \mathbb{E}I_i^{l_k}\left(\frac{1-\rho}{1-\rho_{iH}}\right)^{m_1} \prod_{j=2}^{k}\left(\lambda_{iH}(1-\rho)^j \mathbb{E}\theta_{iH}^j\right)^{m_j}$$

$$= \sum_{\mathbf{m}^{(k)} \in S_k} (-1)^{k+l_k}(1-\rho)^{k-l_k} c_k\left(\mathbf{m}^{(k)}\right) \mathbb{E}\big((1-\rho)I_i\big)^{l_k}\left(\frac{1}{1-\rho_{iH}}\right)^{m_1} \prod_{j=2}^{k}\left(\lambda_{iH}\mathbb{E}\theta_{iH}^j\right)^{m_j}.$$

Setting $\rho \to 1$ in the above equation yields

$$\lim_{\rho \to 1} \mathbb{E}\big((1-\rho)T_{I_i, iH}\big)^k$$

$$= \lim_{\rho \to 1} \sum_{\substack{\mathbf{m}^{(k)} \in S_k \\ l_k = k}} c_k\left(\mathbf{m}^{(k)}\right) \mathbb{E}\big((1-\rho)I_i\big)^{l_k}\left(\frac{1}{1-\rho_{iH}}\right)^{m_1} \prod_{j=2}^{k}\left(\lambda_{iH}\mathbb{E}\theta_{iH}^j\right)^{m_j}$$

$$= \lim_{\rho \to 1} \mathbb{E}\left(\frac{(1-\rho)I_i}{1-\rho_{iH}}\right)^k = (1-p_i)\frac{\prod_{j=0}^{k-1}(\alpha+j)}{[\nu_i(1-\hat{\rho}_{iH})]^k},$$

where the second equation follows since $l_k = k$ holds iff $\mathbf{m}^{(k)} = (k, 0, \ldots, 0)$. By Lemma 1, $(1-\rho)T_{I_i, iH}$ obeys a mixture distribution of a Gamma distribution with a point distribution.

Now we present the main theorem concerning the limiting distributions of the scaled delay.

Theorem 2. *For $i = 1, 2$, $k = 1, 2, \ldots, K_i$, the LSTs of the limiting scaled delays are given by*

$$\lim_{\rho \to 1} \widetilde{W}_{ik}(s) = (1 - \hat{\rho}_{iL,Hk}) \frac{1 - \hat{\rho}_{iH}}{s(1 - \hat{\rho}_i) \mathbb{E} S^{tot}} \left[1 - \left(\frac{(1 - \hat{\rho}_{iH}) \nu_i}{(1 - \hat{\rho}_{iH}) \nu_i + s} \right)^{\alpha} \right] + \hat{\rho}_{iL,Hk},$$

$$(4)$$

where

$$\hat{\rho}_{iL,Hk} = \frac{\hat{\rho}_{iL}}{1 - \hat{\rho}_{iH} - \hat{\rho}_{ik}},$$

$$\alpha = \frac{2\hat{\rho}_1 \hat{\rho}_2 \mathbb{E} S^{tot}}{\hat{\lambda}_1 \mathbb{E} B_1^2 + \hat{\lambda}_2 \mathbb{E} B_2^2},$$

$$\nu_i = \frac{2\hat{\rho}_i}{\hat{\lambda}_1 \mathbb{E} B_1^2 + \hat{\lambda}_2 \mathbb{E} B_2^2}, \quad i = 1, 2,$$

$$\mathbb{E} S^{tot} = \frac{p_1}{1 - p_1} \mathbb{E} S_{11} + \mathbb{E} S_{12} + \mathbb{E} S_{21} + \frac{p_2}{1 - p_2} \mathbb{E} S_{22} = \frac{2 - p_1 - p_2}{(1 - p_1)(1 - p_2)} \sigma.$$

Equivalently,

$$\lim_{\rho \to 1} Pr(\mathcal{W}_{ik} \le t) = (1 - \hat{\rho}_{iL,Hk}) Pr(U \Gamma_{(\alpha+1,(1-\hat{\rho}_{iH})\nu_i)} \le t) + \hat{\rho}_{iL,Hk},$$

where U and $\Gamma_{(\alpha+1,(1-\hat{\rho}_{iH}))}$ are mutually independent.

Proof. Taking the limit of (2) yields

$$\lim_{\rho \to 1} \widetilde{\mathcal{W}}_{ik}(s) = \lim_{\rho \to 1} \widetilde{W}_{ik}((1 - \rho)s)$$

$$= \lim_{\rho \to 1} \frac{(1 - \rho_{ik,H})(1 - \rho)s}{(1 - \rho)s - \lambda_{ik}\left(1 - \widetilde{B}_{ik,H}((1 - \rho)s)\right)} \left[(1 - \rho_{iL,Hk}) \frac{1 - \widetilde{T}_{I_i,iH}((1 - \rho)s)}{\frac{\sigma}{\pi_i}(1 - \rho_i)s} \right.$$

$$\left. + \rho_{iL,Hk} \frac{1 - \widetilde{T}_{B_iL,iH}((1 - \rho)s)}{(1 - \rho)s \mathbb{E} B_{iL,H}} \right]$$

$$= (1 - \hat{\rho}_{iL,Hk}) \frac{1 - \lim_{\rho \to 1} \widetilde{T}_{I_i,iH}((1 - \rho)s)}{\frac{\sigma}{\pi_i}(1 - \hat{\rho}_i)s} + \hat{\rho}_{iL,Hk}$$

$$= (1 - \hat{\rho}_{iL,Hk}) \frac{(1 - p_i)\pi_i}{\sigma(1 - \hat{\rho}_i)s} \left[1 - \left(\frac{\nu_i}{\nu_i + s} \right)^{\alpha} \right] + \hat{\rho}_{iL,Hk},$$

where the last equation follows from Proposition 1. Substituting $\frac{(1-p_i)\pi_i}{\sigma} = \frac{1}{\mathbb{E} S^{tot}}$ into the above equation immediately yields (4).

Corollary 1. *For non-priority model, the LSTs of the limiting scaled delays are given by*

$$\lim_{\rho \to 1} \widetilde{\mathcal{W}}_{i,FCFS}(s) = \frac{1}{s(1-\hat{\rho}_i)\mathbb{E}S^{tot}} \left[1 - \left(\frac{\nu_i}{\nu_i + s} \right)^{\alpha} \right].$$

Equivalently,

$$\lim_{\rho \to 1} Pr(\mathcal{W}_{i,FCFS} \le t) = Pr(U\Gamma_{(\alpha+1,\nu_i)} \le t),$$

where U and $\Gamma_{(\alpha+1,\nu_i)}$ are mutually independent ($i = 1,2$).

*Remark 2 (**HTAP (Heavy Traffic Averaging Principle)**).* By utilizing the averaging principle in [2,3], we give an intuitive interpretation of the heavy traffic behavior. For simplicity, we denote $T_{I_i,iH} = (1 - \rho)T_{I_i,iH}$ and set $T^*_{I_i,iH}$ as the length-biased random variable of $T_{I_i,iH}$ with p.d.f $f_{T^*_{I_i,iH}}(x) = x f_{T_{I_i,iH}}(x)/\mathbb{E}T_{I_i,iH}$. By Proposition 1, we have

$$T^*_{I_i,iH} \to_d \Gamma_{(\alpha+1,(1-\hat{\rho}_{iH})\nu_i)}, \quad \text{as } \rho \to 1, \quad i = 1,2.$$

According to the averaging principle, the total scaled workloads keep constant during a cycle, whereas the workloads of an individual queue change much faster and can be modeled as a fluid system. Therefore, we regard the system in heavy traffic as a fluid model under exhaustive service policy.

From Subsect. 3.1, an arbitrary type-ik customer arrives during a $T_{B_{iL},iHk}$-cycle with probability of $1 - \hat{\rho}_{iL,Hk}$ and during a $T_{I_i,iHk}$-cycle with probability of $\hat{\rho}_{iL,Hk}$ in the heavy traffic. Since the scaled length of the delay $T_{B_{iL},iH}$ approaches zero, the tagged type-ik customer within a $T_{B_{iL},iHk}$-cycle would get service immediately. If he arrives within a $T_{I_i,iHk}$-cycle, we consider the following two cases: when the tagged type-ik customer arrives,

1. If the server is within the initial delay $T_{I_i,iH}$ and $UT^*_{I_i,iH}$ time units have passed, then there will be $[(1-U)+\hat{\rho}_{ik,H}U]T^*_{I_i,iH}$ works served ahead, which has a uniformly distribution on $[\hat{\rho}_{ik,H}T^*_{I_i,iH}, T^*_{I_i,iH}]$.

2. Otherwise, if $U\frac{\hat{\rho}_{ik,H}}{1-\hat{\rho}_{ik,H}}T^*_{I_i,iH}$ time units within the visit time (the period of the standard $M/G/1$ queue within an $T_{I_i,iHk}$-cycle) have elapsed when it arrives, then the delay will equal $[U\frac{\hat{\rho}_{ik,H}}{1-\hat{\rho}_{ik,H}}+1]T^*_{I_i,iH}\hat{\rho}_{ik,H} - U\frac{\hat{\rho}_{ik,H}}{1-\hat{\rho}_{ik,H}}T^*_{I_i,iH} = \hat{\rho}_{ik,H}T^*_{I_i,iH}$ (the work that arrived ahead minus that which has already been served), which is a uniform distribution on $[0, \hat{\rho}_{ik,H}T^*_{I_i,iH}]$.

To sum up, the waiting time of the tagged type-ik customer can be expressed by

$$\lim_{\rho \to 1} Pr(\mathcal{W}_{ik} \le t) = (1 - \hat{\rho}_{iL,Hk})Pr(UT^*_{I_i,iH} \le t) + \hat{\rho}_{iL,Hk},$$

which is equivalent to (4).

*Remark 3 (**Comparison with non-priority case**).* Compared to the non-priority case, Theorem 2 illustrates that, under the heavy traffic scaling, for an arbitrary type-ik customer,

- By preempting the service of type-iL customers within $T_{B_{iL},iHk}$-cycle, the scaled delay has a point distribution of $\hat{\rho}_{iL,Hk} = \frac{\hat{\rho}_{iL}}{1-\hat{\rho}_{iH}-\hat{\rho}_{ik}}$ at 0.
- Being preempted by type-iH customers, The scaled delay is delayed with rate parameter of $1 - \hat{\rho}_{iH}$ times.
- The impact of priority policy on the mean waiting time within Q_i remains unchanged, i.e. giving priority to customers with smaller service times could reduce the mean response time.

5 Numerical Examples

In this section we give a few examples to illustrate the impact of priority policies. For simplicity, we consider a polling system with three-type customers at Q_1 described in Table 1 with exponentially distributed service time and exponentially distributed switch-over times. Note that the first three parameters in the rows of Ratio of arriving rate and Service rate in Table 1 are parameters belonging to the three-type customers at Q_1. We set type-11 customers of the highest priority, closely followed by type-12 customers and then type-13 customers. Hereafter, we mainly focus on the waiting times in the following three cases:

- FCFS: FCFS service policy within Q_1;
- Priority: $\mu_{11} > \mu_{12} > \mu_{13}$: $\mu_{11} = 2, \mu_{12} = 1, \mu_{13} = 0.5$;
- Priority: $\mu_{11} < \mu_{12} < \mu_{13}$: $\mu_{11} = 0.5, \mu_{12} = 1, \mu_{13} = 2$.

Table 1. Parameter values of the investigated polling system

Parameter	Parameter values
Ratio of arriving rate	$2 : 1 : 2 : 5$
Service rate	$2, 1, 0.5, 2$
Mean service time	$\mathbb{E}B = 0.85$
Probability of transitions	$p_1 = 0.4, p_2 = 0.3$
Switch-over times	$\mathbb{E}S_{11} = 3, \mathbb{E}S_{12} = 2, \mathbb{E}S_{21} = 2.5, \mathbb{E}S_{22} = 1.5$
Average switch-over time	$\sigma = 2.3077$

To test the validity of the limiting distributions of the scaled delay in the heavy traffic and the interpolation approximations, we utilize the SimEvents toolbox of Matlab to undertake the simulations. SimEvents provides a discrete-event simulation engine and component library for Simulink. We can model event-driven communication between components to analyze and optimize end-to-end latencies, throughput and other performance characteristics. Libraries of predefined blocks, such as queues, servers, and switches, enable us to accurately represent queueing system and customize routing, processing delays, prioritization, and other operations.

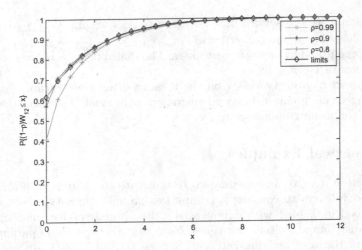

Fig. 2. The CDFs of scaled delay of type-12 customers in heavy traffic

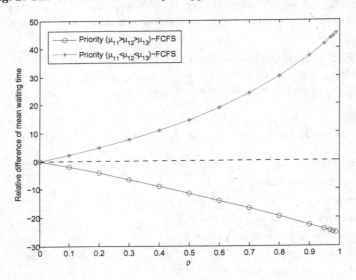

Fig. 3. Relative difference of the mean waiting times under different priority policy

Regarding the heavy traffic behavior, we take $\rho = 0.8, 0.9, 0.99$ to describe the procedure of $\rho \to 1$ while the ratio of the arriving rate is fixed. In this example, we consider the Priority case of $\mu_{11} > \mu_{12} > \mu_{13}$ and take type-12 customers for example. The empirical CDF (Cumulative Distribution Function) of the waiting times are plotted in Fig. 2. The asymptotic distribution of the scaled delay is also plotted. Each simulation runs for time of at least 10^7 units and during which at least 10^7 customers are served. The first 10^5 customers served are discarded due to the warm-up phase. From Fig. 2, it readily shows that the approximation performs very well when ρ is close to 1.

In Fig. 3, we plot the relative difference of mean waiting times between the priority cases and non-priority case, which is calculated by $\frac{W_{1,priority}-W_{1,FCFS}}{W_{1,FCFS}} \times$ 100% in priority cases of $\mu_{11} > \mu_{12} > \mu_{13}$ and $\mu_{11} < \mu_{12} < \mu_{13}$ respectively. Depicted in Fig. 3, the mean waiting time in priority case of $\mu_{11} > \mu_{12} > \mu_{13}$ decreases by 25% in comparison with FCFS when ρ is up to 1, while in priority case of $\mu_{11} < \mu_{12} < \mu_{13}$, it increases by up to 45% in comparison with FCFS. It illustrates that, the performance will get much better for higher load ρ if the smaller jobs are prioritized, otherwise, it will get much worse.

6   Conclusions

In this paper, we have introduced the priority policy into a 2-queue Markovian polling system and presented the LSTs and means of waiting times of each type customers and their limiting behaviors under the heavy traffic scaling. All the works allow us to present the impact of priority policy on the waiting times. It has illustrated that prioritizing jobs with smaller mean service time could reduce the mean response time. In particular, the impact could be dramatic in heavy traffic regime. This conclusion can be well applied in the traffic on campus. For example, it could disperse the congestion much faster by giving priorities to students with bicycles over students on foot during the rush hours.

Although we have made some achievements of the Markovian priority polling system, it leaves many extension works. Here we discuss some topics for further research.

1. *Other priority policy*. We have depicted that the impact of priority policy could be dramatic, which refutes the assertion that the priority policy within a queue only has a minor effect on overall performance. Hence, it is essential to exploit other priority policies in polling systems to improve the system performance without purchasing additional resources.
2. *Other service disciplines*. Although the explicit performance expressions might be hard to obtain in polling systems with gated or other branching-type service discipline, we could focus on the interpolation approximations by utilizing the heavy traffic limit and light traffic limit. Then the question remains to exploit the light traffic limits. The exhaustiveness (see [16]) of a branching-type service discipline may be helpful.

Acknowledgements. This work is partially supported by the National Natural Science Foundation of China (11671404) and by the Fundamental Research Funds for the Central Universities (WUT:2017IVA069). The authors also gratefully acknowledge the helpful comments and suggestions of the reviewers.

References

1. Boxma, O.J., Weststrate, J.A.: Waiting times in polling systems with Markovian server routing. In: Stiege, G., Lie, J.S. (eds.) Messung, Modellierung und Bewertung von Rechensystemen und Netzen. INFORMATIK, vol. 218, pp. 89–104. Springer, Heidelberg (1989). https://doi.org/10.1007/978-3-642-75079-3_8
2. Coffman Jr., E., Puhalskii, A., Reiman, M.: Polling systems with zero switchover times: a heavy-traffic averaging principle. Ann. Appl. Probab. **5**(3), 681–719 (1995)
3. Coffman Jr., E., Puhalskii, A., Reiman, M.: Polling systems in heavy traffic: a bessel process limit. Math. Oper. Res. **23**(2), 257–304 (1998)
4. Dorsman, J.P.L., Borst, S.C., Boxma, O.J., Vlasiou, M.: Markovian polling systems with an application to wireless random-access networks. Perform. Eval. **85–86**(1), 33–51 (2015)
5. Dorsman, J.P.L., Boxma, O.J., van der Mei, R.D.: On two-queue Markovian polling systems with exhaustive service. Queueing Syst. **78**(4), 287–311 (2014)
6. Fiems, D., Altman, E.: Gated polling with stationary ergodic walking times, Markovian routing and random feedback. Ann. Oper. Res. **198**(1), 145–164 (2012)
7. Foss, S., Last, G., et al.: Stability of polling systems with exhaustive service policies and state-dependent routing. Ann. Appl. Probab. **6**(1), 116–137 (1996)
8. Fuhrmann, S., Cooper, R.B.: Stochastic decompositions in the M/G/1 queue with generalized vacations. Oper. Res. **33**(5), 1117–1129 (1985)
9. Kella, O., Yechiali, U.: Priorities in M/G/1 queue with server vacations. Nav. Res. Logist. **35**(1), 23–34 (1988)
10. Kleinrock, L., Levy, H.: The analysis of random polling systems. Oper. Res. **36**(5), 716–732 (1988)
11. Liu, Z., Chu, Y., Wu, J.: Heavy-traffic asymptotics of priority polling system with threshold service policy. Comput. Oper. Res. **65**, 19–28 (2016)
12. van der Mei, R.D.: Distribution of the delay in polling systems in heavy traffic. Perform. Eval. **38**(2), 133–148 (1999)
13. van der Mei, R.D.: Polling systems with switch-over times under heavy load: moments of the delay. Queueing Syst. **36**(4), 381–404 (2000)
14. Srinivasan, M.M.: Nondeterministic polling systems. Manag. Sci. **37**(6), 667–681 (1991)
15. Wierman, A., Winands, E.M.M., Boxma, O.J.: Scheduling in polling systems. Perform. Eval. **64**(9), 1009–1028 (2007)
16. Winands, E.M.M.: Polling, production & priorities, Ph.D. thesis. Technische Universiteit Eindhoven (2007)

Customer Equilibrium and Social Optimization in a Markovian Queue with Fuzzy Parameters

Gang Chen, Zaiming Liu, and Jingchuan Zhang[✉]

School of Mathematics and Statistics,
Central South University, Changsha 410083, China
chengmathcsu@163.com, {math_lzm,math_zjc}@csu.edu.cn

Abstract. This paper analyzes the customers' equilibrium strategy and optimal social benefit in a Markovian queueing system, in which the arrival rate, service rate of customers, as well as the reward and holding cost are all fuzzy numbers. Based on Zadeh's extension principle, we investigate the membership functions of the optimal and equilibrium strategies in both observable and unobservable cases. Furthermore, by applying the α-cut approach, the family of crisp strategy is described by formulating a pair of parametric nonlinear programs, through which the membership functions of the strategy can be derived. Finally, numerical examples are solved successfully to illustrate the validity of the proposed approach and to show the relationship of these strategies and social benefits. Our main contribution is showing that the value of equilibrium and optimal strategies have no deterministic relationship, which are different from the results in the corresponding crisp queues. Moreover, the successful extension of queue game to fuzzy environments can provide more precise information to the system managers.

Keywords: Fuzzy queue · Equilibrium strategies · Social benefit
Membership functions · Parametric nonlinear programming

1 Introduction

Queueing literature is recently devoting an increasing attention to the economic analysis of queueing systems. During the last decades, there is an emerging tendency to study queueing systems from an economic viewpoint. Admittedly, customers' decentralized behavior and socially optimal control of arrivals have gained an ascending attention. Such an economic analysis of queueing systems was pioneered by Naor [20]. Hassin and Haviv [12] and Stidham [25] summarized the main approaches and several results about various models with extensive bibliographical references [1,9,13,18,22,24]. Most of them are forced on the ideas in three directions. One such direction is the consideration of fluid queues and non-Markovian models [7,9,15]. The second direction is the study of two-dimensional Markovian models and the third direction concerns the effect of the level of information on the customer behavior [10,11,26].

© Springer International Publishing AG 2017
W. Yue et al. (Eds.): QTNA 2017, LNCS 10591, pp. 297–311, 2017.
https://doi.org/10.1007/978-3-319-68520-5_18

In the existing literature of queueing game, the inter-arrival times and service times of customers are required to follow certain probability distributions, that is the system parameters are all assumed to be crisp numbers. However, in many real-world applications, the statistical information may be obtained subjectively. Actually, the arrival pattern, service pattern are better described by linguistic terms such as 'the mean arrival rate is approximately 5', 'the mean service rate is approximately 10', rather than by probability distributions. In other words, these system parameters are more possibilistic as well as probabilistic. In order to face the fuzzy environment factor, fuzzy queues are potentially useful and are much more realistic than the commonly used crisp queues [17]. Recently, in fuzzy logic literature, fuzzy queues are largely studied (see Wang et al. [27], Munoz and Ruspini [19], Jolai et al. [14], Chen [5] etc.). A comprehensive discussion on fuzzy queueing systems can be found in the survey papers of Buckley et al. [2,3]. In real life situations, the decision maker faces the major difficulty to forecast demand, which is due to lack of knowledge or inherent vagueness, which imply respectively randomness and fuzziness. As we all know, for the management of the queueing systems it is more important to consider the fuzzy environment, which can provide more precise and comprehensive information about the decision policy for the system manager. When the equilibrium analysis of customer behaviors in the usual crisp queues can be extended to fuzzy queues, these queueing models will have wider applications. To the best of our knowledge, no previous study has considered the customers' equilibrium strategy and social optimization in fuzzy queues.

In this paper, based on the results of corresponding crisp queue, we analyze the membership functions of the equilibrium strategy and optimal social benefit in fuzzy queues through the Zadeh's extension principle and α-cuts [21]. The benefit and significance of such fuzzy strategies lie in the fact that it completely maintains the fuzziness of the input information. As the value of α varies, the family of crisp strategy and social benefit are then described and solved by parametric nonlinear programming (NLP). Based on Zadeh's extension principle and NLP, we construct an approach to solve the membership functions of the equilibrium strategy and optimal social benefit. Moreover, we present some numerical examples to demonstrate how the proposed approach can be applied to this model and make a comprehensive analysis of the numerical results.

The goal of this work is to analyze the optimal and equilibrium balking strategies in a fuzzy queue under both observable and unobservable cases. The novelties of this study include three main aspects. First, our paper is the first to apply the fuzzy set theory to study the equilibrium strategy of customers and optimal social benefits in the fuzzy queueing system. The important feature of the present work is to consider both fuzziness and randomness in a queue for the customers strategy. Second, a parametric nonlinear programming (NLP) analysis method is proposed in this paper to derive the membership functions of the optimal and equilibrium strategies and social benefits. This study fills a gap in the rich literatures on the analysis of the strategic behavior of customers in queues. Finally, the study of the model with fuzzy parameters may provide more precise information to managers and improve decision-making. So that we should take advantage of fuzzy queuing systems to study the customers' behaviors and optimal social benefit decisions. Our finding reveals that different from the strategy in corresponding crisp queueing system, the optimal strategies are not always larger than the equilibrium strategies (threshold and arrival rate) in the present fuzzy queues.

The rest of the paper is structured as follows. In Sect. 2, we briefly describe the fuzzy queue and give the summary of known results in the corresponding crisp queues. A mathematical programming approach is developed for deriving the membership functions of the equilibrium optimal strategies and social benefits for the observable and unobservable cases in Sect. 3. Section 4 outlines a solution procedure to numerically construct the membership function of the optimal and equilibrium balking strategies in the system and some numerical examples are provided. This paper ends with Sect. 5 where conclusions and future scope are given.

2   Model Description

We consider a single-server FCFS queueing system model where customers arrive according to a Poisson process with rate Λ. The service times of all customers are independent and exponentially distributed with parameter μ. Upon arrival, customers are allowed to decide whether to join or balk. Every customer receives a reward of R units after service. This reward R may represent his satisfaction or the added value of the customer obtains from being served. On the other hand, there also exists a waiting cost of C per unit time when the customers remain in the system. We assume that customers are risk-neutral, and choose to join the queue if and only if the expected cost of waiting is less than or equal to R. Customers are risk neutral and they want to maximize their expected net payoffs. The relative traffic intensity for the system is denoted by $\rho = \frac{\Lambda}{\mu}$.

We represent the state of the system at time t by $N(t)$, which denotes the number of customers in the system (including the one being served) at time t. It is clear that the process $\{N(t), t \geq 0\}$ is a continuous time Markov chain with the state space $E = \{0, 1, 2, \ldots\}$. In this paper, we consider two cases depending on the levels of system information available to customers at their arrival instants, before the decision is made: observable case (customers observe the queue length $N(t)$) and unobservable case (customers do not observe the queue length $N(t)$).

In the case of observable system, it can be shown that customers who wish to maximize their individual welfare will follow a pure threshold strategy [8]. This means that there exists an integer n_e such that newly-arrived customers will join the queue if and only if the number of other customers already present in the system is smaller than n_e. For the social optimization, one aims to find the integer n_* which maximizes the overall social welfare $S_{ob}(n) = \frac{1-\rho^n}{1-\rho^{n+1}} - \frac{1}{\nu_s}\left[\frac{1}{1-\rho} - \frac{(n+1)\rho^{(n+1)}}{1-\rho^{(n+1)}}\right]$. Furthermore, we denote $\nu_s = \frac{R\mu}{C}$, and ν satisfies $\nu_s = \frac{\nu(1-\rho)-\rho(1-\rho^\nu)}{(1-\rho)^2}$.

In the case of unobservable system, every customer has the same information upon entering the system. It is therefore reasonable to assume that all customers adopt the same randomized strategy for deciding whether to join or balk. Specifically, the common strategy of customers can be represented by a value $p \in [0,1]$ such that a customer will choose to join the queue with probability p. It follows that the rate at which customers join the queue is $\lambda = p\Lambda$, and this represents a stationary Poisson process of its own. For the social optimization, one aims to control the effective arrival rate λ_* which maximizes the overall social welfare $S_{un}(\lambda) = R\lambda - \frac{C\lambda}{\mu - \lambda}$, $\lambda \in [0, \Lambda]$. The following table give the known results of the relationship of the strategies in $M/M/1$ queues [23].

In the queueing model, we assume that customer arrival rate Λ, service rate μ, reward R and holding cost C are fuzzy numbers which are approximately known. Then, we represent these fuzzy numbers as follows:

$$\tilde{\Lambda} = \{(x, \eta_{\tilde{\Lambda}}(x)) | x \in X\}, \quad \tilde{\mu} = \{(y, \eta_{\tilde{\mu}}(y)) | y \in Y\}.$$

$$\tilde{R} = \{(r, \eta_{\tilde{R}}(r)) | r \in G\}, \quad \tilde{C} = \{(c, \eta_{\tilde{C}}(c)) | c \in H\}.$$

where X, Y, G, H are the crisp universal sets of x, y, r, c respectively, and $\eta_{\tilde{\Lambda}}(x)$, $\eta_{\tilde{\mu}}(y)$, $\eta_{\tilde{R}}(r)$ and $\eta_{\tilde{C}}(c)$ are the corresponding membership functions. Because the arrival rate $\tilde{\Lambda}$, service rate $\tilde{\mu}$, reward $\tilde{R}$ and holding cost $\tilde{C}$ are fuzzy, the steady-state probabilities and the performance measures for this system in the steady state must also be fuzzy.

Any convex normalized fuzzy subset $\tilde{A}$ on $\mathbb{R}$ (where $\mathbb{R}$ is the set of real numbers) with membership function $\eta_{\tilde{A}}$ is called a fuzzy number (Dubois [6]). The fuzzy number $\tilde{A}$ is said to be a trapezoidal fuzzy number if it is determined by the crisp numbers $[a_1, a_2, a_3, a_4]$, where $a_1 < a_2 < a_3 < a_4$, with membership function of the form

$$\eta_{\tilde{A}}(a) = \begin{cases} (a - a_1)/(a_2 - a_1) & a_1 \leq a \leq a_2, \\ 1 & a_2 \leq a \leq a_3, \\ (a_4 - a)/(a_4 - a_3) & a_3 \leq a \leq a_4. \end{cases}$$

Let $\tilde{A} = [a_1, a_2, a_3, a_4]$ and $\tilde{B} = [b_1, b_2, b_3, b_4]$ be two trapezoidal fuzzy numbers. Some fuzzy arithmetic operations under the functional principle (Chen [4]) for trapezoidal fuzzy numbers are given below:

(1) Non-negative fuzzy number $\tilde{A} \geq 0$, i.e. $a_1, a_2, a_3, a_4 \geq 0$.
(2) Subtraction $\ominus \tilde{B} = (-b_4, -b_3, -b_2, -b_1)$, $\tilde{A} \ominus \tilde{B} = (a_1 - b_4, a_2 - b_3, a_3 - b_2, a_4 - b_1)$.
(3) Division (If $a_1, b_1, a_2, b_2, a_3, b_3, a_4$ and b_4 are all positive real numbers)

$$\oslash \tilde{B} = \tilde{B}^{-1} = (\frac{1}{b_4}, \frac{1}{b_3}, \frac{1}{b_2}, \frac{1}{b_1}), \quad \tilde{A} \oslash \tilde{B} = (\frac{a_1}{b_4}, \frac{a_2}{b_3}, \frac{a_3}{b_2}, \frac{a_4}{b_1})$$

In this paper, we just consider the models with trapezoidal fuzzy parameters, which allow us derive the analysis results and numerical examples. From now on we assume that $\tilde{R} - \frac{\tilde{C}}{\tilde{\mu}} \geq 0$, otherwise no customers will ever enter. Furthermore, we assume that $\tilde{\rho} = \frac{\tilde{\Lambda}}{\tilde{\mu}}$ and $1 - \tilde{\rho} > 0$ for the unobservable case.

For the fuzzy queueing systems, we are interested in the optimal and equilibrium strategies which include the equilibrium and optimal thresholds for the observable case, optimal and equilibrium arrival rates for the unobservable case. Because the system parameters $\tilde{\Lambda}$, $\tilde{\mu}$, $\tilde{R}$ and $\tilde{C}$ are fuzzy, the equilibrium thresholds, arrival rates and social benefits must also be fuzzy. They are denoted by $\tilde{n}_e$, $\tilde{n}_*$, $\tilde{\lambda}_e$, $\tilde{\lambda}_*$, $\tilde{S}_{ob}$, $\tilde{S}_{un}$ respectively. For the optimal social strategies, the objective functions are designed to determine the optimal threshold and arrival rate that maximize the fuzzy expected social benefits in both observable and unobservable cases as follows:

$$\tilde{S}_{ob}(n) = \frac{1 - \tilde{\rho}^n}{1 - \tilde{\rho}^{n+1}} - \frac{\tilde{C}}{\tilde{R}\tilde{\mu}}\left[\frac{1}{1 - \tilde{\rho}} - \frac{(n + 1)\tilde{\rho}^{(n+1)}}{1 - \tilde{\rho}^{(n+1)}}\right],$$

$$\tilde{S}_{un}(\lambda) = \tilde{R}\lambda - \frac{\tilde{C}\lambda}{\tilde{\mu} - \lambda}.$$

Note that the maximal expected social benefit per unit time is a fuzzy number, not a crisp number. Consequently, the optimum cannot be obtained directly; instead, we wish to derive its membership function. We denote $\Omega = \{\lfloor \nu \rfloor \mid \frac{ry}{c} = \frac{\nu(1-x/y) - (x/y)(1-(x/y)^\nu)}{(1-x/y)^2}\}$. By Zadeh's extension principle (Zimmermann [28]), the membership functions of the objective values $\tilde{n}_e$, $\tilde{n}_*$, $\tilde{\lambda}_e$, $\tilde{\lambda}_*$, $\tilde{S}_{ob}$, $\tilde{S}_{un}$ are defined as

$$\eta_{\tilde{n}_e}(z) = \sup_{x \in X, y \in Y, r \in G, c \in H} \min\{\eta_{\tilde{\Lambda}}(x), \eta_{\tilde{\mu}}(y), \eta_{\tilde{R}}(r), \eta_{\tilde{C}}(c) \mid z = n_e(x, y, r, c)\},$$

$$(1)$$

$$\eta_{\tilde{n}_*}(z) = \sup_{x \in X, y \in Y, r \in G, c \in H} \min\{\eta_{\tilde{\Lambda}}(x), \eta_{\tilde{\mu}}(y), \eta_{\tilde{R}}(r), \eta_{\tilde{C}}(c) \mid z = n_*(x, y, r, c)\},$$

$$(2)$$

$$\eta_{\tilde{\lambda}_e}(z) = \sup_{x \in X, y \in Y, r \in G, c \in H} \min\{\eta_{\tilde{\Lambda}}(x), \eta_{\tilde{\mu}}(y), \eta_{\tilde{R}}(r), \eta_{\tilde{C}}(c) \mid z = \lambda_e(x, y, r, c)\},$$

$$(3)$$

$$\eta_{\tilde{\lambda}_*}(z) = \sup_{x \in X, y \in Y, r \in G, c \in H} \min\{\eta_{\tilde{\Lambda}}(x), \eta_{\tilde{\mu}}(y), \eta_{\tilde{R}}(r), \eta_{\tilde{C}}(c) \mid z = \lambda_*(x, y, r, c)\},$$

$$(4)$$

$$\eta_{\tilde{S}_{ob}}(z) = \sup_{x \in X, y \in Y, r \in G, c \in H} \min\{\eta_{\tilde{\Lambda}}(x), \eta_{\tilde{\mu}}(y), \eta_{\tilde{R}}(r), \eta_{\tilde{C}}(c) \mid z = \max_{n \geq 0} S_{ob}(n)\},$$

$$(5)$$

$$\eta_{\tilde{S}_{un}}(z) = \sup_{x \in X, y \in Y, r \in G, c \in H} \min\{\eta_{\tilde{\Lambda}}(x), \eta_{\tilde{\mu}}(y), \eta_{\tilde{R}}(r), \eta_{\tilde{C}}(c) \mid z = \max_{0 \leq \lambda \leq \Lambda} S_{un}(\lambda)\},$$

$$(6)$$

In this paper, the system characteristics of interest are the optimal and equilibrium strategies in $M/M/1$ queueing system for both observable and unobservable cases. From Table 1, we have the results as follows:

Table 1. Summary of known results about optimal and equilibrium strategy in $M/M/1$ queues

Cases	Optimal strategy		Relationship	Social welfare
	Selfish optimal	Social optimal		
Observable	$n_e = \lfloor \nu_s \rfloor$	$n_* = \lfloor \nu \rfloor$	$n_e \leq n_*$	$S_{ob}(n_*)$
Unobservable	$\lambda_e = \min\{\mu - \frac{C}{R}, \Lambda\}$	$\lambda_* = \min\{\mu(1 - \sqrt{\frac{C}{R\mu}}), \Lambda\}$	$\lambda_e \leq \lambda_*$	$S_{un}(\lambda_*)$

$$n_e(x, y, r, c) = \lfloor \frac{ry}{c} \rfloor,$$

$$n_*(x, y, r, c) = \lfloor \nu \rfloor, \nu \in \Omega,$$

$$\lambda_e(x, y, r, c) = \min\{y - \frac{c}{r}, x\},$$

$$\lambda_*(x, y, r, c) = \min\{y(1 - \sqrt{\frac{c}{ry}}), x\},$$

$$S_{ob}(n) = \frac{1 - (x/y)^n}{1 - (x/y)^{n+1}} - \frac{c}{ry}\left[\frac{1}{1 - x/y} - \frac{(n+1)(x/y)^{(n+1)}}{1 - (x/y)^{(n+1)}}\right],$$

$$S_{un}(\lambda) = r\lambda - \frac{c\lambda}{y - \lambda}, 0 \leq \lambda \leq x.$$

The membership functions above are not in the usual forms for practical use making it very difficult to imagine their shapes. In this paper we approach the problem using a mathematical programming technique. These parametric nonlinear programs are developed to find the α-cuts of equilibrium and optimal strategies based on the Zadeh's extension principle.

3   The Parametric Nonlinear Programming Approach

In most cases, the values of the membership functions of the optimal strategies cannot be solved analytically. Consequently, a closed-form membership function for these strategies and social benefits cannot be obtained. However, the numerical solutions for $\eta_{\tilde{n}_e}(z), \eta_{\tilde{n}_*}(z), \eta_{\tilde{\lambda}_e}(z), \eta_{\tilde{\lambda}_*}(z), \eta_{\tilde{S}_{ob}}(z), \eta_{\tilde{S}_{un}}(z)$ at different possibility levels can be collected to approximate the shapes of the membership functions. That is, the set of crisp intervals reveals the shape of the fuzzy optimal strategies. The α-cuts of the fuzzy parameters $\tilde{\Lambda}, \tilde{\mu}, R$ and C are defined as follows.

$$\Lambda(\alpha) = \{x \in X \mid \eta_{\tilde{\Lambda}}(x) \geq \alpha\}, \quad \mu(\alpha) = \{y \in Y \mid \eta_{\tilde{\mu}}(y) \geq \alpha\},$$
$$R(\alpha) = \{r \in G \mid \eta_{\tilde{R}}(r) \geq \alpha\}, \quad C(\alpha) = \{c \in H \mid \eta_{\tilde{C}}(c) \geq \alpha\}.$$

Notably, $\Lambda(\alpha)$, $\mu(\alpha)$, $R(\alpha)$ and $C(\alpha)$ are crisp sets rather than fuzzy sets. Using α-cuts, $\tilde{\Lambda}, \tilde{\mu}, \tilde{R}$ and $\tilde{C}$ can be represented by different levels of confidence intervals

(Klir and Yuan [16]). Because $\tilde{\Lambda}$, $\tilde{\mu}$, $\tilde{R}$ and $\tilde{C}$ are assumed to be fuzzy numbers, their cuts, as defined above, are crisp intervals that can be expressed in the following alternate forms:

$$\Lambda(\alpha) = [\min_x \{x \in X \mid \eta_{\tilde{\Lambda}}(x) \geq \alpha\}, \quad \max_x \{x \in X \mid \eta_{\tilde{\Lambda}}(x) \geq \alpha\}] = [X_\alpha^L, X_\alpha^U],$$

$$\mu(\alpha) = [\min_y \{y \in Y \mid \eta_{\tilde{\mu}}(y) \geq \alpha\}, \quad \max_y \{y \in Y \mid \eta_{\tilde{\mu}}(y) \geq \alpha\}] = [Y_\alpha^L, Y_\alpha^U],$$

$$R(\alpha) = [\min_r \{r \in G \mid \eta_{\tilde{R}}(r) \geq \alpha\}, \quad \max_r \{r \in G \mid \eta_{\tilde{R}}(r) \geq \alpha\}] = [G_\alpha^L, G_\alpha^U],$$

$$C(\alpha) = [\min_c \{c \in H \mid \eta_{\tilde{C}}(c) \geq \alpha\}, \quad \max_c \{c \in H \mid \eta_{\tilde{C}}(c) \geq \alpha\}] = [H_\alpha^L, H_\alpha^U].$$

We denote $f = n_e, \lambda_e, n_*, \lambda_*, S_{ob}, S_{un}$. Clearly, the membership functions defined in (1)–(6) are also parameterized by α. Thus, using Zadeh's extension principle, we can use the α-cuts of $\tilde{f}$ to construct these membership functions. The first step is to determine the lower bound $f^L(\alpha)$ and the upper bound $f^U(\alpha)$ on the α-cut of $\tilde{f}$. According to (1)–(6), $\eta_{\tilde{f}}(z)$ is the minimum of several other membership functions $\eta_{\tilde{\Lambda}}(x)$, $\eta_{\tilde{\mu}}(y)$, $\eta_{\tilde{R}}(r)$ and $\eta_{\tilde{C}}(c)$. To ensure that the obtained interval for the membership value is indeed the α-cut of $\tilde{f}$, at least one of x, y, r, c must lie on the boundary of its α-cut to satisfy $\eta_{\tilde{f}}(z) = \alpha$. In other words, to satisfy $\eta_{\tilde{f}}(z) = \alpha$, it is required that $\eta_{\tilde{\Lambda}}(x) \geq \alpha$, $\eta_{\tilde{\mu}}(y) \geq \alpha$, $\eta_{\tilde{R}}(r) \geq \alpha$ and $\eta_{\tilde{C}}(c) \geq \alpha$, and at least one of these four inequalities should be active such that $z = n_e, \lambda_e, n_*, \lambda_*, S_{ob}^*, S_{un}^*$. This outcome can be accomplished via parametric programming. Because $f^L(\alpha)$ and $f^U(\alpha)$ are the minimum and maximum of f, these terms can be expressed as follows.

(i) For the membership function $\eta_{\tilde{n}_e}(z)$ of fuzzy number $\tilde{n}_e$, we have

$$n_e^L(\alpha) = \min\{\lfloor \frac{ry}{c} \rfloor \mid Y_\alpha^L \leq y \leq Y_\alpha^U, G_\alpha^L \leq r \leq G_\alpha^U, H_\alpha^L \leq c \leq H_\alpha^U\}$$

$$n_e^U(\alpha) = \max\{\lfloor \frac{ry}{c} \rfloor \mid Y_\alpha^L \leq y \leq Y_\alpha^U, G_\alpha^L \leq r \leq G_\alpha^U, H_\alpha^L \leq c \leq H_\alpha^U\}.$$

which can be reformulated as follows:

$$n_e^L(\alpha) = \lfloor \frac{G_\alpha^L Y_\alpha^L}{H_\alpha^U} \rfloor, \quad n_e^U(\alpha) = \lfloor \frac{G_\alpha^U Y_\alpha^U}{H_\alpha^L} \rfloor, \tag{7}$$

(ii) For the membership function $\eta_{\tilde{n}_*}(z)$ of fuzzy number $\tilde{n}_*$, we have

$$n_*^L(\alpha) = \min\{\lfloor \nu \rfloor \mid \nu \in \Omega, X_\alpha^L \leq x \leq X_\alpha^U, Y_\alpha^L \leq y \leq Y_\alpha^U, G_\alpha^L \leq r \leq G_\alpha^U, H_\alpha^L \leq c \leq H_\alpha^U\}$$

$$n_*^U(\alpha) = \max\{\lfloor \nu \rfloor \mid \nu \in \Omega, X_\alpha^L \leq x \leq X_\alpha^U, Y_\alpha^L \leq y \leq Y_\alpha^U, G_\alpha^L \leq r \leq G_\alpha^U, H_\alpha^L \leq c \leq H_\alpha^U\}.$$

which can be reformulated as follows:

$$n_*^L(\alpha) = \min\{\lfloor \nu \rfloor \mid \nu \in \Omega\}, \quad n_*^U(\alpha) = \max\{\lfloor \nu \rfloor \mid \nu \in \Omega\},$$

$$s.t. \; X_\alpha^L \leq x \leq X_\alpha^U, \; Y_\alpha^L \leq y \leq Y_\alpha^U, G_\alpha^L \leq r \leq G_\alpha^U, H_\alpha^L \leq c \leq H_\alpha^U. \tag{8}$$

(iii) For the membership function $\eta_{\tilde{\lambda}_e}(z)$ of fuzzy number $\tilde{\lambda}_e$, we have

$$\lambda_e^L(\alpha) = \min\{\min\{y - \frac{c}{r}, x\} \mid X_\alpha^L \leq x \leq X_\alpha^U, Y_\alpha^L \leq y \leq Y_\alpha^U, G_\alpha^L \leq r \leq G_\alpha^U, H_\alpha^L \leq c \leq H_\alpha^U\},$$

$$\lambda_e^U(\alpha) = \max\{\min\{y - \frac{c}{r}, x\} \mid X_\alpha^L \leq x \leq X_\alpha^U, Y_\alpha^L \leq y \leq Y_\alpha^U, G_\alpha^L \leq r \leq G_\alpha^U, H_\alpha^L \leq c \leq H_\alpha^U\},$$

which can be reformulated as follows:

$$\lambda_e^L(\alpha) = \min\{Y_\alpha^L - \frac{H_\alpha^U}{G_\alpha^L}, X_\alpha^L\}, \quad \lambda_e^U(\alpha) = \min\{Y_\alpha^U - \frac{H_\alpha^L}{G_\alpha^U}, X_\alpha^U\}. \tag{9}$$

(iv) For the membership function $\eta_{\tilde{\lambda}_*}(z)$ of fuzzy number $\tilde{\lambda}_*$, we have

$$\lambda_*^L(\alpha) = \min\{\min\{y(1 - \sqrt{\frac{c}{ry}}), x\} \mid X_\alpha^L \le x \le X_\alpha^U, Y_\alpha^L \le y \le Y_\alpha^U, G_\alpha^L \le r$$
$$\le G_\alpha^U, H_\alpha^L \le c \le H_\alpha^U\},$$

$$\lambda_*^U(\alpha) = \max\{\min\{y(1 - \sqrt{\frac{c}{ry}}), x\} \mid X_\alpha^L \le x \le X_\alpha^U, Y_\alpha^L \le y \le Y_\alpha^U, G_\alpha^L$$
$$\le r \le G_\alpha^U, H_\alpha^L \le c \le H_\alpha^U\},$$

which can be reformulated as follows:

$$\lambda_*^L(\alpha) = \min\{Y_\alpha^L(1 - \sqrt{\frac{H_\alpha^U}{G_\alpha^L Y_\alpha^L}}), X_\alpha^L\}, \quad \lambda_*^U(\alpha) = \min\{Y_\alpha^U(1 - \sqrt{\frac{H_\alpha^L}{G_\alpha^U Y_\alpha^U}}), X_\alpha^U\}. \tag{10}$$

(v) For the membership function $\eta_{\tilde{S}_{ob}}$ of fuzzy number $\tilde{S}_{ob}$, we have

$$S_{ob}^L(\alpha) = \min\{S_{ob}^*(n) \mid X_\alpha^L \le x \le X_\alpha^U, Y_\alpha^L \le y \le Y_\alpha^U, G_\alpha^L \le r \le G_\alpha^U, H_\alpha^L \le c \le H_\alpha^U\},$$

$$S_{ob}^U(\alpha) = \max\{S_{ob}^*(n) \mid X_\alpha^L \le x \le X_\alpha^U, Y_\alpha^L \le y \le Y_\alpha^U, G_\alpha^L \le r \le G_\alpha^U, H_\alpha^L \le c \le H_\alpha^U\},$$

which can be reformulated as follows:

$$S_{ob}^L(\alpha) = \min_{n \in \Omega}\{S_{ob}(n, x, y) \mid r = G_\alpha^L, c = H_\alpha^U\},$$
$$S_{ob}^U(\alpha) = \max_{n \ge 0}\{S_{ob}(n, x, y) \mid r = G_\alpha^U, c = H_\alpha^L\}, \tag{11}$$
$$s.t. \quad X_\alpha^L \le x \le X_\alpha^U, \quad Y_\alpha^L \le y \le Y_\alpha^U.$$

(vi) For the membership function $\eta_{\tilde{S}_{un}}$ of fuzzy number $\tilde{S}_{un}$, we have

$$S_{un}^L(\alpha) = \min\{S_{un}^*(\lambda) \mid X_\alpha^L \le x \le X_\alpha^U, Y_\alpha^L \le y \le Y_\alpha^U, G_\alpha^L \le r \le G_\alpha^U, H_\alpha^L \le c \le H_\alpha^U\},$$

$$S_{un}^U(\alpha) = \max\{S_{un}^*(\lambda) \mid X_\alpha^L \le x \le X_\alpha^U, Y_\alpha^L \le y \le Y_\alpha^U, G_\alpha^L \le r \le G_\alpha^U, H_\alpha^L \le c \le H_\alpha^U\},$$

which can be reformulated as follows:

$$S_{un}^L(\alpha) = \min\{S_{un}(x, y) \mid \lambda = \min\{y(1 - \sqrt{\frac{c}{ry}}), x\}, y = Y_\alpha^U, r = G_\alpha^U, c = H_\alpha^L\},$$

$$S_{un}^U(\alpha) = \max_{0 \le \lambda \le x}\{S_{un}(\lambda, x, y) \mid y = Y_\alpha^L, r = G_\alpha^L, c = H_\alpha^U\}, \tag{12}$$

$$s.t. \quad X_\alpha^L \le x \le X_\alpha^U.$$

where $S_{ob}^* = \max_{n \geq 0}\{S_{ob}(n)\}$ and $S_{un}^* = \max_{0 \leq \lambda \leq \Lambda}\{S_{un}(\lambda)\}$. This can be accomplished by the parametric NLP techniques. If both $f^L(\alpha)$ and $f^U(\alpha)$ of $\tilde{f}$ are invertible with respect to α, then a left shape function $L(z)$ and a right shape function $R(z)$ can be obtained. In most cases, the values of z_i cannot be solved analytically. Consequently, a closed-form membership function for $\tilde{f}$ cannot be obtained. However, the numerical solutions for z_i at different possibility level α can be collected to approximate the shapes of $L(z)$ and $R(z)$. In the following section, we present an efficient solution algorithm to compute the membership values of f at different possibility level α.

4   Solution Algorithm and Numerical Examples

Let the arrival rate and service rate be trapezoidal fuzzy number represented by $[x_1, x_2, x_3, x_4]$, $[y_1, y_2, y_3, y_4]$, $[r_1, r_2, r_3, r_4]$ and $[c_1, c_2, c_3, c_4]$ per unit time, respectively, where $x_4 > x_3 > x_2 > x_1$, $y_4 > y_3 > y_2 > y_1$, $r_4 > r_3 > r_2 > r_1$ and $c_4 > c_3 > c_2 > c_1$. The system manager wants to evaluate the optimal and equilibrium strategies in both unobservable and observable cases. Using the proposed approach stated in Sect. 3, it is easy to obtain the α-cut sets of $\tilde{\Lambda}$, $\tilde{\mu}$, $\tilde{R}$ and $\tilde{C}$:

$$[X_\alpha^L, X_\alpha^U] = [(x_2 - x_1)\alpha + x_1, x_4 - (x_4 - x_3)\alpha],$$
$$[Y_\alpha^L, Y_\alpha^U] = [(y_2 - y_1)\alpha + y_1, y_4 - (y_4 - y_3)\alpha],$$
$$[G_\alpha^L, G_\alpha^U] = [(r_2 - r_1)\alpha + r_1, r_4 - (r_4 - r_3)\alpha],$$
$$[H_\alpha^L, H_\alpha^U] = [(c_2 - c_1)\alpha + c_1, c_4 - (c_4 - c_3)\alpha].$$

Thus, following (7)–(12), a set of parametric nonlinear programs for deriving the membership function of fuzzy numbers $\tilde{n}_e$, $\tilde{n}_*$, $\tilde{\lambda}_e$, $\tilde{\lambda}_*$, $\tilde{S}_{ob}$, $\tilde{S}_{un}$ for $M/M/1$ system are given as follows.

For the equilibrium threshold strategy, we have

$$n_e^L(\alpha) = \left\lfloor \frac{[(r_2 - r_1)\alpha + r_1][(y_2 - y_1)\alpha + y_1]}{c_4 - (c_4 - c_3)\alpha} \right\rfloor,$$
$$n_e^U(\alpha) = \left\lfloor \frac{[r_4 - (r_4 - r_3)\alpha][y_4 - (y_4 - y_3)\alpha]}{(c_2 - c_1)\alpha + c_1} \right\rfloor.$$

For the optimal threshold strategy of the social welfare, we have

$$n_*^L(\alpha) = \min\{\lfloor \nu \rfloor \mid \nu \in \Omega\}, \quad n_*^U(\alpha) = \max\{\lfloor \nu \rfloor \mid \nu \in \Omega\},$$
$$s.t. \quad (x_2 - x_1)\alpha + x_1 \leq x \leq x_4 - (x_4 - x_3)\alpha, \quad (y_2 - y_1)\alpha + y_1 \leq y \leq y_4 - (y_4 - y_3)\alpha,$$
$$(r_2 - r_1)\alpha + r_1 \leq r \leq r_4 - (r_4 - r_3)\alpha, \quad (c_2 - c_1)\alpha + c_1 \leq c \leq c_4 - (c_4 - c_3)\alpha.$$

For the equilibrium arrival rate strategy, we have

$$\lambda_e^L(\alpha) = \min\left\{(y_2 - y_1)\alpha + y_1 - \frac{c_4 - (c_4 - c_3)\alpha}{(r_2 - r_1)\alpha + r_1}, \quad (x_2 - x_1)\alpha + x_1\right\},$$
$$\lambda_e^U(\alpha) = \min\left\{y_4 - (y_4 - y_3)\alpha - \frac{(c_2 - c_1)\alpha + c_1}{r_4 - (r_4 - r_3)\alpha}, \quad x_4 - (x_4 - x_3)\alpha\right\}.$$

For the optimal arrival rate strategy of the social welfare, we have

$$\lambda_*^L(\alpha) = \min\{[(y_2 - y_1)\alpha + y_1](1 - \sqrt{\frac{c_4 - (c_4 - c_3)\alpha}{[(r_2 - r_1)\alpha + r_1][(y_2 - y_1)\alpha + y_1]}}),$$

$$(x_2 - x_1)\alpha + x_1\},$$

$$\lambda_*^U(\alpha) = \min\{[y_4 - (y_4 - y_3)\alpha](1 - \sqrt{\frac{(c_2 - c_1)\alpha + c_1}{[r_4 - (r_4 - r_3)\alpha][y_4 - (y_4 - y_3)\alpha]}}),$$

$$x_4 - (x_4 - x_3)\alpha\}.$$

For the optimal social warfare in observable case, we have

$$S_{ob}^L(\alpha) = \min_{n \in \Omega}\{S_{ob}(n, x, y) \mid r = (r_2 - r_1)\alpha + r_1, c = c_4 - (c_4 - c_3)\alpha\},$$

$$S_{ob}^U(\alpha) = \max_{n \geq 0}\{S_{ob}(n, x, y) \mid r = r_4 - (r_4 - r_3)\alpha, c = (c_2 - c_1)\alpha + c_1\},$$

$$s.t. \quad (x_2 - x_1)\alpha + x_1 \leq x \leq x_4 - (x_4 - x_3)\alpha, \quad (y_2 - y_1)\alpha + y_1 \leq y \leq y_4 - (y_4 - y_3)\alpha.$$

For the optimal social warfare in unobservable case, we have

$$S_{un}^L(\alpha) = \min_{0 \leq \lambda \leq x}\{S_{un}(\lambda, x) \mid y = (y_2 - y_1)\alpha + y_1,$$

$$r = (r_2 - r_1)\alpha + r_1, c = c_4 - (c_4 - c_3)\alpha\},$$

$$S_{un}^U(\alpha) = \max_{0 \leq \lambda \leq x}\{S_{un}(\lambda, x) \mid y = y_4 - (y_4 - y_3)\alpha, r = r_4 - (r_4 - r_3)\alpha,$$

$$c = (c_2 - c_1)\alpha + c_1\},$$

$$s.t. \quad (x_2 - x_1)\alpha + x_1 \leq x \leq x_4 - (x_4 - x_3)\alpha.$$

The proposed approach is based on Zadeh's extension principle and parametric nonlinear programming. Therefore, we can summarize the solution procedure into the following algorithm:

Step 1: Input the arrival rate, service rate, reward and holding cost, which are trapezoidal fuzzy number represented by $[x_1, x_2, x_3, x_4]$, $[y_1, y_2, y_3, y_4]$, $[r_1, r_2, r_3, r_4]$ and $[c_1, c_2, c_3, c_4]$.

Step 2: Output the numbers $X_\alpha^L, X_\alpha^U, Y_\alpha^L, Y_\alpha^U, G_\alpha^L, G_\alpha^U$ and H_α^L, H_α^U.

Step 3: According to (i)–(vi), calculate the crisp interval $[f_\alpha^L, f_\alpha^U]$ where $f = n_e, n_*, \lambda_e, \lambda_*, S_{ob}, S_{un}$ of α-cut level by the parametric nonlinear programming method.

Next, to illustrate the validity and suitability of the proposed model, several numerical experiments are considered and their results are included in this section. Due to the complexity of four fuzzy variables, it is impossible to determine the analytical solution of the crisp interval $[S_e^L(\alpha), S_e^U(\alpha)]$ in terms of α. Consequently, it is very difficult to obtain a closed-form membership function of f. Instead, a software Maltab version 6.0 for windows is used to solve the mathematical programs and then the shape of f can be found. Here we enumerate 101 values of α: $0, 0.01, 0.02, \ldots, 1.00$. The figures depict the rough shape from these values. In the following examples, we assume that the arrival rate, service rate, reward and holding cost rate are trapezoidal fuzzy numbers represented by $[1, 2, 3, 4]$, $[5, 6, 7, 8]$, $[10, 15, 20, 25]$, $[2, 4, 6, 8]$ per unit time, respectively. As is shown in the figures, we can make the following observations.

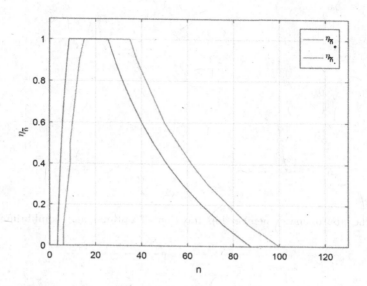

Fig. 1. The approximate membership function of optimal and equilibrium threshold n.

In the numerical experiments, we focus on exploring the membership functions of the equilibrium and optimal strategies in the observable and unobservable cases. Figures 1, 2, 3 show the MATLAB output results of the analysis for each input parameter of our problem in fuzzy queueing system. As can be seen in these figures, the membership functions of the strategies and social benefits are approximately trapezoidal, which are result from the assumption that the four fuzzy parameters in this model are trapezoidal fuzzy numbers.

In Fig. 1, we describe the characteristics of the membership function of the optimal and equilibrium thresholds for observable case. The range of the equilibrium and optimal thresholds are $n_e \in [3, 87]$ and $n_* \in [6, 100]$ respectively, which indicates that the equilibrium and optimal thresholds in the fuzzy queueing system belong to a large interval rather than by crisp values, so that more information is provided to the system designers. Moreover, another thing worth noting is the relationship of the derived membership functions. For example, for $\alpha = 0.6$, it is easy to see that $n_e^\alpha \in [3, 87]$ and $n_*^\alpha \in [6, 100]$ so that $n_e^\alpha \bigcap n_*^\alpha \neq \emptyset$. Therefore, in fuzzy queueing system, there is no exact relationship between the values of n_e and n_*, which is totally different from the results (see Table 1) in the corresponding crisp queueing system. Figure 2 illustrates the equilibrium and optimal customer arrival rate in unobservable case. Taking the same analysis method for Fig. 1, as illustrated in Fig. 2, $\lambda_e \in [1.0, 4.0]$ and $\lambda_* \in [1.0, 7.2]$, it is to

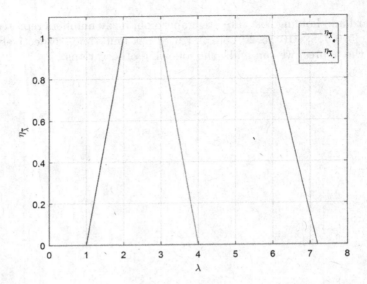

Fig. 2. The approximate membership function of optimal and equilibrium arrival rate λ.

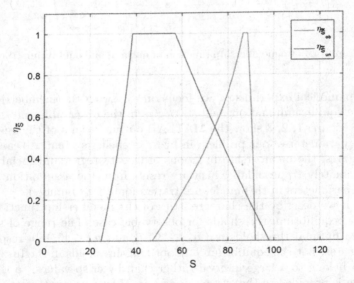

Fig. 3. The approximate membership function of optimal social benefit S.

easy to see $\lambda_e \bigcap \lambda_* \neq \emptyset$. Therefore, similar results can be obtained that no explicitly relationship between optimal arrival rate λ_* and equilibrium arrival rate λ_e exists, which is also different from the corresponding crisp queueing system. The above information will be very useful for designing a queueing system.

Figure 3 is concerned with the fuzzy expected social welfare between observable case and unobservable case. Based on the equilibrium analysis, we examine the impact of information for the system. Such as, for $\alpha = 1$, we have $S_{ob} > S_{un}$, which means that the social benefit in unobservable case is small. So more information can significantly promote the system performance. However, when change the value of $\eta_{\tilde{S}}$ in the range of $[0, 1]$, the result may not be valid. The value of S_{ob} is in the range $[47.06, 92.00]$ in observable case, while in unobservable case the value of S_{un} will never exceed 98.00 or fall below 24.71. The two curves intersect in two points in the interval of $[60, 100]$, the values of S_{un} and S_{ob} are difficult to compare in Fig. 3, but it is obviously that the range of S_{un} is lager than that of S_{ob}. From this point of view, we can find that the impact of expected social benefit with fuzzy factors in unobservable case is greater. Therefore, to avoid the loss of the system, system designers should take more fuzzy factors into consideration. The above results will certainly be useful and significant for system managers. In addition, we find a consequence of the fact that there is no definite relationship between S_{ob} and S_{un}. As the performance measures are expressed by membership functions rather than by crisp values, more information provided to the system designers and decision makers may be helpful to improve the existing systems. Since the fuzzy performance measures of fuzzy queues derived from the proposed approach maintain the fuzziness of input information; therefore the derived results can be used to represent the real time systems as fuzzy system more accurately.

5   Conclusions

In this paper, we analyzed the Nash equilibrium behavior of customers in Markovian queueing systems with fuzzy parameters. For the observable and unobservable cases, we investigated the membership functions of the customer's equilibrium balking strategy and optimal social benefits. We presented an efficient algorithm to construct the membership functions of the strategies and social benefits. Numerical solutions for different α values were calculated to approximate the membership functions by NLP. Moreover, some numerical examples are provided, which allow us to find some interesting results differing from the crisp queueing system. In contrast to existing studies, the results derived from the proposed solution procedure conserve the fuzziness of the input information.

From the above results there arise some interesting extensions of the model which we may study in the near future. One possible change is to consider the systems where the customer arrival time and service time distributions are general distributions. It is not worthy that instead of considering fuzzy triangular numbers we can also considered different types of fuzzy numbers such as trapezoidal, exponential, bell-shaped fuzzy etc. Furthermore, the method of our paper also can be promoted to some different queueing systems. Another way to generalize the model is to study some models with different strategies which can be used in more practical applications, such as the queueing systems with impatient customers, multiple priority customers and vacation policy. In addition, further extensions would be researched about the equilibrium strategy and dynamic pricing problem in the queueing system with fuzzy parameters or fuzzy states.

Acknowledgements. This work is partially supported by the National Natural Science Foundation of China (11671404), and the Fundamental Research Funds for the Central Universities of Central South University (2017zzts061, 2017zzts386). The authors also gratefully acknowledge the helpful comments and suggestions of the reviewers.

References

1. Bountali, O., Economou, A.: Equilibrium joining strategies in batch service queueing systems. Eur. J. Oper. Res. **260**(3), 1142–1151 (2017)
2. Buckley, J.J.: Fuzzy queuing theory. Fuzzy Probabilities. Studies in Fuzziness and Soft Computing, vol. 115, pp. 61–69. Springer, Heidelberg (2003). https://doi.org/10.1007/978-3-642-86786-6_5
3. Buckley, J.J., Feuring, T., Hayashi, Y.: Fuzzy queueing theory revisited. Int. J. Uncertain. Fuzziness Knowl.-Based Syst. **9**(05), 527–537 (2001)
4. Chen, S.H.: Ranking fuzzy numbers with maximizing set and minimizing set. Fuzzy Sets Syst. **17**(2), 113–129 (1985)
5. Chen, S.P.: Time value of delays in unreliable production systems with mixed uncertainties of fuzziness and randomness. Eur. J. Oper. Res. **255**(3), 834–844 (2016)
6. Dubois, D.J.: Fuzzy Sets and Systems: Theory and Applications, vol. 144. Academic press, Cambridge (1980)
7. Economou, A., Gómez-Corral, A., Kanta, S.: Optimal balking strategies in single-server queues with general service and vacation times. Perform. Eval. **68**(10), 967–982 (2011)
8. Economou, A., Kanta, S.: Equilibrium balking strategies in the observable single-server queue with breakdowns and repairs. Oper. Res. Lett. **36**(6), 696–699 (2008)
9. Economou, A., Manou, A.: Strategic behavior in an observable fluid queue with an alternating service process. Eur. J. Oper. Res. **254**, 148–160 (2016)
10. Guo, P., Hassin, R.: Strategic behavior and social optimization in Markovian vacation queues: the case of heterogeneous customers. Eur. J. Oper. Res. **222**(2), 278–286 (2012)
11. Guo, P., Sun, W., Wang, Y.: Equilibrium and optimal strategies to join a queue with partial information on service times. Eur. J. Oper. Res. **214**(2), 284–297 (2011)
12. Hassin, R., Haviv, M.: To Queue or Not to Queue: Equilibrium Behavior in Queueing Systems, vol. 59. Springer Science & Business Media, Heidelberg (2003). https://doi.org/10.1007/978-1-4615-0359-0
13. Hassin, R., Snitkovsky, R.I.: Strategic customer behavior in a queueing system with a loss subsystem. Queueing Syst. **86**, 361–387 (2017)
14. Jolai, F., Asadzadeh, S.M., Ghodsi, R., Bagheri-Marani, S.: A multi-objective fuzzy queuing priority assignment model. Appl. Math. Model. **40**(21), 9500–9513 (2016)
15. Kerner, Y.: Equilibrium joining probabilities for an M/G/1 queue. Games Econ. Behav. **71**(2), 521–526 (2011)
16. Klir, G., Yuan, B.: Fuzzy Sets and Fuzzy Logic, vol. 4. Prentice Hall, Upper Saddle River (1995)

17. Li, R.J., Lee, E.: Analysis of fuzzy queues. Comput. Math. Appl. **17**(7), 1143–1147 (1989)
18. Ma, Y., Liu, Z., Zhang, Z.G.: Equilibrium in vacation queueing system with complementary services. Qual. Technol. Quant. Manag. **14**(1), 114–127 (2017)
19. Muñoz, E., Ruspini, E.H.: Simulation of fuzzy queueing systems with a variable number of servers, arrival rate, and service rate. IEEE Trans. Fuzzy Syst. **22**(4), 892–903 (2014)
20. Naor, P.: The regulation of queue size by levying tolls. Econom.: J. Econom. Soc. **37**, 15–24 (1969)
21. Negi, D., Lee, E.: Analysis and simulation of fuzzy queues. Fuzzy Sets Syst. **46**(3), 321–330 (1992)
22. Panda, G., Goswami, V., Banik, A.D.: Equilibrium and socially optimal balking strategies in Markovian queues with vacations and sequential abandonment. Asia-Pac. J. Oper. Res. **33**(05), 1650036 (2016)
23. Shone, R., Knight, V.A., Williams, J.E.: Comparisons between observable and unobservable M/M/1 queues with respect to optimal customer behavior. Eur. J. Oper. Res. **227**(1), 133–141 (2013)
24. Simhon, E., Hayel, Y., Starobinski, D., Zhu, Q.: Optimal information disclosure policies in strategic queueing games. Oper. Res. Lett. **44**(1), 109–113 (2016)
25. Stidham Jr., S.: Optimal Design of Queueing Systems. CRC Press, Boca Raton (2009)
26. Wang, J., Zhang, F.: Strategic joining in M/M/1 retrial queues. Eur. J. Oper. Res. **230**(1), 76–87 (2013)
27. Wang, T.Y., Yang, D.Y., Li, M.J.: Fuzzy analysis for the N-policy queues with infinite capacity. Int. J. Inf. Manag. Sci. **21**, 41–56 (2010)
28. Zimmermann, H.J.: Fuzzy Set Theory and Its Applications. Springer Science & Business Media, Heidelberg (2011)

Trilogy on Computing Maximal Eigenpair

Mu-Fa Chen$^{(\boxtimes)}$

School of Mathematical Sciences, Beijing Normal University, Beijing 100875, China
mfchen@bnu.edu.cn

Abstract. The eigenpair here means the twins of eigenvalue and corresponding eigenvector. The talk introduces the three steps of our study on computing the maximal eigenpair. In the first two steps, we construct efficient initials for a known but dangerous algorithm, first for tridiagonal matrices and then for the irreducible matrices, having nonnegative off-diagonal elements. In the third step, we present two global algorithms which are still efficient and work well for a quite large class of matrices, even complex for instance.

Keywords: Maximal eigenpair · Efficient initial · Tridiagonal matrix Global algorithm

1 Introduction

This paper is a continuation of [4]. For the reader's convenience, we review shortly the first part of [4], especially the story of the proportion of 1000 and 2 of iterations for two different algorithms.

The most famous result on the maximal eigenpair should be the Perron-Frobenius theorem. For nonnegative (pointwise) and irreducible A, if Trace$(A) > 0$, then the theorem says there exists uniquely a maximal eigenvalue $\rho(A) > 0$ with positive left-eigenvector u and positive right-eigenvector g:

$$uA = \lambda u, \qquad Ag = \lambda g, \qquad \lambda = \rho(A).$$

These eigenvectors are also unique up to a constant. Before going to the main body of the paper, let us make two remarks.

(1) We need to study the right-eigenvector g only. Otherwise, use the transpose A^* instead of A.

(2) The matrix A is required to be irreducible with nonnegative off-diagonal elements, its diagonal elements can be arbitrary. Otherwise, use a shift $A + mI$ for large m:

$$(A + mI)g = \lambda g \Longleftrightarrow Ag = (\lambda - m)g, \tag{1}$$

their eigenvector remains the same but the maximal eigenvalues are shifted to each other.

W. Yue et al. (Eds.): QTNA 2017, LNCS 10591, pp. 312–329, 2017.
https://doi.org/10.1007/978-3-319-68520-5_19

Consider the following matrix.

$$
Q = \begin{pmatrix}
-1^2 & 1^2 & 0 & 0 & \cdots \\
1^2 & -1^2 - 2^2 & 2^2 & 0 & \cdots \\
0 & 2^2 & -2^2 - 3^2 & 3^2 & \cdots \\
\vdots & \vdots & \ddots & \ddots & \ddots \\
0 & 0 & 0 & N^2 & -N^2 - (N+1)^2
\end{pmatrix}. \tag{2}
$$

The main character of the matrix is the sequence $\{k^2\}$. The sum of each row equals zero except the last row. Actually, this matrix is truncated from the corresponding infinite one, in which case we have known that the maximal eigenvalue is $-1/4$ (refer to ([2]; Example 3.6)).

Example 1. *Let $N = 7$. Then the maximal eigenvalue is -0.525268 with eigenvector:*

$$
g \approx (55.878,\ 26.5271,\ 15.7059,\ 9.97983,\ 6.43129,\ 4.0251,\ 2.2954,\ 1)^*,
$$

where the vector $v^* =$ the transpose of v.

We now want to practice the standard algorithms in matrix eigenvalue computation. The first method in computing the maximal eigenpair is the *Power Iteration*, introduced in 1929. Starting from a vector v_0 having a nonzero component in the direction of g, normalized with respect to a norm $\|\cdot\|$. At the kth step, iterate v_k by the formula

$$
v_k = \frac{Av_{k-1}}{\|Av_{k-1}\|}, \quad z_k = \|Av_k\|, \quad k \geqslant 1. \tag{3}
$$

Then we have the convergence: $v_k \to g$ (first pointwise and then uniformly) and $z_k \to \rho(Q)$ as $k \to \infty$. If we rewrite v_k as

$$
v_k = \frac{A^k v_0}{\|A^k v_0\|},
$$

one sees where the name "power" comes from. For our example, to use the Power Iteration, we adopt the ℓ^1-norm and choose $v_0 = \tilde{v}_0/\|\tilde{v}_0\|$, where

$$
\tilde{v}_0 = (1,\ 0.587624,\ 0.426178,\ 0.329975,\ 0.260701,\ 0.204394, 0.153593, 0.101142)^*.
$$

This initial comes from a formula to be given in the next section. In Fig. 1 below, the upper curve is g, the lower one is modified from $\tilde{v}_0$, renormalized so that its last component becomes one. Clearly, these two functions are quite different, one may worry about the effectiveness of the choice of v_0. Anyhow, having the experience of computing its eigensystem, I expect to finish the computation in a few of seconds. Unexpectly, I got a difficult time to compute the maximal eigenpair for this simple example. Altogether, I computed it for 180 times, not in one day, using 1000 iterations. The printed pdf-file of the outputs has 64 pages. Figure 2 give us the outputs.

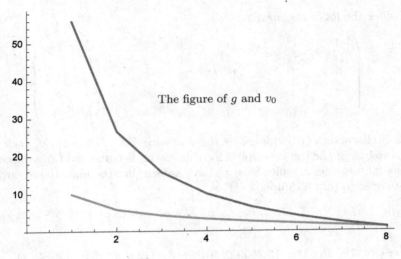

The figure of g and v_0

Fig. 1. g and v_0

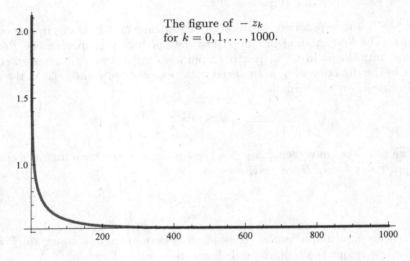

The figure of $-z_k$
for $k = 0, 1, \ldots, 1000$.

Fig. 2. $-z_k$ for $k = 0, 1, \ldots, 1000$

The figure shows that the convergence of z_k goes quickly at the beginning of the iterations. This means that our initial v_0 is good enough. Then the convergence goes very slow which means that the Power Iteration Algorithm converges very slowly.

Let us have a look at the convergence of the power iteration. Suppose that the eigenvalues are all different for simplicity. Denote by (λ_j, g_j) the eigenpairs with maximal one (λ_0, g_0). Write $v_0 = \sum_{j=0}^{N} c_j g_j$ for some constants (c_j). Then $c_0 \neq 0$ by assumption and

$$A^k v_0 = \sum_{j=0}^{N} c_j \lambda_j^k g_j = c_0 \lambda_0^k \left[g_0 + \sum_{j=1}^{N} \frac{c_j}{c_0} \left(\frac{\lambda_j}{\lambda_0} \right)^k g_j \right].$$

Since $|\lambda_j/\lambda_0| < 1$ for each $j \geqslant 1$ and $\|g_0\| = 1$, we have

$$\frac{A^k v_0}{\|A^k v_0\|} = \frac{c_0}{|c_0|} g_0 + O\left(\left| \frac{\lambda_1}{\lambda_0} \right|^k \right) \qquad \text{as } k \to \infty,$$

where $|\lambda_1| := \max\{|\lambda_j| : j > 0\}$. Since $|\lambda_1/\lambda_0|$ can be very closed to 1, this explains the reason why the convergence of the method can be very slow.

Before moving further, let us mention that the power method can be also used to compute the minimal eigenvalue $\lambda_{\min}(A)$, simply replace A by A^{-1}. That is the *Inverse Iteration* introduced in 1944:

$$v_k = \frac{A^{-1} v_{k-1}}{\|A^{-1} v_{k-1}\|} \iff v_k = \frac{A^{-k} v_0}{\|A^{-k} v_0\|}. \tag{4}$$

It is interesting to note that the equivalent assertion on the right-hand side is exactly the input-output method in economy.

To come back to compute the maximal $\rho(A)$ rather than $\lambda_{\min}(A)$, we add a shift z to A: replacing A by $A - zI$. Actually, it is even better to replace the last one by $zI - A$ since we will often use $z > \rho(A)$ rather than $z < \rho(A)$, the details will be explained at the beginning of Sect. 4 below. When z is close enough to $\rho(A)$, the leading eigenvalue of $(zI - A)^{-1}$ becomes $(z - \rho(A))^{-1}$. Furthermore, we can even use a variant shift $z_{k-1}I$ to accelerate the convergence speed. Thus, we have arrived at the second algorithm in computing the maximal eigenpair, the *Rayleigh Quotient Iteration* (RQI), a variant of the *Inverse Iteration*. From now on, unless otherwise stated, we often use the ℓ^2-norm. Starting from an approximating pair (z_0, v_0) of the maximal one $(\rho(A), g)$ with $v_0^* v_0 = 1$, use the following iteration.

$$v_k = \frac{(z_{k-1}I - A)^{-1} v_{k-1}}{\|(z_{k-1}I - A)^{-1} v_{k-1}\|}, \qquad z_k = v_k^* A v_k, \qquad k \geqslant 1. \tag{5}$$

If (z_0, v_0) is close enough to $(\rho(A), g)$, then

$$v_k \to g \quad \text{and} \quad z_k \to \rho(A) \quad \text{as } k \to \infty.$$

Since for each $k \geqslant 1$, $v_k^* v_k = 1$, we have $z_k = v_k^* A v_k / (v_k^* v_k)$. That is where the name "Rayleigh Quotient" comes from. Unless otherwise stated, z_0 is setting to be $v_0^* A v_0$.

Having the hard time spent in the first algorithm, I wondered how many iterations are required using this algorithm. Of course, I can no longer bear 1000 iterations. To be honest, I hope to finish the computation within 100 iterations. What happens now?

Example 2. *For the same matrix Q and $\tilde{v}_0$ as in Example 1.1, by RQI, we need two iterations only:*

$$z_1 \approx -0.528215, \quad z_2 \approx -0.525268.$$

The result came to me, not enough to say surprisingly, I was shocked indeed. This shows not only the power of the second method but also the effectiveness of my initial v_0. From the examples above, we have seen the story of the proportion of 1000 and 2.

For simplicity, from now on, we often write $\lambda_j := \lambda_j(-Q)$. In particular $\lambda_0 = -\rho(Q) > 0$. Instead of our previous v_0, we adopt the uniform distribution:

$$v_0 = (1, 1, 1, 1, 1, 1, 1, 1)^* / \sqrt{8}.$$

This is somehow fair since we usually have no knowledge about g in advance.

Example 3. *Let Q be the same as above. Use the uniform distribution v_0 and set $z_0 = v_0^*(-Q)v_0$. Then*

$$(z_1, z_2, z_3, \mathbf{z_4}) \approx (4.78557, 5.67061, 5.91766, \mathbf{5.91867}).$$
$$(\lambda_0, \lambda_1, \mathbf{\lambda_2}) \approx (0.525268, 2.00758, \mathbf{5.91867}).$$

The computation becomes stable at the 4th iteration. Unfortunately, it is not what we want λ_0 but λ_2. In other words, the algorithm converges to a pitfall. Very often, there are $n - 1$ pitfalls for a matrix having n eigenvalues. This shows once again our initial $\tilde{v}_0$ is efficient and the RQI is quite dangerous.

Hopefully, everyone here has heard the name *Google's PageRank*. In other words, the Google's search is based on the maximal left-eigenvector. On this topic, the book [8] was published 11 years ago. In this book, the Power Iteration is included but not the RQI. It should be clear that for PageRank, we need to consider not only large system, but also fast algorithm.

It may the correct position to mention a part of the motivations for the present study.

- Google's search–PageRank.
- Input–output method in economy. In this and the previous cases, the computation of the maximal eigenvector is required.
- Stability speed of stochastic systems. Here, for the stationary distribution of a Markov chain, we need to compute the eigenvector; and for the stability rate, we need to study the maximal (or the fist nontrivial) eigenvalue.
- Principal component analysis for BigData. One choice is to study the so-called five-diagonal matrices. The second approach is using the maximal eigenvector to analysis the role played by the components, somehow similar to the PageRank.
- For image recognition, one often uses Poisson or Toeplitz matrices, which are more or less the same as the Quasi-birth-death matrices studied in queueing theory. The discrete difference equations of elliptic partial differential equations are included in this class: the block-tridiagonal matrices.
- The effectiveness of random algorithm, say Markov Chain Monte Carlo for instance, is described by the convergence speed. This is also related to the algorithms for machine learning.

– As in the last item, a mathematical tool to describe the phase transitions is the first nontrivial eigenvalue (the next eigenpair in general). This is the original place where the author was attracted to the topic.

Since the wide range of the applications of the topic, there is a large number of publications. The author is unable to present a carefully chosen list of references here, what instead are two random selected references: [8,11].

Up to now, we have discussed only a small size $8 \times 8 \, (N = 7)$ matrix. How about large N? In computational mathematics, one often expects the number of iterations grows in a polynomial way N^α for α greater or equal to 1. In our efficient case, since $2 = 8^{1/3}$, we expect to have $10000^{1/3} \approx 22$ iterations for $N + 1 = 10^4$. The next table subverts completely my imagination.

Here z_0 is defined by

$$z_0 = 7/(8\delta_1) + v_0^*(-Q)v_0/8,$$

where v_0 and δ_1 are computed by our general formulas to be defined in the next section. We compute the matrices of order $8, 100, \ldots, 10^4$ by using MatLab in a notebook, in no more than 30 s, the iterations finish at the second step. This means that the outputs starting from z_2 are the same and coincide with λ_0. See the first row for instance, which becomes stable at the first step indeed. We do not believe such a result for some days, so we checked it in different ways. First, since $\lambda_0 = 1/4$ when $N = \infty$, the answers of λ_0 given in the fourth column are reasonable. More essentially, by using the output v_2, we can deduce upper and lower bounds of λ_0 (using ([2]; Theorem 2.4 (3))), and then the ratio upper/ lower is presented in the last column. In each case, the algorithm is significant up to 6 digits. For the large scale matrices here and in Sect. 4, the computations are completed by Yue-Shuang Li.

2   Efficient Initials: Tridiagonal Case

It is the position to write down the formulas of v_0 and δ_1. Then our initial z_0 used in Table 1 is a little modification of δ_1^{-1}: a convex combination of δ_1^{-1} and $v_0^*(-Q)v_0$.

Table 1. Comparison of RQI for different N.

$N + 1$	z_0	z_1	$z_2 = \lambda_0$	Upper/lower
8	0.523309	0.525268	0.525268	$1 + 10^{-11}$
100	0.387333	0.376393	0.376383	$1 + 10^{-8}$
500	0.349147	0.338342	0.338329	$1 + 10^{-7}$
1000	0.338027	0.327254	0.32724	$1 + 10^{-7}$
5000	0.319895	0.30855	0.308529	$1 + 10^{-7}$
7500	0.316529	0.304942	0.304918	$1 + 10^{-7}$
10^4	0.31437	0.302586	0.302561	$1 + 10^{-7}$

Let us consider the tridiagonal matrix (cf. ([3]; Sect. 3) and ([6]; Subsect. 4.4)). Fix $N \geqslant 1$, denote by $E = \{0, 1, \ldots, N\}$ the set of indices. By a shift if necessary, we may reduce A to Q with negative diagonals: $Q^c = A - mI$, $m := \max_{i \in E} \sum_{j \in E} a_{ij}$,

$$
Q^c = \begin{pmatrix}
-b_0 - c_0 & b_0 & 0 & 0 & \cdots \\
a_1 & -a_1 - b_1 - c_1 & b_1 & 0 & \cdots \\
0 & a_2 & -a_2 - b_2 - c_2 & b_2 & \cdots \\
\vdots & \vdots & & \ddots & \ddots & \ddots \\
0 & 0 & 0 & a_N & -a_N - c_N
\end{pmatrix}.
$$

Thus, we have three sequences $\{a_i > 0\}$, $\{b_i > 0\}$, and $\{c_i \geqslant 0\}$. Our main assumption here is that the first two sequences are positive and $c_i \not\equiv 0$. In order to define our initials, we need three new sequences, $\{h_k\}$, $\{\mu_k\}$, and $\{\varphi_k\}$.

First, we define the sequence $\{h_k\}$:

$$
h_0 = 1, \quad h_n = h_{n-1} r_{n-1}, \qquad 1 \leqslant n \leqslant N; \tag{6}
$$

here we need another sequence $\{r_k\}$:

$$
r_0 = 1 + \frac{c_0}{b_0}, \quad r_n = 1 + \frac{a_n + c_n}{b_n} - \frac{a_n}{b_n r_{n-1}}, \qquad 1 \leqslant n < N.
$$

Here and in what follows, our iterations are often of one-step. Note that if $c_k = 0$ for every $k < N$, then we do not need the sequence $\{h_k\}$, simply set $h_k \equiv 1$. An easier way to remember this (h_i) is as follows. It is nearly harmonic of Q^c except at the last point N:

$$
Q^{c \setminus \text{the last row}} h = 0, \tag{7}
$$

where $B^{\setminus \text{the last row}}$ means the matrix modified from B by removing its last low.

We now use H-transform, it is designed to remove the sequence (c_i):

$$
\widetilde{Q} = \text{Diag}\,(h_i)^{-1} Q^c \, \text{Diag}\,(h_i).
$$

Then

$$
\widetilde{Q} = \begin{pmatrix}
-b_0 & b_0 & 0 & 0 & \cdots \\
a_1 & -a_1 - b_1 & b_1 & 0 & \cdots \\
0 & a_2 & -a_2 - b_2 & b_2 & \cdots \\
\vdots & \vdots & & \ddots & \ddots & \ddots \\
0 & 0 & 0 & a_N & -a_N - c_N
\end{pmatrix}
$$

for some modified $\{a_i > 0\}$, $\{b_i > 0\}$, and $c_N > 0$. Of course, Q^c and $\widetilde{Q}$ have the same spectrum. In particular, under the H-transform,

$$
(\lambda_{\min}(-Q^c),\, g) \to (\lambda_{\min}(-\widetilde{Q}) = \lambda_{\min}(-Q^c),\, \text{Diag}\,(h_i)^{-1} g).
$$

From now on, for simplicity, we denote by Q the matrix replacing c_N by b_N in $\widetilde{Q}$.

Next, we define the second sequence $\{\mu_k\}$:

$$\mu_0 = 1, \quad \mu_n = \mu_{n-1}\frac{b_{n-1}}{a_n}, \qquad 1 \leqslant n \leqslant N. \tag{8}$$

And then define the third one $\{\varphi_k\}$ as follows.

$$\varphi_n = \sum_{k=n}^{N} \frac{1}{\mu_k b_k}, \qquad 0 \leqslant n \leqslant N. \tag{9}$$

We are now ready to define v_0 and δ_1 (or z_0) using the sequences (μ_i) and (φ_i).

$$\tilde{v}_0(i) = \sqrt{\varphi_i}, \ i \leqslant N; \qquad v_0 = \tilde{v}_0/\|\tilde{v}_0\|; \quad \|\cdot\| := \|\cdot\|_{L^2(\mu)} \tag{10}$$

$$\delta_1 = \max_{0\leqslant n\leqslant N}\left[\sqrt{\varphi_n}\sum_{k=0}^{n}\mu_k\sqrt{\varphi_k} + \frac{1}{\sqrt{\varphi_n}}\sum_{n+1\leqslant j\leqslant N}\mu_j\varphi_j^{3/2}\right] =: z_0^{-1} \tag{11}$$

with a convention $\sum_\emptyset = 0$.

Finally, having constructed the initials (v_0, z_0), the RQI goes as follows. Solve w_k:

$$(-Q - z_{k-1}I)w_k = v_{k-1}, \qquad k \geqslant 1; \tag{12}$$

and define

$$v_k = w_k/\|w_k\|, \qquad z_k = (v_k, -Qv_k)_{L^2(\mu)}.$$

Then

$$v_k \to g \quad \text{and} \quad z_k \to \lambda_0 \qquad \text{as } k \to \infty.$$

Before moving further, let us mention that there is an explicit representation of the solution (w_i) to Eq. (12). Assume that we are given $v := v_{k-1}$ and $z := z_{k-1}$. Set

$$M_{sj} = \mu_j \sum_{k=j}^{s} \frac{1}{\mu_k b_k}, \qquad 0 \leqslant j \leqslant s \leqslant N. \tag{13}$$

Define two independent sequences $\{A(s)\}$ and $\{B(s)\}$, recurrently:

$$\begin{cases} A(s) = -\sum_{0\leqslant j\leqslant s-1} M_{s-1,j}(v(j) + zA(j)), \\ B(s) = 1 - z\sum_{0\leqslant j\leqslant s-1} M_{s-1,j}B(j), \qquad 0 \leqslant s \leqslant N. \end{cases} \tag{14}$$

Set

$$x = \frac{\sum_{j=0}^{N}\mu_j(v(j) + zA(j)) - \mu_N b_N A(N)}{\mu_N b_N B(N) - z\sum_{j=0}^{N}\mu_j B(j)}. \tag{15}$$

Then the required solution $w_k := \{w(s) : s \in E\}$ can be expressed as $w(s) = A(s) + xB(s)\,(s \in E)$.

To finish the algorithm, we return to the estimates of $\left(\lambda_{\min}(-Q^c), g(Q^c)\right)$ $(g(Q^c) = g(-Q^c))$ or further $(\rho(A), g(A))$ if necessary, where $g(A)$, for instance, denotes the maximal eigenvector of A. Suppose that the iterations are stopped at $k = k_0$ and set $(\bar{z}, \bar{v}) = (z_{k_0}, v_{k_0})$ for simplicity. Then, we have

$$\left(\lambda_{\min}\left(-Q^c\right), \operatorname{Diag}(h_i)^{-1} g(Q^c)\right) = \left(\lambda_{\min}\left(-\widetilde{Q}\right), g(\widetilde{Q})\right) \approx (\bar{z}, \bar{v}),$$

and so

$$\left(\lambda_{\min}(-Q^c), g(Q^c)\right) \approx \left(\bar{z}, \operatorname{Diag}(h_i) \bar{v}\right). \tag{16}$$

Because $\lambda_{\min}(-Q^c) = m - \rho(A)$, we obtain

$$\left(\rho(A), g(A)\right) \approx \left(m - \bar{z}, \operatorname{Diag}(h_i) \bar{v}\right). \tag{17}$$

Now, the question is the possibility from the tridiagonal case to the general one.

3   Efficient Initials: The General Case (([3]; Subsect. 4.2) and ([6]; Subsect. 4.5))

When we first look at the question just mentioned, it seems quite a long distance to go from the special tridiagonal case to the general one. However, in the eigenvalue computation theory, there is the so-called Lanczos tridiagonalization procedure to handle the job, as discussed in ([3]; Appendix of Sect. 3). Nevertheless, what we adopted in ([3]; Sect. 4) is a completely different approach. Here is our main idea. Note that the initials v_0 and δ_1 constructed in the last section are explicitly expressed by the new sequences. In other words, we have used three new sequences $\{h_k\}$, $\{\mu_k\}$, and $\{\varphi_k\}$ instead of the original three $\{a_i\}$, $\{b_i\}$, and $\{c_i\}$ to describe our initials. Very fortunately, the former three sequences do have clearly the probabilistic meaning, which then leads us a way to go to the general setup. Shortly, we construct these sequences by solving three linear equations (usually, we do not have explicit solution in such a general setup). Then use them to construct the initials and further apply the RQI-algorithm.

Let $A = (a_{ij} : i, j \in E)$ be the same as given at the beginning of the paper. Set $A_i = \sum_{j \in E} a_{ij}$ and define

$$Q^c = A - \left(\max_{i \in E} A_i\right) I.$$

We can now state the probabilistic/analytic meaning of the required three sequences (h_i), (μ_i), and (φ_i).

- (h_i) is the harmonic function of Q^c except at the right endpoint N, as mentioned in the last section.
- (μ_i) is the invariant measure (stationary distribution) of the matrix Q^c removing the sequence (c_i).
- (φ_i) is the tail related to the transiency series, refer to ([3]; Lemma 24 and its proof).

We now begin with our construction. Let $h = (h_0, h_1, \ldots, h_N)^*$ (with $h_0 = 1$) solve the equation

$$Q^{c \,\backslash\text{the last row}} h = 0$$

and define

$$\widetilde{Q} = \text{Diag}(h_i)^{-1} Q^c \,\text{Diag}(h_i).$$

Then for which we have

$$c_0 = \ldots = c_{N-1} = 0, \quad c_N =: q_{N,N+1} > 0.$$

This is very much similar to the tridiagonal case.

Next, let $Q = \widetilde{Q}$. Let $\varphi = (\varphi_0, \varphi_1, \ldots, \varphi_N)^*$ (with $\varphi_0 = 1$) solve the equation

$$\varphi^{\,\backslash\text{the first row}} = P^{\,\backslash\text{the first row}} \varphi,$$

where

$$P = \text{Diag}\left((-q_{ii})^{-1}\right) Q + I.$$

Thirdly, assume that $\mu := (\mu_0, \mu_1, \ldots, \mu_N)$ with $\mu_0 = 1$ solves the equation

$$Q^{* \,\backslash\text{the last row}} \mu^* = 0.$$

Having these sequences at hand, we can define the initials

$$\tilde{v}_0(i) = \sqrt{\varphi_i}, \quad i \leqslant N; \quad v_0 = \tilde{v}_0 / \|\tilde{v}_0\|_\mu; \quad z_0 = (v_0, -Qv_0)_\mu.$$

Then, go to the RQI as usual. For $k \geqslant 1$, let w_k solve the equation

$$(-Q - z_{k-1}I) w_k = v_{k-1}$$

and set

$$v_k = w_k / \|w_k\|_\mu, \qquad z_k = (v_k, -Qv_k)_\mu.$$

Then $(z_k, v_k) \to (\lambda_0, g)$ as $k \to \infty$.

We remark that there is an alternative choice (more safe) of z_0:

$$z_0^{-1} = \frac{1}{1 - \varphi_1} \max_{0 \leqslant n \leqslant N} \left[\sqrt{\varphi_n} \sum_{k=0}^{n} \mu_k \sqrt{\varphi_k} + \frac{1}{\sqrt{\varphi_n}} \sum_{n+1 \leqslant j \leqslant N} \mu_j \varphi_j^{3/2} \right]$$

which is almost a copy of the one used in the last section.

The procedure for returning to the estimates of $\left(\lambda_{\min}(-Q^c), g(Q^c)\right)$ or further $(\rho(A), g(A))$ is very much the same as in the last section.

To conclude this section, we introduce two examples to illustrate the efficiency of the extended initials for tridiagonally dominant matrices. The next two examples were computed by Xu Zhu, a master student in Shanghai.

Example 4 (Block-tridiagonal matrix). *Consider the matrix*

$$
Q = \begin{pmatrix}
A_0 & B_0 & 0 & 0 & \cdots \\
C_1 & A_1 & B_1 & 0 & \cdots \\
0 & C_2 & A_2 & B_2 & \cdots \\
\vdots & \vdots & \ddots & \ddots & \ddots \\
0 & 0 & 0 & C_N & A_N
\end{pmatrix},
$$

where A_k, B_k, C_k are 40×40-matrices, $B's$ and $C's$ are identity matrices, and $A's$ are tridiagonal matrices. For this model, two iterations are enough to arrive at the required results (Table 2).

Table 2. Outputs for Poisson matrix.

$N+1$	z_0	z_1	$z_2 = \lambda_0$
1600	7.985026	7.988219	7.988263
3600	7.993232	7.994676	7.994696
6400	7.996161	7.988256	7.987972

Example 5 (Toeplitz matrix). *Consider the matrix*

$$
A = \begin{pmatrix}
1 & 2 & 3 & \cdots & n-1 & n \\
2 & 1 & 2 & \cdots & n-2 & n-1 \\
\vdots & \vdots & \vdots & \ddots & \vdots & \vdots \\
n-1 & n-2 & n-3 & \cdots & 1 & 2 \\
n & n-1 & n-2 & \cdots & 2 & 1
\end{pmatrix}.
$$

For this model, three iterations are enough to arrive at the required results (Table 3).

Table 3. Outputs for Toeplitz matrix.

$N+1$	$z_0 \times 10^6$	$z_1 \times 10^6$	$z_2 \times 10^6$	$z_3 = \lambda_0$
1600	0.156992	0.451326	0.390252	0.389890
3600	0.157398	2.30731	1.97816	1.97591
6400	0.157450	7.32791	6.25506	6.24718

As mentioned before, the extended algorithm should be powerful for the tridiagonally dominant matrices. How about more general case? Two questions are often asked to me by specialists in computational mathematics: do you allow more negative off-diagonal elements? How about complex matrices? My answer is: they are too far away from me, since those matrices can not be a generator of a Markov chain, I do not have a tool to handle them. Alternatively, I have studied some more general matrices than the tridiagonal ones: the block-tridiagonal matrices, the lower triangular plus upper-diagonal, the upper triangular plus lower-diagonal, and so on. Certainly, we can do a lot case by case, but this seems still a long way to achieve a global algorithm. So we do need a different idea.

4   Global Algorithms

Several months ago, AlphaGo came to my attention. From which I learnt the subject of machine learning. After some days, I suddenly thought, since we are doing the computational mathematics, why can not let the computer help us to find a high efficiency initial value? Why can not we leave this hard task to the computer? If so, then we can start from a relatively simple and common initial value, let the computer help us to gradually improve it.

The first step is easy, simply choose the uniform distribution as our initial v_0:

$$v_0 = (1, 1, \cdots, 1)^* / \sqrt{N + 1}.$$

As mentioned before, this initial vector is fair and universal. One may feel strange at the first look at "global" in the title of this section. However, with this universal v_0, the power iteration is already a global algorithm. Unfortunately, the convergence of this method is too slow, and hence is often not practical. To quicken the speed, we should add a shift which now has a very heavy duty for our algorithm. The main trouble is that the usual Rayleigh quotient $v_0^* A v_0 / (v_0^* v_0)$ can not be used as z_0, otherwise, it will often lead to a pitfall, as illustrated by Example 1.3. The main reason is that our v_0 is too rough and so z_0 deduced from it is also too rough. Now, how to choose z_0 and further z_n?

Clearly, for avoiding the pitfalls, we have to choose z_0 from the outside of the spectrum of A (denoted by $\mathrm{Sp}(A)$), and as close to $\rho(A)$ as possible to quicken the convergence speed. For nonnegative A, $\mathrm{Sp}(A)$ is located in a circle with radius $\rho(A)$ in the complex plane. Thus, the safe region should be on the outside of $\mathrm{Sp}(A)$. Since $\rho(A)$ is located at the boundary on the right-hand side of the circle, the effective area should be on the real axis on the right-hand side of, but a little away from, $\rho(A)$.

For the matrix Q used in this paper, since $\rho(Q) < 0$, its spectrum $Sp(Q)$ is located on the left-hand side of the origin. Then, one can simply choose $z_0 = 0$ as an initial. See Fig. 3.

Having these idea in mind, we can now state two of our global algorithms. Each of them uses the same initials:

$$v_0 = \text{uniform distribution}, \qquad z_0 = \max_{0 \leqslant i \leqslant N} \frac{A v_0}{v_0}(i),$$

where for two vectors f and g, $(f/g)(i) = f_i / g_i$.

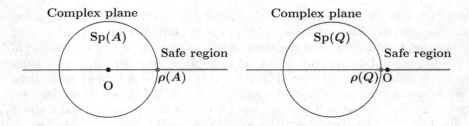

Fig. 3. Safe region in complex plane.

Algorithm 1 (Specific Rayleigh quotient iteration). *At step $k \geqslant 1$, for given $v := v_{k-1}$ and $z := z_{k-1}$, let w solve the equation*

$$(zI - A)w = v.$$

Set $v_k = w/\|w\|$ and let $z_k = v_k^* A v_k$.

This algorithm goes back to [3], Subsect. 4.1 with Choice I.

Algorithm 2 (Shifted inverse iteration). *Everything is the same as in Algorithm 1, except redefine z_k as follows:*

$$z_k = \max_{0 \leqslant i \leqslant N} \frac{A v_k}{v_k}(i)$$

for $k \geqslant 1$ (or equivalently, $k \geqslant 0$).

The comparison of these algorithms are the following: with unknown small probability, Algorithm 1 is less safe than Algorithm 2, but the former one has a faster convergence speed than the latter one with possibility 1/5 for instance.

With the worrying on the safety and convergence speed in mind, we examine two examples which are non-symmetric.

The first example below is a lower triangular plus the upper-diagonal. It is far away from the tridiagonal one, we want to see what can be happened.

Example 6 ([6]; Example 7). *Let*

$$Q = \begin{pmatrix} -1 & 1 & 0 & 0 \cdots\cdots & 0 & 0 \\ a_1 & -a_1-2 & 2 & 0 \cdots\cdots & 0 & 0 \\ a_2 & 0 & -a_2-3 & 3 \cdots\cdots & 0 & 0 \\ \vdots & \vdots & \vdots & \vdots \cdots\cdots & N-1 & 0 \\ a_{N-1} & 0 & 0 & 0 \cdots & -a_{N-1}-N & N \\ a_N & 0 & 0 & 0 \cdots\cdots & 0 & -a_N-N-1 \end{pmatrix}. \quad (18)$$

For this matrix, we have computed several cases:

$$a_k = 1/(k+1), \quad a_k \equiv 1, \quad a_k = k, \quad a_k = k^2.$$

Among them, the first one is the hardest and is hence presented below.
For different N, the outputs of our algorithm are given in Table 4.

Table 4. The outputs for different N by our algorithm.

$N+1$	z_1	z_2	z_3	z_4	z_5	z_6
8	0.276727	0.427307	0.451902	0.452339		
16	0.222132	0.367827	0.399959	0.400910		
32	0.187826	0.329646	0.370364	0.372308	0.372311	
50	0.171657	0.311197	0.357814	0.360776	0.360784	
100	0.152106	0.287996	0.343847	0.349166	0.349197	
500	0.121403	0.247450	0.321751	0.336811	0.337186	
1000	0.111879	0.233257	0.313274	0.334155	0.335009	0.335010
5000	0.0947429	0.205212	0.293025	0.328961	0.332609	0.332635
10^4	0.0888963	0.194859	0.284064	0.326285	0.332113	0.332188

The next example is upper triangular plus lower-diagonal. It is motivated from the classical branching process. Denote by $(p_k : k \geqslant 0)$ a given probability measure with $p_1 = 0$. Let

$$Q = \begin{pmatrix} -1 & p_2 & p_3 & p_4 & \cdots\cdots & p_{N-1} & \sum_{k \geqslant N} p_k \\ 2p_0 & -2 & 2p_2 & 2p_3 & \cdots\cdots & 2p_{N-2} & 2\sum_{k \geqslant N-1} p_k \\ 0 & 3p_0 & -3 & 3p_2 & \cdots & 3p_{N-3} & 3\sum_{k \geqslant N-2} p_k \\ \vdots & \vdots & \vdots & \ddots & \ddots & \ddots & \ddots \\ \vdots & \vdots & \vdots & \ddots & \ddots & -N+1 & (N-1)\sum_{k \geqslant 2} p_k \\ 0 & 0 & 0 & 0 & \cdots\cdots & Np_0 & -Np_0 \end{pmatrix},$$

The matrix is defined on $E := \{1, 2, \ldots, N\}$. Set $M_1 = \sum_{k \in E} k p_k$. When $N = \infty$, it is subcritical iff $M_1 < 1$, to which the maximal eigenvalue should be positive. Otherwise, the convergence rate should be zero.

Now, we fix

$$p_0 = \alpha/2, \ p_1 = 0, \ p_2 = (2-\alpha)/2^2, \ \ldots p_n = (2-\alpha)/2^n, \cdots, \qquad \alpha \in (0, 2).$$

Then $M_1 = 3(2-\alpha)/2$ and hence we are in the subcritical case iff $\alpha \in (4/3, 2)$.

Example 7 ([6]; Example 9). *Set $\alpha = 7/4$. We want to know how fast the local $(N < \infty)$ maximal eigenvalue becomes stable (i.e., close enough to the converge rate at $N = \infty$). Up to $N = 10^4$, the steps of the iterations we need are no more than 6. To quicken the convergence, we adopt an improved algorithm. Then the outputs of the approximation of the minimal eigenvalue of $-Q$ for different N are given in Table 5.*

The computation in each case costs no more than one minute. Besides, starting from $N = 50$, the final outputs are all the same: 0.625, which then can be regarded as a very good approximation of $\lambda_{\min}(-Q)$ at infinity $N = \infty$.

It is the position to compare our global algorithm with that given in the last section. At the first look, here in the two examples above, we need about

Table 5. The outputs in the subcritical case.

$N+1$	z_1	z_2	z_3	z_4
8	0.637800	0.638153		
16	0.621430	0.625490	0.625539	
50	0.609976	0.624052	0.624997	0.625000
100	0.606948	0.623377	0.624991	0.625000
500	0.604409	0.622116	0.624962	0.625000
1000	0.604082	0.621688	0.624944	0.625000
5000	0.603817	0.620838	0.62489	0.625000
10^4	0.603784	0.620511	0.624861	0.625000

6 iterations, double of the ones given in the last section. Note that for the initials of the algorithm in the last section, we need solve three additional linear equations, which are more or less the same as three additional iterations. Hence the efficiency of these two algorithms are very close to each other. Actually, the computation time used for the algorithm in the last section is much more than the new one here.

It is quite surprising that our new algorithms work for a much general class of matrices, out of the scope of [3]. Here we consider the maximal eigenpair only. The example below allows partially negative off-diagonal elements.

Example 8 (([9]; Example (7)), ([6]; Example 12)). *Let*

$$A = \begin{pmatrix} -1 & 8 & -1 \\ 8 & 8 & 8 \\ -1 & 8 & 8 \end{pmatrix}.$$

Then the eigenvalues of A are as follows.

$$17.5124, \quad -7.4675, \quad 4.95513.$$

The corresponding maximal eigenvector is

$$(0.486078, 1.24981, 1)^*$$

which is positive.

Here are the outputs of our algorithms. Both algorithms are started at $z_0 = 24$ (Table 6).

Furthermore, we can even consider some complex matrices.

Example 9 (([10]; Example 2.1), ([6]; Example 15)). *Let*

$$A = \begin{pmatrix} 0.75 - 1.125\,i & 0.5882 - 0.1471\,i & 1.0735 + 1.4191\,i \\ -0.5 - i & 2.1765 + 0.7059\,i & 2.1471 - 0.4118\,i \\ 2.75 - 0.125\,i & 0.5882 - 0.1471\,i & -0.9265 + 0.4191\,i \end{pmatrix},$$

Table 6. The outputs for a matrix with more negative elements.

n	z_n: Algorithm 1	z_n: Algorithm 2
1	17.3772	18.5316
2	17.5124	17.5416
3		17.5124

where the coefficients are all accurate, to four decimal digits. Then A has eigen-values

$$3, \quad -2 - i, \quad 1 + i$$

with maximal eigenvector

$$(0.408237, \quad 0.816507, \quad 0.408237)^*.$$

The outputs (y_n) (but not (z_n)) of ([6]; Algorithm 14), a variant of Algorithm 2, are as follows (Table 7).

Table 7. The outputs for a complex matrix.

y_1	y_2	y_3
$3.03949 - 0.0451599\,i$	$3.00471 - 0.0015769\,i$	3

We mention that a simple sufficient condition for the use of our algorithms is the following:

$$\text{Re}\,(A^n) > 0 \text{ for large enough } n, \text{ up to a shift } mI. \tag{19}$$

Then we have the Perron–Frobenius property: there exists the maximal eigen-value $\rho(A) > 0$ having simple left- and right-eigenvectors.

Hopefully, the reader would now be accept the use of "global" here for our new algorithms. They are very much efficient indeed. One may ask about the convergence speed of the algorithms. Even though we do not have a universal estimate for each model in such a general setup, it is known however that the shifted inverse algorithm is a fast cubic one, and hence should be fast enough in practice. This explains the reason why our algorithms are fast enough in the general setup. Certainly, in the tridiagonal dominate case, one can use the algorithms presented in the previous sections. Especially, in the tridiagonal situ-ation, we have analytically basic estimates which guarantee the efficiency of the algorithms. See [4] for a long way to reach the present level.

When talking about the eigenvalues, the first reaction for many people (at least for me, 30 years ago) is that well, we have known a great deal about the subject. However, it is not the trues. One may ask himself that for eigenvalues, how large matrix have you computed by hand? As far as I know, 2×2 only in analytic computation by hand. It is not so easy to compute them for a 3×3 matrix, except using computer. Even I have worked on the topic for about 30 years, I have not been brave enough to compute the maximal eigenvector, we use its mimic only to estimate the maximal eigenvalue (or more generally the first nontrivial eigenvalue). The first paper I wrote on the numerical computation is [3]. It is known that the most algorithms in computational mathematics are local, the Newton algorithm (which is a quadratic algorithm) for instance. Hence, our global algorithms are somehow unusual.

About three years ago, I heard a lecture that dealt with a circuit board optimization problem. The author uses the Newton method. I said it was too dangerous and could fall into the trap. The speaker answered me that yes, it is dangerous, but no one in the world can solve this problem. Can we try annealing algorithm? I asked. He replied that it was too slow. We all know that in the global optimization, a big problem (not yet cracked) is how to escape from the local traps. The story we are talking about today seems to have opened a small hole for algorithms and optimization problems, and perhaps you will be here to create a new field.

Acknowledgments. This paper is based on a series of talks: Central South University (2017/6), 2017 IMS-China ICSP (2017/6), Summer School on Stochastic Processes at BNU (2017/7), the 9th Summer Camp for Excellent College Students at BNU (2017/7), Sichun University (2017/7), the 12th International Conference on Queueing Theory and Network Applications (2017/8), the 2nd Sino-Russian Seminar on Asymptotic Methods in Probability Theory and Mathematical Statistics & the 10th Probability Limit Theory and Statistic Large Sample Theory Seminar (2017/9). The author thanks professors Zhen-Ting Hou, Zai-Ming Liu, Zhen-Qing Chen, Elton P. Hsu, Jing Yang, Xiao-Jing Xu, An-Min Li, Lian-Gang Peng, Quan-Lin Li, Zhi-Dong Bai, Ning-Zhong Shi, Jian-Hua Guo, Zheng-Yan Lin for their invitations and hospitality. The author also thanks Ms Jing-Yu Ma for the help in editing the paper. Research supported in part by National Natural Science Foundation of China (No. 11131003 and 11626245) the "985" project from the Ministry of Education in China, and the Project Funded by the Priority Academic Program Development of Jiangsu Higher Education Institutions.

References

1. Chen, M.F.: Eigenvalues, Inequalities, and Ergodic Theory. Springer, Heidelberg (2005). https://doi.org/10.1007/b138904
2. Chen, M.F.: Speed of stability for birth-death processes. Front. Math. China **5**, 379–515 (2010)
3. Chen, M.F.: Efficient initials for computing the maximal eigenpair. Front. Math. China **11**, 1379–1418 (2016)
4. Chen, M.F.: The charming leading eigenpair. To appear in Advances in Mathematics (China) (2017)
5. Chen, M.F.: Efficient Algorithm for Principal Eigenpair of Discrete p-Laplacian. Preprint
6. Chen, M.F.: Global Algorithms for Maximal Eigenpair. Preprint

7. Golub, G.H., van der Vorst, H.A.: Eigenvalue computation in the 20th century. J. Comput. Appl. Math. **123**, 35–65 (2000)
8. Langville, A.N., Meyer, C.D.: Google's PageRank and Beyond: The Science of Search Engine Rankings. Princeton University, Princeton (2006)
9. Noutsos, D.: Perron Frobenius Theory and Some Extensions (2008). http://www.pdfdrive.net/perron-frobenius-theory-and-some-extensions-e10082439.html
10. Noutsos, D., Varga, R.S.: On the Perron-Frobenius theory for complex matrices. Linear Algebra Appl. **473**, 1071–1088 (2012)
11. Solomon, J.: Numerical Algorithms: Methods for Computer Vision, Machine Learning, and Graphics. CRC Press, Boca Raton (2015)

Author Index

Printed in the United States
By Bookmasters